GEOMETRY AND INTERPOLATION
OF CURVES AND SURFACES

This text takes a practical, step-by-step approach to algebraic curves and surface interpolation motivated by the understanding of the many practical applications in engineering analysis, approximation, and curve-plotting problems. Because of its usefulness for computing, the algebraic approach is the main theme, but a brief discussion of the synthetic approach is also presented as a way of gaining additional insight before proceeding with the algebraic manipulation.

The authors start with simple interpolation, including splines. In an intuitive fashion they extend these simple procedures to the production of conic sections. They then introduce projective coordinates as tools for dealing with higher-order curves and such important concepts as singular points. They present many applications and concrete examples, including an analysis of the rational and polynomial cubics, parabolic interpolation, geometric approximation, and the numerical solution of trajectory problems. In the final chapter they apply the basic theory to the construction of finite-element basis functions and surface interpolants over nonregular shapes and discuss the simple cases of the Steiner surface and the cubic surface.

Professionals, students, and researchers in applied mathematics, solid modeling, graphics, robotics, and engineering design and analysis will find this a useful reference.

Robin J. Y. McLeod is President and Chief Executive Officer of Saltire Software Inc. in Beaverton Oregon.

M. Louisa Baart is Associate Professor of Mathematics at Pochefstroom University, South Africa.

GEOMETRY AND INTERPOLATION OF CURVES AND SURFACES

ROBIN J. Y. McLEOD   M. LOUISA BAART

CAMBRIDGE
UNIVERSITY PRESS

CAMBRIDGE UNIVERSITY PRESS
Cambridge, New York, Melbourne, Madrid, Cape Town, Singapore,
São Paulo, Delhi, Dubai, Tokyo, Mexico City

Cambridge University Press
The Edinburgh Building, Cambridge CB2 8RU, UK

Published in the United States of America by Cambridge University Press, New York

www.cambridge.org
Information on this title: www.cambridge.org/9780521159395

© Cambridge University Press 1998

First published 1998
First paperback edition 2010

A catalogue record for this publication is available from the British Library

Library of Congress Cataloguing in Publication Data
McLeod, Robin J. Y.
Geometry and interpolation of curves and surfaces/Robin, J. Y.
McLeod, M. Louisa Baart.
p.   cm.
Includes bibliographical references.
ISBN 0-521-32153-0 (hb)
1. Curves, Algebraic.   2. Surfaces, Algebraic;   3. Interpolation.
4. Geometry, Algebraic.   I. Baart, M. Louisa (Maria Louisa)
II. Title.
QA565.M39   1998
516.3'52-dc21   97-43729
                                                         CIP

ISBN 978-0-521-32153-2 Hardback
ISBN 978-0-521-15939-5 Paperback

Contents

v

Preface

Background

The text we present here is the outcome of many years of research and study, by the authors, on a variety of applied problems. Our primary training was in numerical analysis, and from that viewpoint we pursued some problems in finite-element methods, an important area of engineering analysis. Soon we began to suspect that some aspects of geometry might be useful in our research, but we did not know which ones. Geometry seemed rather confusing, and there were many different kinds and much material. We were overwhelmed. Each branch of the subject seemed an enormous study, and we wanted to cut down our work to the areas that, we hoped, would be most relevant to our applied interests. We hoped for a simple program that would yield quick results in only a few days of study. It was not that simple.

However, as we studied we discovered some remarkable things. We found that the subject was much more beautiful than we had imagined, and we were drawn towards it. We also found that, as we explored, we stumbled across answers to applied problems that had defeated mathematicians for years. We even found that the geometry made significant contributions to areas we had not initially set out to explore, and we were drawn into research on curve interpolation, geometrical approximation, computer graphics, and finite-difference methods.

This book is the story of our journey, though we have organized it in a well-structured way and not as the meandering network of pathways that we actually trod.

The book could serve as a complete text for a year-long course on algebraic geometry and applications, but shorter courses, using only selected parts of the text, are equally possible. We will make some specific suggestions later in the Preface, but first, let us set the scene by giving an example.

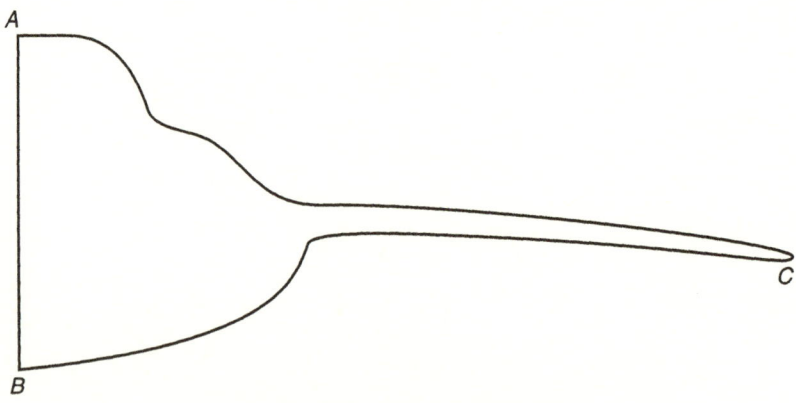

Figure 0.1. Aircraft cross section.

A Real-Life Example

Let us imagine that we have been posed the problem of giving a mathematical description of a curve like the one represented in Figure 0.1. There are two common situations in which such a problem may arise. The first situation is what we shall refer to as the *design problem*. In this situation we are not given any curve, but have to create one. The criteria determining whether or not the curve we created is a "good" one are concerned with constraints on the entire problem. For example, in aircraft design these criteria may be connected with loading factors, lift, strength, and so on. In other environments, such as graphic art designs for the motion-picture industry or as used by clothing manufacturers, the criteria may be nothing more than the creation of esthetically pleasing curves. The second situation, which we shall refer to as the *approximation problem*, is the case where we are provided with a specific curve – perhaps a draughtsman's sketch, perhaps a set of points – and we have to produce a curve that is "close" to the given curve. Many problems involve aspects of both design and approximation. In any of the above situations it will be necessary to produce some curve and to know what curve has been produced. It may not be necessary to have a closed-form mathematical expression for the equation of the curve, provided that there is an algorithm for reproducing the points on the curve to within some specified tolerance. In most cases, however, some closed-form expression is indeed sought, even though in many of them, as when machining a part using a numerically controlled machine, the curves are reapproximated and finally represented as a set of discrete points.

Suppose we want to approximate a curve similar to the one shown in Figure 0.1. In addition, we may have been given the distance *AB*, the distance from *C* to the line *AB*, and perhaps a few other important measurements, but otherwise we know nothing about the curve apart from its qualitative behavior as displayed in the drawing, that is, we have visual rather than mathematical information. We must mathematically produce a curve which is, in some sense, close to the given curve and displays the same qualitative behavior. How should we set about solving such a problem? Our first thought might be to choose some function $y = f(x)$, then to select a set of x-values $\{x_i : i = 1, 2, \ldots, n\}$ and force the corresponding y-values $f(x_i)$ to be such that the points $(x_i, f(x_i))$, for $i = 1, 2, \ldots, n$, lie on the curve. But what are x and y? Our drawing was not supplied with coordinate axes, and any imposition of axes will be somewhat arbitrary. This is clearly undesirable, since our approximation would depend not only on the set of points $\{(x_i, f(x_i))\}$, but also on the particular orientation of the axes. Moreover, since functions are single-valued, there is no function that can approximate both the upper part and the lower part of the curve in Figure 0.1. We could use the common, if ponderous, technique of thinking of the curve as being composed of several pieces, each selected in such a way that it does, in fact, represent a function in its own local coordinate system. However, if we divide the curve into pieces, then it may be that a piecewise approximation technique, such as using a conic to approximate each piece, would yield satisfactory results. Such as technique is often used, and we will describe it in Chapter 2. Alternatively, we could abandon the "$y = f(x)$" form of approximation and use a curve representation that permits curves that can indeed "go round corners."

Two such representations come to mind, namely, a curve defined parametrically, that is, in the form $x = X(t)$, $y = Y(t)$, and a curve given implicitly, that is, in the form $f(x, y) = 0$. The simplest parametric form is polynomials. We may think that we are now in an ideal situation, for we now have a simple representation of the approximating curve, although, as we shall discuss in Chapter 5, such curves in general possess loops and cusps and may also display unwanted inflection points. This behavior would clearly be undesirable, and any method of approximation that uses such curves ought to pay due attention to such potentially troublesome qualitative behavior. Furthermore, there is the difficulty, when using parametric curves, of quantifying what is an appropriate meaning for "close." The usual norms are pertinent only to function approximation and are therefore not immediately applicable to the situation of curve approximation using parametric curves. Some of the special techniques that can be used to measure the distance between curves will be discussed in Chapter 6.

Were we to choose an implicit definition for the approximating curve, then the questions of closeness of approximation that arise in parametric approximation still prevail. As to the form of the function $f(x, y)$, it would be reasonable, as a first attempt, to choose polynomials, the simplest nontrivial case being that of a quadratic polynomial, hence giving us a conic approximation. Even a cursory glance at the curve in Figure 0.1 would indicate that no single conic could be expected to give a reasonable approximation to this curve. A disadvantage of using implicitly defined curves is the difficulty in plotting the final curve. With curves that are given as functions or with parametrically defined curves no such problems arise, but the plotting of implicitly defined curves involves, in general, the solution of a nonlinear equation. We shall return to this question in Chapter 6.

Further questions come to mind. Is there a place for approximation with implicitly defined curves other than conics? Is the choice of parameter important in parametric approximation? Is there a connection between parametric approximation and approximation using implicitly defined curves? How can one control curvature? Some of these questions and the problems cited above will be discussed later. In the meantime, we should notice that in each of the above postulated situations our proposed technique would involve some form of interpolation. It is very important, therefore, that we develop a clear understanding of simple interpolation before proceeding to other and more sophisticated techniques, so many of which rely heavily upon it.

The key step in the finite-element method is the use of a local basis whose construction is the solution of an interpolation problem, and most of the curve and surface design in CAD/CAM (computer-aided design/computer-aided manufacture) is, in fact, interpolation, although here, as in some other applications, there are some "nonstandard" interpolation problems. Interpolation of data is the topic under discussion in Chapter 1. Many of the curve interpolation problems will be discussed in Chapter 6, and surface interpolation will be discussed in Chapter 7.

Course Suggestions

Elementary Geometry

Many students have met conics in high school or during a first-year college math course. For such students a course starting with Chapter 2 might be suitable. If the instructor desires some links with interpolation, Sections 1.6.1 through 1.7.2 from Chapter 1 will be instructive.

At this point the instructor must make another choice. If some elementary synthetic geometry is desired, then Chapter 3 should be studied before proceeding with Chapter 4. However, Chapter 3 can be omitted and the student can proceed directly to Chapter 4. The latter chapter is a basic introduction to algebraic projective geometry and should not be omitted. The approach is, essentially, a transformational approach and is consistent with the current way of approaching coordinate geometry in American high schools. The material, however, goes well beyond that covered in high schools.

This outline will suffice for a simple course on algebraic geometry at an elementary level. If time permits, some examples from Chapter 6 (perhaps from Section 6.7 or 6.8) might provide an interesting applications section to the course.

Algebraic Curves

For the more advanced student or the student specifically interested in algebraic curves and with some pertinent previous background, Chapter 2 can be covered quickly, but it is not recommended that it be ignored. Once more, Chapter 3, though adding another important viewpoint, is not essential. Chapter 4 is essential, and Chapter 5 is the central chapter on algebraic curves and would be the climax of such a course. A course purely on the theory could stop there, and this material could be covered easily in a semester. However, many students nowadays are interested in the utility of the mathematics they study, and for such students, Chapter 6 is strongly recommended. If time permits, the student can then jump to Section 7.8 to get an introduction to one type of algebraic surface.

Interpolation and Applied Geometry (Introductory)

This would be a course for applied mathematicians, scientists, computer scientists, and engineers. For this course Chapters 1 and 2 are essential, Chapter 3 could be omitted but Chapter 4 should be carefully studied, Chapter 5 should be omitted but the examples and applications in Chapter 6 are very important. The examples in Section 7.5 would give worthwhile exposure to two-dimensional problems and would provide an excellent introduction to the rational surfaces of Section 7.8. Once more, this material could be covered in a semester.

Applied Algebraic Geometry

For the senior student who already has a good mathematical background, a course concentrating on the geometry and applications could be constructed as follows. Most of Chapter 1 could be omitted, but Sections 1.6.1 through 1.6.3 should be studied. Chapter 2 could be glossed over and Chapter 3 omitted. Chapter 4 is also likely to be review. Chapters 5, 6, and 7 would form the main part of the course.

Applied Projective Geometry (a Year Course)

For the serious student wanting to understand the critical points, the entire book is recommended, and it is recommended that it be studied in the sequence presented. It is further recommended that certain things definitely not be omitted.

The finite linear interpolation problem of Chapter 1 is important because it is simple and complete and serves as a striking counterpoint to the nonlinear geometric case. Haar's theorem is the eye-opener.

Chapter 2 takes two steps forward and one back. The geometrical arguments are powerful, but the techniques break down and underline the need for more sophisticated mathematical tools. It is important to appreciate the necessity for more general methods.

Chapter 3 should be stressed. Synthetic methods are often ignored, but it is these very methods that provide insight and are often much more incisive than the algebraic counterparts. Often algebraic solution is impossible without the prior insight gained by the synthetic method.

Chapter 5 gets to the heart of the matter. The theory here is critical. Without it a student cannot fully understand rational curves, and it is these curves that give us the main applications.

1

Interpolation

1.1 Introduction

In this chapter we describe what we mean by an interpolation problem, outline the basic theory, show both its generality and its limitations, highlight two-dimensional polynomial interpolation, and find the relationship between this problem and its geometrical counterpart.

The reader is probably familiar with the words "interpolate" and "extrapolate" in the context of finding an approximation to some function value not already tabulated. If the value sought lay between two values already existing in the table then the process of finding the approximation was referred to as "interpolation." Occasionally one sought to extend the table beyond its last entry or add a value before the already existing first value. This process was called "extrapolation." In this book, however, we use the word "interpolation" to refer to a process of passing some curve through given points. When we refer to an *interpolant* we mean a function (or a curve, surface or hypersurface) whose graph passes through a given set of points. We call these points the *interpolation points*. Interpolation is thus distinguished from an *approximation* method, which may not agree with the given function on any set of points but is expected to be "close" to the given function over some interval.

1.2 Polynomial and Rational Interpolation

Let us discuss some examples before we begin to study the theory and applications of the interpolation of data by functions.

1.2.1 Examples

Example 1.1 (*Quadratic polynomial interpolation*) *Find a quadratic polynomial function that assumes the values* 7 *at* $x = -1$, -1 *at* $x = 1$ *and* 1 *at* $x = 2$.

1

We seek a function $p(x) = ax^2 + bx + c$ such that $p(-1) = 7$, $p(1) = -1$, and $p(2) = 1$. Performing the evaluations, we obtain

$$a - b + c = 7,$$

$$a + b + c = -1,$$

$$4a + 2b + c = 1,$$

or, in matrix notation,

$$\begin{pmatrix} 1 & -1 & 1 \\ 1 & 1 & 1 \\ 4 & 2 & 1 \end{pmatrix} \begin{pmatrix} a \\ b \\ c \end{pmatrix} = \begin{pmatrix} 7 \\ -1 \\ 1 \end{pmatrix}.$$

The determinant of the matrix on the left of the equation has the value -6; hence a unique solution exists (because the determinant is nonzero), namely

$$\begin{pmatrix} a \\ b \\ c \end{pmatrix} = -\frac{1}{6} \begin{pmatrix} -1 & 3 & -2 \\ 3 & -3 & 0 \\ -2 & -6 & 2 \end{pmatrix} \begin{pmatrix} 7 \\ -1 \\ 1 \end{pmatrix} = \begin{pmatrix} 2 \\ -4 \\ 1 \end{pmatrix}.$$

Therefore, the required polynomial is $p(x) = 2x^2 - 4x + 1$. This interpolating parabola is shown in Figure 1.1.

Example 1.2 *(Rational-function interpolation)* *Find a function of the form*

$$r(x) = \frac{ax + b}{x + c}$$

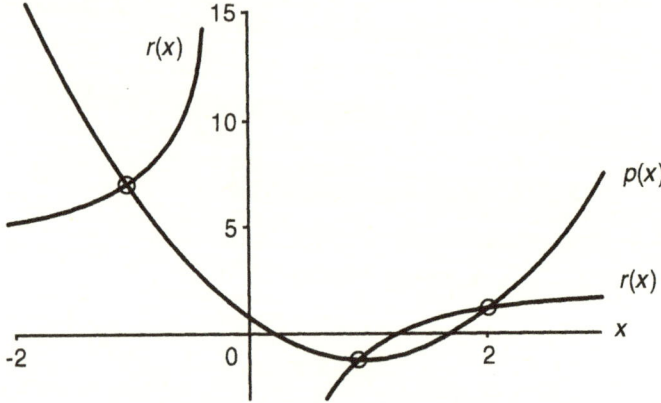

Figure 1.1. Two interpolants through three points.

that assumes the same values at the same points as the polynomial in the previous example, namely $r(-1) = 7, r(1) = -1,$ *and* $r(2) = 1.$

The interpolation constraints give us

$$\frac{-a + b}{-1 + c} = 7, \qquad \frac{a + b}{1 + c} = -1, \qquad \frac{2a + b}{2 + c} = 1,$$

which in matrix notation (after cross multiplication and rearrangement) is equivalent to

$$\begin{pmatrix} -1 & 1 & -7 \\ 1 & 1 & 1 \\ 2 & 1 & -1 \end{pmatrix} \begin{pmatrix} a \\ b \\ c \end{pmatrix} = \begin{pmatrix} -7 \\ -1 \\ 2 \end{pmatrix}.$$

Again the matrix on the left is nonsingular, and the unique solution is given by

$$\begin{pmatrix} a \\ b \\ c \end{pmatrix} = \frac{1}{12} \begin{pmatrix} -2 & -6 & 8 \\ 3 & 15 & -6 \\ -1 & 3 & -2 \end{pmatrix} \begin{pmatrix} -7 \\ -1 \\ 2 \end{pmatrix} = \begin{pmatrix} 3 \\ -4 \\ 0 \end{pmatrix}.$$

Hence the required rational function, shown in Figure 1.1, is

$$r(x) = \frac{3x - 4}{x}.$$

1.2.2 Discussion

Both these examples involved a function with three parameters—the a, b, and c in each case—and the corresponding function had to satisfy three conditions. However, in the first example there will be a unique solution regardless of what values of $p(x)$ we were interpolating, whereas in the second example there are infinitely many sets of values for $r(x)$ for which there is no solution. For example, there is no $r(x)$ of the given form that satisfies the conditions $r(-1) = -1, r(1) = 3,$ and $r(2) = 5.$ In this chapter we will concentrate on polynomial interpolation and similar problems, leaving rational interpolation for later chapters.

We note that for Example 1.1 the existence and uniqueness of the solution do not depend on the specific values interpolated, i.e., $7, -1,$ and $1,$ since the matrix on the left of the first matrix equation is nonsingular and does not depend on these values. Similarly, as long as the three x-values used for the interpolation are distinct, the matrix concerned is nonsingular and a unique solution exists. This result, stated in more general form, is proved in Example 1.6.

1.3 Definitions

Let us give a few definitions and then proceed with our main discussion of the finite linear interpolation problem.

A *field* F is a set of elements called scalars together with two binary laws of operation on F. These laws are called addition (written $+$) and multiplication (written $\times$). By convention, if no symbol appears between two members of F then the multiplication sign is assumed. Thus, for $\alpha, \beta \in F$, $\alpha\beta$ is taken to mean $\alpha \times \beta$. For any $\alpha, \beta, \gamma \in F$ the following properties must be satisfied:

1. $\alpha + \beta \in F$. F is closed under addition.
2. $\alpha + \beta = \beta + \alpha$. Addition is commutative.
3. $\alpha + (\beta + \gamma) = (\alpha + \beta) + \gamma$. Addition is associative.
4. There exists an element written $0 \in F$ such that $\alpha + 0 = \alpha$.
5. There exists an element called the additive inverse and written $-\alpha$ such that $\alpha + (-\alpha) = 0$. We usually write $\alpha - \alpha = 0$ rather than $\alpha + (-\alpha) = 0$.
6. $\alpha\beta \in F$. Closed under multiplication.
7. $\alpha\beta = \beta\alpha$. Multiplication is commutative.
8. $\alpha(\beta\gamma) = (\alpha\beta)\gamma$. Multiplication is associative.
9. There exists an element written $1 \in F$ such that $\alpha 1 = \alpha$.
10. For $\alpha \neq 0$ there exists an element called the multiplicative inverse and written α^{-1} such that $\alpha\alpha^{-1} = 1$.
11. $\alpha(\beta + \gamma) = \alpha\beta + \alpha\gamma$. Multiplication is distributive with respect to addition.

We will usually work over the field of real numbers R and write $1/\alpha$ or $\frac{1}{\alpha}$ for α^{-1}.

A *vector space* V over a field F is a set of elements called vectors together with two binary laws of operation. One of these laws is called vector addition (or simply, addition) and operates on two vectors. The second binary law, called scalar multiplication, operates on a scalar and a vector. The vector addition is written $+$. If $\alpha \in F$ and $f \in V$ then αf is taken to mean scalar multiplication between the scalar α and the vector f. We do not define a multiplication between two vectors. For any $f, g, h \in V$ and $\alpha, \beta \in F$, the following properties must be satisfied:

1. $f + g \in V$. V is closed under addition.
2. $f + g = g + f$. Addition is commutative.
3. $f + (g + h) = (f + g) + h$. Addition is associative.
4. There exists an element written 0 and belonging to V such that $f + 0 = f$.
5. For each $f \in V$ there exists an element of V called the additive inverse and

written $-f$ such that $f + (-f) = 0$. We usually write $f - f = 0$ rather than $f + (-f) = 0$.

6. $\alpha f \in V$. V is closed under scalar multiplication.

7. $1f = f$.

8. $\alpha(f + g) = \alpha f + \alpha g$. Scalar multiplication is distributive with respect to vector addition.

9. $(\alpha + \beta)f = \alpha f + \beta f$. Scalar multiplication is distributive with respect to addition in the field.

When the field is the field of real numbers, we will refer to the vector space as a *real vector space*.

A subset S of a vector space V is said to be a *subspace* if it is also a vector space over the same field as V. A subset S will be a subspace if it is closed under addition and scalar multiplication and contains the zero vector (Exercise 1.12). For example, the set $S = \{(x, x) : x \in R\}$ is a subspace of the two-dimensional real vector space $R_2 = \{(x, y) : x, y \in R\}$ over the field R, with the usual laws for scalar multiplication and vector addition.

Let $f_i \in V, i = 1, 2, \ldots, n$.

The set $\{f_i : i = 1, 2, \ldots, n\}$ is said to *span* V if, for each $f \in V$, there exists a set of scalars $\alpha_i \in F$ such that

$$f = \sum_{i=1}^{n} \alpha_i f_i.$$

The set $\{f_i : i = 1, 2, \ldots, n\}$, is said to form a *linearly independent* set if the equation

$$\sum_{i=1}^{n} \beta_i f_i = 0, \qquad \beta_i \in F, \quad i = 1, 2, \ldots, n,$$

implies $\beta_i = 0, i = 1, 2, \ldots, n$.

If a set of vectors is not linearly independent, then it is said to be *linearly dependent*. We often omit the qualifier "linearly" and refer to a set of vectors as being dependent or independent.

If a set of linearly independent vectors $\{f_i : i = 1, 2, \ldots, n\}$ spans V, then the set forms a *basis* for V and the *dimension* of V is n. In this situation each coefficient $\alpha_i, i = 1, 2, \ldots, n$, referred to above is unique.

When the vector space is a space of real-valued functions over some subset S of reals, then the statement

$$\sum_{i=1}^{n} \beta_i f_i = 0$$

may be written $\sum_{i=1}^{n} \beta_i f_i(x) = 0$, and is taken to mean

$$\sum_{i=1}^{n} \beta_i f_i(x) = 0 \qquad \text{for all} \quad x \in S \subseteq R,$$

where S is the particular subset of the reals that is under consideration.

Example 1.3 *Examine the dependence of the vectors $f_1 = (1\ 0\ 1\ 0)^T$, $f_2 = (0\ 1\ 0\ 1)^T$, $f_3 = (0\ 1\ 1\ 0)^T$, and $f_4 = (1\ 0\ 0\ 1)^T$ in R_4, where the superscript T denotes the transpose of the (row) vector.*

We set

$$\beta_1 \begin{pmatrix} 1 \\ 0 \\ 1 \\ 0 \end{pmatrix} + \beta_2 \begin{pmatrix} 0 \\ 1 \\ 0 \\ 1 \end{pmatrix} + \beta_3 \begin{pmatrix} 0 \\ 1 \\ 1 \\ 0 \end{pmatrix} + \beta_4 \begin{pmatrix} 1 \\ 0 \\ 0 \\ 1 \end{pmatrix} = \begin{pmatrix} 0 \\ 0 \\ 0 \\ 0 \end{pmatrix},$$

that is,

$$\begin{pmatrix} 1 & 0 & 0 & 1 \\ 0 & 1 & 1 & 0 \\ 1 & 0 & 1 & 0 \\ 0 & 1 & 0 & 1 \end{pmatrix} \begin{pmatrix} \beta_1 \\ \beta_2 \\ \beta_3 \\ \beta_4 \end{pmatrix} = \begin{pmatrix} 0 \\ 0 \\ 0 \\ 0 \end{pmatrix}.$$

Proceeding by Gaussian elimination, we get

$$\begin{pmatrix} 1 & 0 & 0 & 1 \\ 0 & 1 & 1 & 0 \\ 0 & 0 & 1 & -1 \\ 0 & 0 & 0 & 0 \end{pmatrix} \begin{pmatrix} \beta_1 \\ \beta_2 \\ \beta_3 \\ \beta_4 \end{pmatrix} = \begin{pmatrix} 0 \\ 0 \\ 0 \\ 0 \end{pmatrix}.$$

We can choose β_4 arbitrarily; then $\beta_3 = \beta_4$, $\beta_2 = -\beta_4$, and $\beta_1 = -\beta_4$. Hence $\sum_{i=1}^{n} \beta_i f_i(x) = 0$ does not imply that $\beta_i = 0$ for $i = 1, 2, 3, 4$, and the given vectors are dependent.

Example 1.4 *Let $V = P_3(x)$, the space of polynomials of degree three in a single variable x. Let $S_1 = \{0, \frac{1}{2}, 1\}$ and $S_2 = \{x : 0 \leq x \leq 1\}$. Show that $f_1 = 1$, $f_2 = x$, $f_3 = x^2$, and $f_4 = x^3$ are dependent on S_1 but independent on S_2.*

Since in this example $\sum_{i=1}^{4} \beta_i f_i$ is a polynomial of degree three, we seek a polynomial of degree three that is zero on S_1. The polynomial $p(x) = \alpha x(x - 1)(2x - 1)$ satisfies this condition. Hence $\sum_{i=1}^{4} \beta_i f_i = 0$ if $\beta_1 = 0$, $\beta_2 = \alpha$,

$\beta_3 = -3\alpha$, and $\beta_4 = 2\alpha$. Therefore, the functions $f_i, i = 1, \ldots, 4$, are dependent on S_1.

To show independence on S_2, select any four (or more) distinct points, for example $0, \frac{1}{2}, 1$, and x_4, where $x_4 \in S_2$ but $x_4 \neq 0, \frac{1}{2}$, or 1. A cubic which is zero at $x = 0, \frac{1}{2}, 1$ must be of the form $p(x) = \alpha x(x - 1)(2x - 1)$. Then $p(x_4) = 0$ implies that $\alpha x_4(x_4 - 1)(2x_4 - 1) = 0$, so that $\alpha = 0$. Therefore the f_i are independent on S_2.

Example 1.5 *Show that x and $\tan x$ are linearly independent solutions of the differential equation*

$$y''(\tan x - x \sec^2 x) + 2(y'x - y) \tan x \sec^2 x = 0, \qquad x \in (0, \pi/2).$$

Let $y_1 = x$; then $y_1' = 1$ and $y_1'' = 0$. Substitution shows that $y_1 = x$ satisfies the differential equation and hence x is a solution.

Let $y_2 = \tan x$; then $y_2' = \sec^2 x$ and $y_2'' = 2 \sec^2 x \tan x$. Substitution shows that $y_2 = \tan x$ satisfies the differential equation.

Suppose $ax + b \tan x = 0$ for all $x \in (0, \pi/2)$. Then choosing $x = \pi/4$ and $x = \pi/3$, we get

$$\begin{pmatrix} \pi/4 & 1 \\ \pi/3 & \sqrt{3} \end{pmatrix} \begin{pmatrix} a \\ b \end{pmatrix} = \begin{pmatrix} 0 \\ 0 \end{pmatrix},$$

which implies $a = b = 0$. Hence x and $\tan x$ are linearly independent on $x \in (0, \pi/2)$.

We introduce a few more definitions.

A *functional L* over a vector space V is a mapping from V to the field F that associates with each vector f of V a unique scalar in F designated by $L(f)$. The functional is said to be a *linear functional* if, in addition, we have

$$L(\alpha f + \beta g) = \alpha L(f) + \beta L(g) \qquad \text{for} \quad f, g \in V \text{ and } \alpha, \beta \in F.$$

For example, consider the two-dimensional vector space R_2 over the field of reals. We define a functional $L((x, y)) \equiv x + y \in R$. Then

$$\begin{aligned} L(\alpha(x_1, y_1) + \beta(x_2, y_2)) &= L((\alpha x_1, \alpha y_1) + (\beta x_2, \beta y_2)) \\ &= L((\alpha x_1 + \beta x_2, \alpha y_1 + \beta y_2)) \\ &= (\alpha x_1 + \beta x_2) + (\alpha y_1 + \beta y_2) \\ &= \alpha(x_1 + y_1) + \beta(x_2 + y_2) \\ &= \alpha L((x_1, y_1)) + \beta L((x_2, y_2)), \end{aligned}$$

so that our functional is linear.

1.3.1 The Dual Space

For two linear functionals L_1 and L_2 and a scalar α, we can define an addition between the linear functionals and a scalar multiplication between a scalar and a linear functional in the following way:

$$(L_1 + L_2)(f) = L_1(f) + L_2(f), \qquad f \in V,$$

$$(\alpha L_1)(f) = \alpha L_1(f), \qquad f \in V, \quad \alpha \in F.$$

With such definitions of addition and scalar multiplication the set of linear functionals over V forms a vector space called the *dual space* (or conjugate space) and denoted by V^*. The zero linear functional is the functional that maps each vector of V onto the zero of the field. We have the following result.

Theorem 1.1 *If V has dimension n then so does V^*.*

Proof. Let $\{v_i : i = 1, 2, \ldots, n\}$ be a basis for V. Then

$$f \in V \;\Rightarrow\; f = \sum_{i=1}^{n} \alpha_i v_i,$$

where, given f, the α_i are unique scalars. Define $\{L_i : i = 1, 2, \ldots, n\}$ by

$$L_i(f) = \alpha_i.$$

Then the $L_i i = 1, \ldots, n$, are linear functionals (Exercise 1.12). To show that they are linearly independent, suppose that $\sum_{i=1}^{n} \beta_i L_i = 0$. Then

$$\sum_{i=1}^{n} \beta_i L_i(f) = 0 \qquad \text{for all} \quad f \in V.$$

In particular,

$$\sum_{i=1}^{n} \beta_i L_i(v_j) = 0, \qquad j = 1, 2, \ldots, n.$$

By definition of the L_i we have $L_i(v_j) = \delta_{ij}$. Therefore $\beta_j = 0$, $j = 1, 2, \ldots, n$, and hence the L_i are linearly independent. The dimension of V^* is therefore at least n.

Let us suppose that V^* has dimension greater than n. Then there must exist a set of $n + 1$ linearly independent linear functionals $L_1, L_2, \ldots, L_{n+1}$. The set of equations

$$\sum_{i=1}^{n+1} \beta_i L_i(v_j) = 0, \qquad j = 1, 2, \ldots, n,$$

is underdetermined and thus possesses a nontrivial solution. For these solution values β_i (not all zero) we have

$$\sum_{i=1}^{n+1} \beta_i L_i(f) = \sum_{i=1}^{n+1} \beta_i L_i \left(\sum_{j=1}^{n} \alpha_j v_j \right)$$

$$= \sum_{j=1}^{n} \alpha_j \left(\sum_{i=1}^{n+1} \beta_i L_i(v_j) \right)$$

$$= 0.$$

Therefore, $\sum_{i=1}^{n+1} \beta_i L_i$ is the zero linear functional, and hence the set $\{L_i : i = 1, 2, \ldots, n+1\}$ is linearly dependent. This contradicts our assumption that the dimension of V^* is greater than n. Therefore, the dimension of V^* is n. ∎

1.4 The Finite Linear Interpolation Problem

We are now in a position to define the simplest kind of interpolation problem, namely the *finite linear interpolation problem*. This problem is stated as follows: Given a vector space V of finite dimension n, a set of linear functionals $L_i, i = 1, 2, \ldots, n$, in the dual space V^* of V, and a set of scalars $\alpha_i, i = 1, 2, \ldots, n$, in the field F, find an $f \in V$ such that $L_i(f) = \alpha_i$ for $i = 1, 2, \ldots, n$.

Let us return to Example 1.1. Polynomials of degree two with the usual addition and multiplication form a real vector space of dimension three, which we denote by $P_2(x)$. Let us define

$$L_1(f) = f(-1), \qquad L_2(f) = f(1), \qquad L_3(f) = f(2),$$

$$\alpha_1 = 7, \quad \alpha_2 = -1, \quad \text{and} \quad \alpha_3 = 1.$$

We can now restate the problem in the terminology of the finite linear interpolation problem, as folllows: find a function $f(x) \in P_2(x)$ such that $L_i(f) = \alpha_i$ for $i = 1, 2, 3$.

We are interested both in determining under what conditions there exists a unique solution to the finite linear interpolation problem and in finding that solution. The solution, when it exists, can always be written as a linear combination of basis vectors. The first step is to determine whether or not the given linear functionals form a basis for V^*.

1.4.1 Matrix Nonsingularity

Theorem 1.2 *Let V be a vector space of dimension n, $\{v_i : i = 1, 2, \ldots, n\}$ a set of vectors in V, and $\{L_i : i = 1, 2, \ldots, n\}$ a set of linear functionals in V^*.*

If the sets $\{v_i\}$ and $\{L_i\}$ are independent in V and V^*, respectively, then the determinant

$$\det(L_i(v_j)) \neq 0.$$

Conversely, if $\det(L_i(v_j)) \neq 0$ and one of the sets $\{v_i\}$ or $\{L_i\}$ is independent, then the other set is also independent.

Proof. We will assume that $\det(L_i(v_j)) = 0$ and show that the $\{L_i\}$ must be dependent, giving us a contradiction. We want, therefore, to show that there exists a set of $\{\alpha_i : i = 1, 2, \ldots, n\}$, not all elements zero, such that $\sum_{i=1}^{n} \alpha_i L_i = 0$. The matrix $(L_i(v_j))$ is singular; therefore there exists a nontrivial solution to the n equations

$$(\alpha_1 \quad \alpha_2 \quad \cdots \quad \alpha_n)(L_i(v_j)) = (0 \quad 0 \quad \cdots \quad 0). \tag{1.1}$$

That is, we have a set of $\{\alpha_i\}$, not all zero, such that

$$\sum_{i=1}^{n} \alpha_i L_i(v_j) = 0, \qquad j = 1, 2, \ldots, n.$$

Therefore, since the v_i, $i = 1, 2, \ldots, n$, are independent, $\sum_{i=1}^{n} \alpha_i L_i$ must be the zero linear functional, and hence the $\{L_i\}$ must be dependent, providing us with our contradiction. Hence $\det(L_i(v_j)) \neq 0$.

Conversely, assume that $\det(L_i(v_j)) \neq 0$. Then the only solution to the system of equations (1.1) is the zero solution. That is, the only linear functional of the form $\sum_{i=1}^{n} \alpha_i L_i$ such that $\sum_{i=1}^{n} \alpha_i L_i(v_j) = 0$, $j = 1, 2, \ldots, n$, is the one given by $\alpha_i = 0$, $i = 1, 2, \ldots, n$. If the set $\{v_i\}$ is independent then this functional will map all vectors in V to zero and hence it is the zero linear functional. Therefore,

$$\sum_{i=1}^{n} \alpha_i L_i = 0 \quad \text{implies} \quad \alpha_i = 0, \quad i = 1, 2, \ldots, n,$$

and hence the $\{L_i\}$ are independent.

If the given $\{L_i\}$ are independent and $\det(L_i(v_j)) \neq 0$, then the only solution of

$$(L_i(v_j)) \begin{pmatrix} \beta_1 \\ \beta_2 \\ \vdots \\ \beta_n \end{pmatrix} = \begin{pmatrix} 0 \\ 0 \\ \vdots \\ 0 \end{pmatrix}$$

is $\beta_i = 0$, $i = 1, 2, \ldots, n$. In this case, since the only vector in V to be mapped onto zero by every member of a basis for V^* must be the zero vector, we find that

$$\sum_{i=1}^{n} \beta_i v_i = 0 \quad \text{implies} \quad \beta_i = 0, \quad i = 1, 2, \ldots, n,$$

and hence the $\{v_i\}$ must be independent. ∎

1.4.2 Uniqueness of Interpolation

We use Theorem 1.2 to prove the basic theorem on the existence and uniqueness of a solution to the finite linear interpolation problem.

Theorem 1.3 *Let V be a vector space of dimension n, and let $\{L_i : i = 1, 2, \ldots, n\}$ be a set of linear functionals in V^*. The finite linear interpolation problem has a unique solution for arbitrary scalars $\{\alpha_i : i = 1, 2, \ldots, n\}$ if and only if the set $\{L_i\}$ is independent in V^*.*

Proof. Let $\{v_i : i = 1, 2, \ldots, n\}$, be a basis for V. By Theorem 1.2, $(L_i(v_j))$ is nonsingular if and only if the set $\{L_i\}$ is independent and hence the system of equations

$$(L_i(v_j)) \begin{pmatrix} \beta_1 \\ \beta_2 \\ \vdots \\ \beta_n \end{pmatrix} = \begin{pmatrix} \alpha_1 \\ \alpha_2 \\ \vdots \\ \alpha_n \end{pmatrix},$$

has a unique solution regardless of the scalars $\{\alpha_i\}$, and $f = \sum_{j=1}^{n} \beta_j v_j$ is the unique solution to the interpolation problem. ∎

1.4.3 The Finite Interpolation Property

If an interpolation problem has a unique solution for *any* set of scalars $\{\alpha_i : i = 1, 2, \ldots, n\}$, we say that it has the *finite interpolation property*. If a good algorithm finds the solution to an interpolation problem for our data points, we can expect that it should be able to find a solution for any other data set. However, even the best algorithm can fail. We shall find that the geometry of the situation will predict failure. Knowing this in advance will enable us to construct methods that will disallow data sets that would lead to failure of a particular method.

The next theorem yields an important technique for establishing whether or not a particular finite linear interpolation problem has the finite interpolation property.

Theorem 1.4 *(Equivalence Theorem) A finite linear interpolation problem has the finite interpolation property if and only if the only solution to the homogenous problem, that is, for $\alpha_i = 0, i = 1, 2, \ldots, n$, is the zero solution.*

Proof. If the only solution to

$$(L_i(v_j)) \begin{pmatrix} \beta_1 \\ \beta_2 \\ \vdots \\ \beta_n \end{pmatrix} = \begin{pmatrix} 0 \\ 0 \\ \vdots \\ 0 \end{pmatrix},$$

where the set $\{v_i\}$ is some basis, is the zero solution, then $(L_i(v_j))$ is nonsingular. Hence the set $\{L_i\}$ is independent and, by Theorem 1.3, the problem has the finite interpolation property. ■

1.4.4 Using the Equivalence Theorem

As an example of the power of this result, let us show that the problem of Example 1.1 has the finite interpolation property. We will do this by the more general method known as *Lagrange polynomial interpolation*, namely interpolation of function values at distinct points. The proof of the existence of a solution is discussed in Example 1.6, and a formula for the solution appears in Example 1.11.

Example 1.6 *Let $V = P_n(x)$, $L_i(f) = f(x_i), i = 0, 1, \ldots, n$, with the points x_i distinct. Show that this Lagrange interpolation problem has the finite interpolation property.*

Examine the homogenous problem $f(x_i) = 0, i = 0, 1, \ldots, n$, and apply the fundamental theorem of algebra, namely that a polynomial of degree n in one variable has exactly n zeros if we count them with the correct multiplicities in the field of complex numbers. The only polynomial of degree n that is zero at $n + 1$ distinct points is the zero polynomial. Hence, by Theorem 1.4, the system has the finite interpolation property.

1.4.5 The Existence of a Normalized Basis

There is another deduction which we can make from Theorem 1.3, which, like Theorem 1.4, is so crucial to our study that we call it a theorem.

Theorem 1.5 *If the finite linear interpolation problem has the finite interpola-tion property, then there exists a unique basis $\{v_i^* : i = 1, 2, \ldots, n\}$ for V that satisfies the condition $L_i(v_j^*) = \delta_{ij}$, and the unique solution to the interpolation problem $L_i(f) = \alpha_i$ is given by $f = \sum_{j=1}^{n} \alpha_j v_j^*$.*

Proof. Since we have a unique solution for each set of $\{\alpha_i\}$, choose n different vectors whose components are given by

$$(\alpha_i = \delta_{ij}, i = 1, 2, \ldots, n) \qquad \text{for} \quad j = i = 1, 2, \ldots, n,$$

and denote the n different solutions by v_j^*. Hence, by construction, $L_i(v_j^*) = \delta_{ij}$. However,

$$f = \sum_{j=1}^{n} \alpha_j v_j^* \in V$$

and

$$L_i(f) = L_i \left(\sum_{j=1}^{n} \alpha_j v_j^* \right) = \sum_{j=1}^{n} \alpha_j L_i(v_j^*) = \alpha_i.$$

Therefore, f is the solution to the problem.

We shall refer to the set $\{v_j^*\}$ as the *normalized basis* (or biorthonormalized basis) with respect to the linear functionals $\{L_i\}$. The normalized basis is inde-pendent of the particular set $\{\alpha_i\}$ occurring in any specific problem, and, using this basis, we have an explicit solution to the interpolation problem without the need for any matrix inversion. ∎

1.4.6 Examples on Using the Basic Theorems

Theorems 1.3, 1.4, and 1.5, together with a few results from geometry that we will develop later in this chapter, provide the basic tools necessary for the construction of two- and three-dimensional interpolation. Let us practice the use of these results by considering the following examples.

Example 1.7 *Let $V = ax + bx^3$ for a and b real, $L_1(f) = f(-1)$ and $L_2(f) = f(1)$. Does this system have the finite interpolation property?*

Certainly x and x^3 form a basis, so let $v_1 = x$ and $v_2 = x^3$. Then

$$\det(L_i(v_j)) = \det \begin{pmatrix} -1 & -1 \\ 1 & 1 \end{pmatrix} = 0,$$

and hence, by Theorem 1.2, the linear functionals are dependent and by Theorem 1.3 we do not have the finite interpolation property. Using the equivalence theorem, Theorem 1.4, we see that any function of the form $\alpha x(1 - x^2)$ is zero at $x = \pm 1$ and lies in the vector space. Theorem 1.4 often provides a very efficient way of examining a particular system. We note that V is a space of odd functions and we cannot make an odd function assume arbitrary values at two points symmetrically placed about the origin.

Example 1.8 *Let*

$$V = P_2(x), \qquad L_1(f) = f'(-1),$$
$$L_2(f) = f(0), \quad and \quad L_3(f) = f'(1).$$

Does this system have the finite interpolation property?

Let $v_1 = 1, v_2 = x, v_3 = x^2$. Then

$$\det(L_i(v_j)) = \det \begin{pmatrix} 0 & 1 & -2 \\ 1 & 0 & 0 \\ 0 & 1 & 2 \end{pmatrix} = 4;$$

hence, by Theorems 1.2 and 1.3, the functionals are independent and we have the finite interpolation property.

A quadratic that has zero slope at two distinct points (the points $x = -1$ and $x = 1$, in this case) must be a constant. Using the second linear functional and zero function value, we see that this constant must be zero. Theorem 1.4 then guarantees the finite interpolation property.

Using Theorem 1.5, we can now easily construct the $v_i^*, i = 1, 2, 3$. They are

$$v_1^* = -\tfrac{1}{4}(x^2 - 2x),$$
$$v_2^* = 1,$$
$$v_3^* = \tfrac{1}{4}(x^2 + 2x).$$

These functions are shown in Figure 1.2. Note that $L_i(v_j^*) = \delta_{ij}$, and if the specific values to be interpolated were $L_1(f) = \alpha_1, L_2(f) = \alpha_2$, and $L_3(f) = \alpha_3$, the unique solution to the problem would be

$$f(x) = -\frac{\alpha_1}{4}(x^2 - 2x) + \alpha_2 + \frac{\alpha_3}{4}(x^2 + 2x).$$

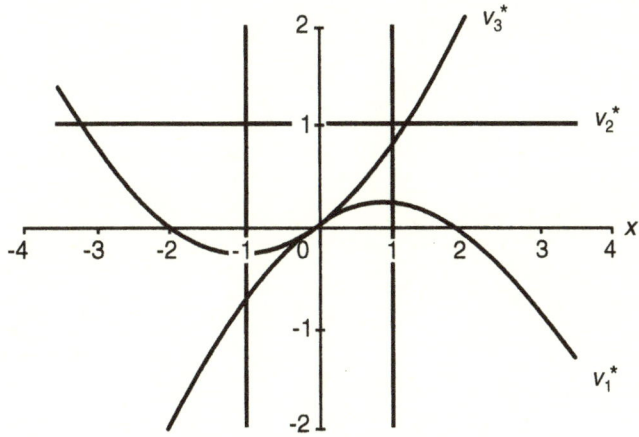

Figure 1.2. Normalized basis functions.

1.4.7 Hermite Interpolation

Charles Hermite posed and solved the interpolation problem of finding a polynomial that interpolates $f^{(i)}(x_j), i = 0, 1, 2, \ldots, n_j, j = 1, 2, \ldots, m$, where the x_j are distinct and $f(x)$ is analytic and without poles in a region containing the set $\{x_j\}$. These problems are therefore referred to as *general Hermite interpolation problems*. If all $n_j = 0$, we have Lagrange interpolation. The case $m = 1, n_1 = n$, gives us our next example.

Example 1.9 *Let $V = P_n(x)$ and $L_i(f) = f^{(i)}(x_0), i = 0, 1, \ldots, n$. Examine this system to determine whether or not it has the finite interpolation property.*

It is easily shown that the only polynomial of degree n that is zero at x_0 and has zero derivatives of up to and including order n at x_0 is the zero polynomial. Theorem 1.4 then assures us of the finite interpolation property. Using Theorem 1.2, we could choose $v_i = x^i, i = 0, 1, \ldots, n$, as a basis for $P_n(x)$. Then

$$\det(L_i(v_j)) = \det \begin{pmatrix} 1 & x_0 & x_0^2 & \cdots & \cdots & x_0^n \\ 0 & 1 & 2x_0 & \cdots & \cdots & nx^{n-1} \\ 0 & 0 & 2 & \cdots & \cdots & n(n-1)x^{n-2} \\ \vdots & \vdots & \vdots & \ddots & & \vdots \\ \vdots & \vdots & \vdots & & \ddots & \vdots \\ 0 & 0 & 0 & \cdots & \cdots & n! \end{pmatrix} \neq 0.$$

A more astute choice of basis would be $v_i = (x - x_0)^i, i = 0, 1, \ldots, n$, in which case

$$\det(L_i(v_j)) = \det \begin{pmatrix} 1 & 0 & 0 & \cdots & \cdots & 0 \\ 0 & 1! & 0 & \cdots & \cdots & 0 \\ 0 & 0 & 2! & \cdots & \cdots & 0 \\ \vdots & \vdots & \vdots & \ddots & & \vdots \\ \vdots & \vdots & \vdots & & \ddots & \vdots \\ 0 & 0 & 0 & \cdots & \cdots & n! \end{pmatrix}.$$

An even more astute choice would be

$$v_i = \frac{(x - x_0)^i}{i!}, \qquad i = 0, 1, \ldots, n.$$

In this case $(L_i(v_j)) = I$, the identity matrix, showing that this last basis is the normalized basis guaranteed to exist by Theorem 1.5.

Example 1.10 *Let V be the vector space spanned by the set of functions $\{\sin i\theta : i = 0, 1, \ldots, n\}$, and let*

$$L_i(f) = \frac{1}{\pi} \int_{-\pi}^{\pi} f(\theta) \sin i\theta \, d\theta.$$

Show that this system has the finite interpolation property.

If we take $v_i = \sin i\theta$, the matrix $(L_i(v_j))$ is the identity matrix. Hence the set $\{L_i\}$ is independent and the required results follow. This problem draws an interesting connection between interpolation and the finite Fourier sine series.

1.4.8 Lagrange Interpolation

One of the most common interpolation problems is Lagrange polynomial interpolation in one dimension. We discussed the existence of solutions to these problems in Example 1.6, so we know that Lagrange interpolation has the finite interpolation property. Let us construct the normalized basis. We will use a special case of a more general type of argument that we shall find useful in the construction of bases for two-dimensional domains (Chapter 7).

Example 1.11 *Construct the normalized basis for Lagrange one-dimensional polynomial interpolation in $P_n(x)$ at $n+1$ distinct points $\{x_i : i = 0, 1, \ldots, n\}$.*

We seek a set of functions $\{l_i(x) : l_i(x) \in P_n(x), i = 0, 1, \ldots, n\}$ such that $L_i(l_j) = \delta_{ij}$. However, in this example $L_i(f) = f(x_i)$. Therefore,

$$l_j(x_i) = \delta_{ij}, \qquad i, j = 0, 1, \ldots, n.$$

We apply a very simple, though important, result. Since $l_j(x_i) = 0$ for $i \neq j$, we know that $x - x_i$ must be a factor of $l_j(x)$. Therefore,

$$l_j(x) = \alpha \prod_{\substack{i=0 \\ i \neq j}}^{n} (x - x_i),$$

where α is some real number yet to be determined. However, $l_j(x_j) = 1$. Hence

$$l_j(x_j) = \alpha \prod_{\substack{i=0 \\ i \neq j}}^{n} (x_j - x_i) = 1,$$

which yields α, and

$$l_j(x) = \prod_{\substack{i=0 \\ i \neq j}}^{n} \frac{x - x_i}{x_j - x_i}, \qquad j = 0, 1, \ldots, n.$$

Since the values x_i are distinct, all the functions $l_j(x)$ are well defined.

In the special case when $n = 3, x_0 = 0, x_1 = 1, x_2 = 2$, and $x_3 = 3$, the four Lagrange basis functions are (Figure 1.3)

$$l_0(x) = -\tfrac{1}{6}(x - 1)(x - 2)(x - 3),$$

$$l_1(x) = \tfrac{1}{2}x(x - 2)(x - 3),$$

$$l_2(x) = -\tfrac{1}{2}x(x - 1)(x - 3),$$

$$l_3(x) = \tfrac{1}{6}x(x - 1)(x - 2).$$

1.4.9 Osculatory Interpolation

Osculatory interpolation is often referred to as Hermite interpolation, and is a special case of the general Hermite interpolation problems. In osculatory interpolation $V = P_{2n+1}(x)$ and the linear functionals are defined by

$$L_i(f) = f(x_i), \qquad M_i(f) = f'(x_i), \qquad i = 0, 1, \ldots, n,$$

with the x_i distinct.

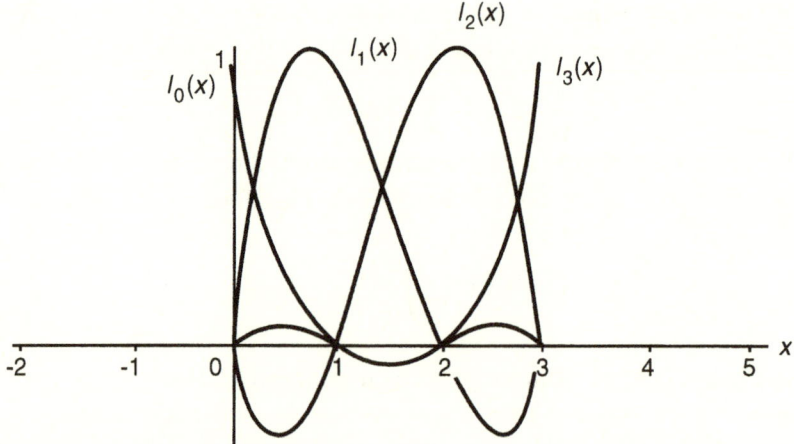

Figure 1.3. Normalized cubic Lagrange basis functions, shown for $x \in [0, 3]$. They are globally defined for all values of x.

Example 1.12 *Assuming that osculatory interpolation has the finite interpolation property, construct the normalized basis.*

Here we are seeking two sets of polynomials $\{u_i(x) \in P_{2n+1}\}$ and $\{v_i(x) \in P_{2n+1}\}$ such that

$$L_i(u_j) = \delta_{ij}, \qquad M_i(u_j) = 0, \qquad L_i(v_j) = 0, \qquad M_i(v_j) = \delta_{ij}.$$

Arguing as in Example 1.11, $x - x_i$ must be a factor of $u_j(x_i)$, since $u_j(x_i) = 0$ for $i \neq j$. Therefore,

$$u_j(x) = p_{n+1}(x) \prod_{\substack{i=0 \\ i \neq j}}^{n} (x - x_i), \qquad p_{n+1}(x) \in P_{n+1}(x).$$

Let us continue this simple argument by noting that, since $u_j(x_i)$ and $u'_j(x_i)$ are both zero for $i \neq j$, then $(x - x_i)^2$ must be a factor of $u_j(x)$. Hence

$$u_j(x) = p_1(x) \prod_{\substack{i=0 \\ i \neq j}}^{n} (x - x_i)^2, \qquad p_1(x) \in P_1(x),$$

which we can write in the form

$$u_j(x) = l_j^2(x)(ax + b),$$

where $l_j(x)$ is defined in Example 1.11. We have only two interpolatory conditions: $u_j(x_j) = 1$ and $u'_j(x_j) = 0$. These conditions will determine a and b,
since $u_j(x_j) = 1$ implies $ax_j + b = 1$, and $u'_j(x_j) = 0$ implies $a + 2l'_j(x_j)$.
Therefore,

$$u_j(x) = l_j^2(x)[1 - 2l'_j(x_j)(x - x_j)], \qquad j = 0, 1, \ldots, n.$$

Similarly, for $v_j(x)$ we see that

$$v_j(x) = (x - x_j)\, l_j^2(x), \qquad j = 0, 1, \ldots, n.$$

The functions $u_j(x)$ and $v_j(x)$, $j = 0, 1, \ldots, n$, are all well defined. The
osculatory interpolant to a differentiable function $f(x)$ at $n + 1$ distinct points
is given by

$$p_{2n+1}(x) = \sum_{i=0}^{n} f(x_i)\, l_i^2(x)[1 - 2l'_i(x_i)(x - x_i)]$$

$$+ \sum_{i=0}^{n} f'(x_i)\, l_i^2(x)(x - x_i).$$

In the special case when $n = 1$, $x_0 = 0$, and $x_1 = 1$, the basis functions for
cubic Hermite interpolation are

$$u_0(x) = (1 - x)^2(1 + 2x),$$
$$u_1(x) = x^2(3 - 2x),$$
$$v_0(x) = x(1 - x)^2,$$
$$v_1(x) = (x - 1)x^2.$$

These basis functions are shown in Figure 1.4 for the restricted interval.

We have now demonstrated three possible ways of establishing the finite
interpolation property in any specific situation. The first was via Theorems
1.2 and 1.3, the second via the equivalence theorem (Theorem 1.4), and the
third, as in Example 1.12, by the actual construction of a well-defined normalized basis that can exist only if the system does have the finite interpolation
property.

1.5 Local and Global Methods

The osculatory interpolation provides an odd-degree polynomial that interpolates function values and first-derivative values at distinct points. Unfortunately,
the degree of the polynomial becomes undesirably large when this interpolation

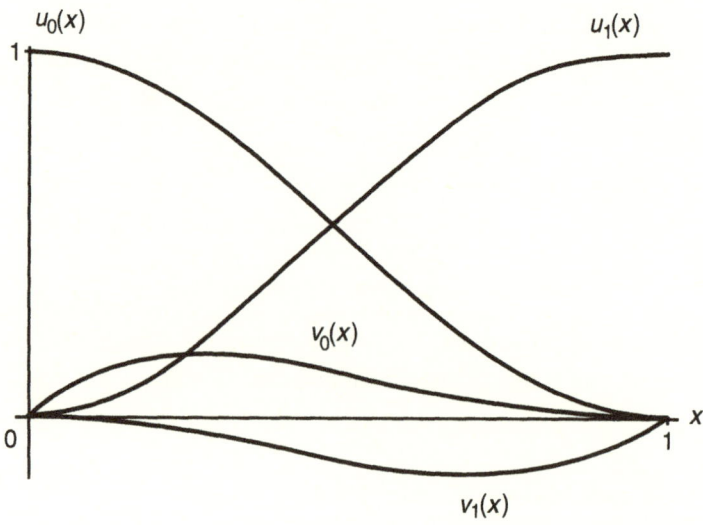

Figure 1.4. Normalized cubic Hermite basis functions, $x \in [0, 1]$.

is done at more than a few points. To avoid this problem it is more usual to per-form osculatory interpolation in a piecewise fashion using a cubic polynomial between each pair of consecutive interpolation points. Such a scheme is often referred to as piecewise cubic Hermite interpolation. Since the first derivative is interpolated at each endpoint of every (local) piece of interpolant, the resulting global interpolant is C^1 continuous. It is not, in general, C^2 continuous. We will call this method *local*, since each segment of the interpolant is constructed from localized information. Local surface interpolation will receive much of our at-tention in Chapter 7. In this section we will discuss some examples of local curve interpolation.

Suppose that we have a curve C_1 that passes through the ordered set of points $\{(x_i, y_i) : i = 1, 2, \ldots, n\}$, the curve having been derived by some interpola-tion method. Suppose that a single ordinate y_j is changed, and let C_2 be the curve produced by the same interpolation procedure applied to the new data set. If the procedure is such that C_1 and C_2 are identical except on the inter-val from the point (x_{j-1}, y_{j-1}) to (x_{j+1}, y_{j+1}), then we say that the method is local, noting that the only change in the curve is in the vicinity of the point which was changed. On the other hand, if the procedure is such that, in gen-eral, the perturbation of any single interpolation point perturbs the resulting curve over its entire length (apart, of course, from the finite number of in-terpolating points, which, by construction, will lie on both curves), then we refer to the method as a *global* method. The method is usually still called local if a perturbation of one point causes a perturbation in a "few," though not all, intervals. If, when using one method, the number of intervals affected is less than that affected by another, we will say that the first method is "more local" than the second. This terminology is somewhat hazy but, we hope, clear enough.

1.5.1 Piecewise Hermite Interpolation

The piecewise cubic Hermite interpolation is therefore as local as possible, since the cubic in each interval is completely determined by the interpolating values at the end points of that particular interval, that is, the function value and the value of the first derivative at each endpoint. When using interpolation as a means of designing or approximating curves it is often preferable to use a local technique, because usually we are satisfied with the fit or shape of the curve over most of its length and wish to make small adjustments to only one part of the curve. The use of a local technique enables us to change the data in this part of the curve, and the resulting approximation will change only in the vicinity of the changed data points.

Another desirable feature of any method of curve design is that it produces a smooth curve. The production of a C^1 curve is almost mandatory. The piecewise cubic Hermite interpolation is therefore a good method from the standpoint of being local and also producing a C^1 curve. This method does, however, require that we know the slopes (first derivatives) at the interpolating points and often this is information which is not available. When this is the case, it is a simple matter to use local data to construct an approximation to the slope at each point in turn. This is equivalent to performing yet another local interpolation. For each point (x_i, y_i) we could define a slope at (x_i, y_i) by taking the slope of the line through the points (x_{i-1}, y_{i-1}) and (x_{i+1}, y_{i+1}). Alternatively, we could use the slope at the point $x = x_i$ of the quadratic Lagrange polynomial interpolant through the points (x_j, y_j) for $j = i - 1, i, i + 1$, or any of a variety of other interpolants. Once we have a set of slopes, we can use the piecewise cubic Hermite interpolation to define a C^1 curve passing through all the given data. Procedures based upon this technique are extremely common and very useful. We could also use local interpolation to construct data sets of second or higher derivatives.

We are discussing here a function approximation or design problem in the sense that the points (x_i, y_i) are ordered, not only in the sense of increasing index but also with the restriction that $x_{i+1} > x_i, i = 0, 1, \ldots, n-1$. This restriction

can be removed by selecting a local coordinate system. It is possible, by
use of local coordinate systems and approximations to derivatives, to produce,
via Hermite interpolation, a piecewise cubic C^1 approximation even to curves
that do not represent functions in any global coordinate system. The piecewise
cubics $y^{[i]}$ in this case will be of the form

$$y^{[i]}\left(x^{[i]}\right) = a_3\left(x^{[i]}\right)^3 + a_2\left(x^{[i]}\right)^2 + a_1 x^{[i]} + a_0,$$

defined between the points (x_{i-1}, y_{i-1}) and (x_i, y_i), where $x^{[i]}$ and $y^{[i]}$ represent some local coordinate system. If each of these systems is derivable from
some global coordinate system by a translation and rotation, then, in global
coordinates x and y, the approximating curve is given by $p_3^{[i]}(x, y) = 0$ between (x_{i-1}, y_{i-1}) and (x_i, y_i), for $i = 1, 2, \ldots, n$, where $p_3^{[i]}(x, y) \in P_3(x, y)$.
However, the $p_3^{[i]}(x, y) = 0$ are rather special cubic curves, for we know,
by construction, that there exists a polynomial parametrization of a particular
form, and, as we shall see in Chapter 5, the general cubic curve is not necessarily parametrizable even in terms of general cubic polynomials. Hence the
$p_3^{[i]}(x, y)$ actually belong to a rather special subset of $P_3(x, y)$.

We can use this rather simple idea of piecewise cubic Hermite interpolation in
a more sophisticated fashion. We could regard our technique of approximating
first derivatives as producing only a first approximation to the slope at each
point and then iteratively refine this approximation in such a way as to achieve
(approximate) continuity of curvature at each point. The resulting curve would
be approximately C^2 continuous, although the initial steps in the algorithm
would not indicate this and not even first derivatives are supplied as input data.
Such an algorithm may sound cumbersome, yet provides what may be the most
common method of producing the required tapes for numerically controlled
machine tools.

1.5.2 Cubic Splines

Another important technique, particularly because of generalizations that permit a relaxation of the restriction $x_{i+1} > x_i$, $i = 0, 1, \ldots, n-1$, is cubic spline
interpolation.

Consider the ordered data as in the piecewise cubic Hermite case (see Example 1.12) and consider a cubic in each interval. Since there are $n + 1$ points
there will be n intervals and we will have n cubic polynomials. These n cubics
will result in our interpolation problem having $4n$ degrees of freedom. The
set of piecewise cubics defined on n intervals forms a vector space over the
real numbers. We define $4n - 2$ of these degrees of freedom in the following
manner.

Let the cubic on the interval $[x_{i-1}, x_i]$ be denoted by $s_i(x)$. We impose the conditions

(1)

$$s_i(x_{i-1}) = y_{i-1}, \, s_i(x_i) = y_i, \qquad i = 1, 2, \ldots, n,$$

(2)

$$s_i'(x_i) = s_{i+1}'(x_i), \qquad i = 1, 2, \ldots, n-1,$$

(3)

$$s_i''(x_i) = s_{i+1}''(x_i), \qquad i = 1, 2, \ldots, n-1.$$

In the notation of the general finite linear interpolation problem, we can define linear functionals in the following manner:

(1)

$$L_i^+(f) = \lim_{x \to x_i^+} f(x), \qquad i = 0, 1, \ldots, n-1,$$

$$L_i^-(f) = \lim_{x \to x_i^-} f(x), \qquad i = 1, 2, \ldots, n,$$

(2)

$$M_i^+(f) = \frac{df}{dx}\bigg|_{x=x_i^+},$$

$$M_i^-(f) = \frac{df}{dx}\bigg|_{x=x_i^-},$$

$$M_i(f) = M_i^+(f) - M_i^-(f), \qquad i = 1, 2, \ldots, n-1,$$

(3)

$$N_i^+(f) = \frac{d^2 f}{dx^2}\bigg|_{x=x_i^+},$$

$$N_i^-(f) = \frac{d^2 f}{dx^2}\bigg|_{x=x_i^-},$$

$$N_i(f) = N_i^+(f) - N_i^-(f), \qquad i = 1, 2, \ldots, n-1,$$

where the notation x^+ or x^- denotes the right-hand or left-hand limits. These linear functionals are defined on piecewise second-order differentiable functions with the only allowable points of discontinuity being the set $\{x_i : i = 0,$

$1, \ldots, n\}$. We can then restate the $4n - 2$ conditions above as

(1)

$$L_i^+(f) = y_i, \qquad i = 0, 1, \ldots, n - 1,$$
$$L_i^-(f) = y_i, \qquad i = 1, 2, \ldots, n,$$

(2)

$$M_i(f) = 0, \qquad i = 1, 2, \ldots, n - 1,$$

(3)

$$N_i(f) = 0, \qquad i = 1, 2, \ldots, n - 1,$$

where f is now a piecewise cubic function. If we include conditions (2) and (3) above in our space of admissible cubics, the resulting vector space is of dimension $2n + 2$.

Two more conditions are necessary in order to specify the problem completely. If these final two conditions are linear functionals, then we will have a finite linear interpolation problem. Common choices for these last two conditions, called *end* conditions, are

(1)

$$f''(x_0) = f''(x_n) = 0,$$

or

(2)

$$f'(x_0) = y_0', \qquad f'(x_n) = y_n'.$$

To use the second of these conditions, we need some estimate of the slope, which we could calculate using differences. Solutions to this interpolation problem using either of these (or other) end conditions are referred to as *cubic splines*. If the first of these end conditions is used, then the solution to the interpolation problem is referred to as the *natural cubic spline*.

Under what conditions do these splines actually exist? In the following we shall have occasion to use Gerschgorin's circle theorem, which gives a method of bounding the eigenvalues of a matrix. This theorem states that, if A is an $n \times n$ matrix and $C_i, i = 1, 2, \ldots, n$, are the disks with centers at a_{ii} and radii $\sum_{j=1, j \neq i}^{n} |a_{ij}|$, then the eigenvalues of A lie within the union of the disks C_i.

Let $s_i''(x_i) = m_i, s_i''(x_{i-1}) = m_{i-1}$, and $h_i = x_i - x_{i-1}, i = 1, 2, \ldots, n$. Each s_i is a cubic, so s_i'' is linear, and, via Lagrange linear interpolation, we have

$$s_i''(x) = \frac{1}{h_i}[m_i(x - x_{i-1}) - m_{i-1}(x - x_i)].$$

Integrating twice, we have

$$s_i(x) = \frac{1}{6h_i}[m_i(x - x_{i-1})^3 - m_{i-1}(x - x_i)^3 + \hat{c}x + \hat{d}],$$

which can be rewritten in the form

$$s_i(x) = \frac{1}{6h_i}[m_i(x - x_{i-1})^3 - m_{i-1}(x - x_i)^3 + c(x - x_i) + d(x - x_i)].$$

Interpolation of the values y_i at the endpoints x_{i-1} and x_i yields

$$s_i(x) = \frac{1}{6h_i}[m_i(x - x_{i-1})^3 - m_{i-1}(x - x_i)^3]$$
$$+ \frac{1}{h_i}\left[(x - x_{i-1})\left(y_i - \frac{m_i h_i^2}{6}\right) - (x - x_i)\left(y_{i-1} - \frac{m_{i-1}h_i^2}{6}\right)\right].$$

We have now incorporated the interpolation of the ordinates together with the continuity of the second derivative. We must still impose continuity in the first derivative. Differentiating s_i and s_{i+1} and demanding continuous first derivatives we get, after some rearranging,

$$\frac{h_i}{6}m_{i-1} + \left(\frac{h_i + h_{i+1}}{3}\right)m_i + \frac{h_{i+1}}{6}m_{i+1} = \frac{y_{i+1} - y_i}{h_{i+1}} - \frac{y_i - y_{i-1}}{h_i}$$

for $i = 1, 2, \ldots, n - 1$. Division by $h_i + h_{i+1}$ and multiplication by 6 changes this to

$$\frac{h_i}{h_i + h_{i+1}}m_{i-1} + 2m_i + \frac{h_{i+1}}{h_i + h_{i+1}}m_{i+1}$$
$$= \frac{6(y_{i+1} - y_i)}{h_{i+1}(h_i + h_{i+1})} - \frac{6(y_i - y_{i-1})}{h_i(h_i + h_{i+1})}$$

for $i = 1, 2, \ldots, n - 1$. Let

$$a_i = \frac{h_i}{h_i + h_{i+1}}, \qquad b_i = \frac{h_{i+1}}{h_i + h_{i+1}}$$
$$c_i = \frac{6}{h_i + h_{i+1}}\left(\frac{y_{i+1} - y_i}{h_{i+1}} - \frac{y_i - y_{i-1}}{h_i}\right),$$

for $i = 1, 2, \ldots, n - 1$. Then we have

$$a_i m_{i-1} + 2m_i + b_i m_{i+1} = c_i, \qquad i = 1, 2, \ldots, n - 1.$$

This is a system of $n-1$ linear equations in the $n+1$ unknowns m_i, $i = 0, 1, \ldots,$ n. We must now impose the two end conditions. Suppose that these are of the

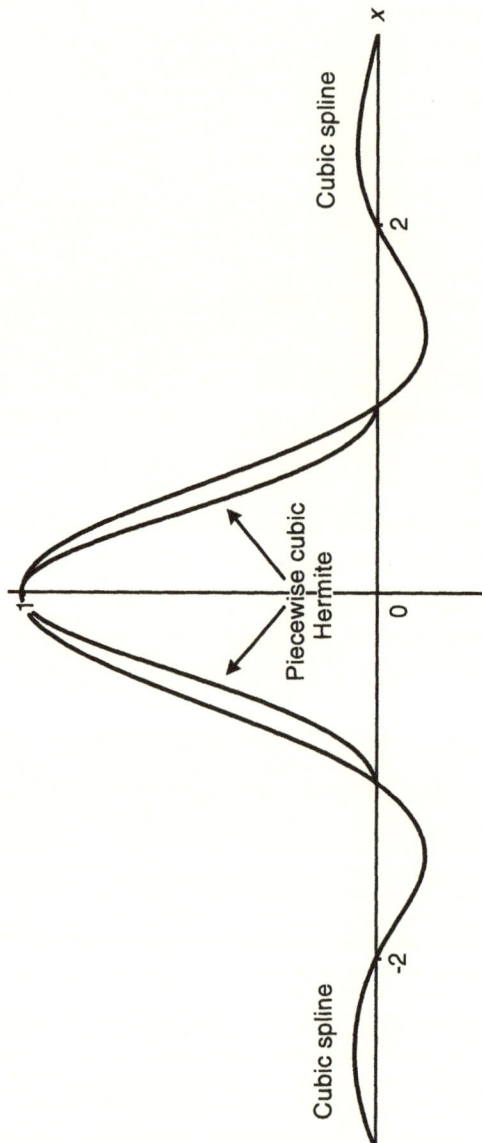

Figure 1.5. Cubic spline and piecewise cubic Hermite basis function.

form

$$2m_0 + b_0 m_1 = c_0,$$

$$a_n m_{n-1} + 2m_n = c_n.$$

Then the system of $n + 1$ equations can be written in matrix form as

$$
\begin{pmatrix}
2 & b_0 & 0 & \cdots & \cdots & \cdots & 0 \\
a_1 & 2 & b_1 & \cdots & \cdots & \cdots & 0 \\
\vdots & & & & & & \vdots \\
0 & \cdots & \cdots & \cdots & a_{n-1} & 2 & b_{n-1} \\
0 & \cdots & \cdots & \cdots & 0 & a_n & 2
\end{pmatrix}
\begin{pmatrix}
m_0 \\ m_1 \\ \vdots \\ \vdots \\ m_n
\end{pmatrix}
=
\begin{pmatrix}
c_0 \\ c_1 \\ \vdots \\ \vdots \\ c_n
\end{pmatrix}.
$$

Now, since our data points were ordered ($x_i > x_{i-1}, i = 1, 2, \ldots, n$), we know that the $h_i, i = 1, 2, \ldots, n$, are all positive. Therefore, $0 < a_i, b_i < 1$ for $i = 1, 2, \ldots, n - 1$, and we know that $a_i + b_i = 1$ for $i = 1, 2, \ldots, n - 1$. If $|b_0|, |a_n| < 2$, we have, by Gerschgorin's circle theorem, $|\lambda - 2| < 2$, where λ is an eigenvalue of the matrix. Hence the matrix is nonsingular and the problem has the finite interpolation property. For the natural spline the first and last equations do not appear, the system reducing to one of order $n - 1$. Therefore, b_0 and a_n do not appear, and we have, as a special case of our analysis, the following theorem.

Theorem 1.6 *The problem of finding the natural cubic spline interpolating the data $(x_i, y_i), i = 0, 1, \ldots, n$, with $x_i > x_{i-1}, i = 1, 2, \ldots, n$, has the finite interpolation property.*

Since the system has the finite interpolation property, we know, from Theorem 1.5, that a normalized basis exists. We have $n+1$ natural cubic splines $S_i(x), i = 0, 1, \ldots, n$, which satisfy the condition

$$S_i(x_j) = \delta_{ij}, \qquad i, j = 0, 1, \ldots, n,$$

and hence the natural spline interpolating $(x_i, y_i), i = 0, 1, \ldots, n$, is given by

$$S(x) = \sum_{i=0}^{n} y_i S_i(x).$$

The functions $S_i(x)$ are called *cardinal splines*. A sketch of one of these basis functions along with the corresponding piecewise cubic Hermite basis function is given in Figure 1.5.

The cardinal spline basis is global in nature, and hence changing a single value y_i will result in changes in the curve at all points apart from the nodal points x_j, $j \neq i$. The construction of the cardinal spline basis guarantees C^2 continuity from the outset.

1.5.3 B-Splines

The unwelcome global nature of the cardinal splines discussed in the previous subsection can be largely circumvented by taking appropriate combinations of the cardinal basis to form a new basis. It would be most convenient to have a cubic spline basis with a completely local support (domain of definition), as in the case of the piecewise cubic Hermite basis (see Sections 1.4.9 and 1.5.1). If we have a cubic spline $S(x)$ that satisfies

$$S(x_j) = 0 \quad \text{for all } j \geq i, \quad \text{and} \quad S'(x_i) = S''(x_i) = 0,$$

then $S(x)$ will be identically zero for all $x \geq x_i$. To see this we need only consider the cubic $s_{i+1}(x)$ on the interval $[x_i, x_{i+1}]$. Since

$$s_{i+1}(x_i) = s'_{i+1}(x_i) = s''_{i+1}(x_i) = 0,$$

it follows that $s_{i+1}(x) = \alpha(x - x_i)^3$. Since $s_{i+1}(x_{i+1}) = 0$, we have $\alpha = 0$ and hence

$$s'_{i+1}(x_{i+1}) = s''_{i+1}(x_{i+1}) = s'_{i+2}(x_{i+1}) = s''_{i+2}(x_{i+1}) = 0.$$

We can now repeat the argument for the subsequent intervals.

A similar argument holds if

$$S(x_j) = 0 \quad \text{for all } j \leq i, \quad \text{and} \quad S'(x_i) = S''(x_i) = 0,$$

showing that $S(x)$ will be identically zero for all $x \leq x_i$.

Ideally then we would seek a cubic spline $S_i(x)$ defined over two subintervals $[x_{i-1}, x_i]$ and $[x_i, x_{i+1}]$ such that

$$S_i(x_j) = \delta_{ij}$$

and

$$S'_i(x_{i-1}) = S''_i(x_{i-1}) = S'_i(x_{i+1}) = S''_i(x_{i+1}) = 0.$$

The symmetry of the situation would indicate that

$$S'_i(x_i) = 0,$$

and hence the cubic $s_{i+1}(x)$ in the interval $[x_i, x_{i+1}]$ would satisfy

$$s_{i+1}(x_{i+1}) = s'_{i+1}(x_{i+1}) = s''_i(x_{i+1}) = 0,$$

$$s'_{i+1}(x_i) = 0, \quad \text{and} \quad s_{i+1}(x_i) = 1.$$

The conditions imposed at $x = x_{i+1}$ imply $s_{i+1}(x) = \alpha(x - x_{i+1})^3$, with only α remaining undetermined, making it impossible to satisfy both the conditions imposed at $x = x_i$. Hence no such cubic spline exists. This also shows that the cardinal spline $S_i(x)$ does not have zero first and second derivatives at x_{i+1}.

Similarly, no cubic spline exists for three intervals (see Exercise 1.32). However, if we extend the cubic spline to four intervals, we can complete the construction. The resulting spline will be called the *B-spline*.

We seek a cubic spline, which we label $B_i(x)$, that satisfies, for points to the right of x_i,

$$B_i(x_i) = 1, \qquad B_i(x_j) = 0, \quad j \geq i + 2,$$

$$B_i'(x_{i+2}) = B_i''(x_{i+2}) = 0,$$

and, for symmetry, we require $B_i'(x_i) = 0$. This implies $B_i(x) \equiv 0$ for all $x \geq x_{i+2}$.

We have two cubic segments (one for the interval $[x_i, x_{i+1}]$, and one for the interval $[x_{i+1}, x_{i+2}]$) that must be adjusted to satisfy these conditions. This gives us a little more freedom. The two cubics are now required to satisfy

$$s_{i+1}(x_i) = 1, \qquad s_{i+1}'(x_i) = 0, \tag{1.2}$$

$$s_{i+1}'(x_{i+1}) = s_{i+2}'(x_{i+1}), \qquad s_{i+1}''(x_{i+1}) = s_{i+2}''(x_{i+1}), \tag{1.3}$$

$$s_{i+2}(x_{i+2}) = s_{i+2}'(x_{i+2}) = s_{i+2}''(x_{i+2}) = 0. \tag{1.4}$$

From the three conditions (1.4) we have immediately that $s_{i+2}(x) = \alpha(x - x_{i+2})^3$. We will satisfy the four conditions (1.2) and (1.3) to define $s_{i+1}(x)$, and then we will impose the condition (so that the global solution will be a spline) $s_{i+1}(x_{i+1}) = s_{i+2}(x_{i+1})$. We first write $s_{i+1}(x)$ in the form

$$s_{i+1}(x) = a_3(x - x_i)^3 + a_2(x - x_i)^2 + a_1(x - x_i) + a_0,$$

and apply the two conditions (1.2) to get $a_0 = 1$ and $a_1 = 0$. The conditions (1.3) then imply

$$\begin{pmatrix} 3h_{i+1}^2 & 2h_{i+1} \\ 6h_{i+1} & 2 \end{pmatrix} \begin{pmatrix} a_3 \\ a_2 \end{pmatrix} = \begin{pmatrix} 3\alpha h_{i+2}^2 \\ -6\alpha h_{i+2} \end{pmatrix}.$$

The matrix on the left is nonsingular, and hence there is always a unique solution to this system of two linear equations. Finally, we put $s_{i+1}(x_{i+1}) = s_{i+2}(x_{i+1})$ and solve for α. In the case of evenly spaced data the h_i are constants h and the solution is

$$s_{i+1}(x) = \frac{1}{4h^3}[3(x - x_{i+1})^3 + 3h(x - x_{i+1})^2 - 3h^2(x - x_{i+1}) + h^3],$$

$$s_{i+2}(x) = -\frac{1}{4h^3}(x - x_{i+2})^3.$$

Similarly, for the two intervals to the left of x_i we have

$$s_i(x) = \frac{1}{4h^3}[3(x_{i-1} - x)^3 + 3h(x_{i-1} - x)^2 - 3h^2(x_{i-1} - x) + h^3],$$

$$s_{i-1}(x) = -\frac{1}{4h^3}(x_{i-2} - x)^3.$$

Thus, the cubic B-spline $B_i(x)$ that satisfies

$$B_i(x_i) = 1, \quad B_i(x_{i\pm1}) = \tfrac{1}{4}, \quad B_i(x_j) = 0, \qquad i = 2, 3, \ldots, n - 2,$$
$$j \neq i, i \pm 1,$$

will be identically zero outside the interval $[x_{i-2}, x_{i+2}]$.

A cubic spline interpolation formula using the B-splines as a basis is referred to as a *B-spline representation*. To extend the basis to $i = 0, 1, n - 1$, and n, we simply imagine additional points x_{-2}, x_{-1}, x_{n+1}, and x_{n+2}, although we use only the part of the spline that is defined in the interval $[x_0, x_n]$. The restriction that $B''(x_{-2}) = B''(x_{n+2}) = 0$ is now of theoretical but not practical significance. The cubic B-spline basis $\{B_i(x)\}$ is then given by

$$B_i(x) = \begin{cases} s_{i-1}(x) & \text{for} \quad x \in [x_{i-2}, x_{i-1}], \\ s_i(x) & \text{for} \quad x \in [x_{i-1}, x_i], \\ s_{i+1}(x) & \text{for} \quad x \in [x_i, x_{i+1}], \\ s_{i+2}(x) & \text{for} \quad x \in [x_{i+1}, x_{i+2}], \\ 0 & \text{otherwise} \end{cases}$$

for $i = 0, 1, \ldots, n$. The support of each cubic B-spline is four intervals, twice as many as the support of the piecewise cubic Hermite. Although the support of each basis function is local, this interpolation procedure is not really "local," because there is no interval in which the B-spline interpolant is not affected by changes outside that interval. We shall return to the question of local interpolation in Chapter 7. Figure 1.6 shows three overlapping B-splines over the interval $x \in [-4, 3]$. The B-spline representation of the interpolant to the points $(x_i, y_i), i = 0, 1, \ldots, n$, is

$$S_B(x) = \sum_{i=0}^{n} a_i B_i(x),$$

where

$$y_i = S_B(x_i) = \tfrac{1}{4}a_{i-1} + a_i + \tfrac{1}{4}a_{i+1}.$$

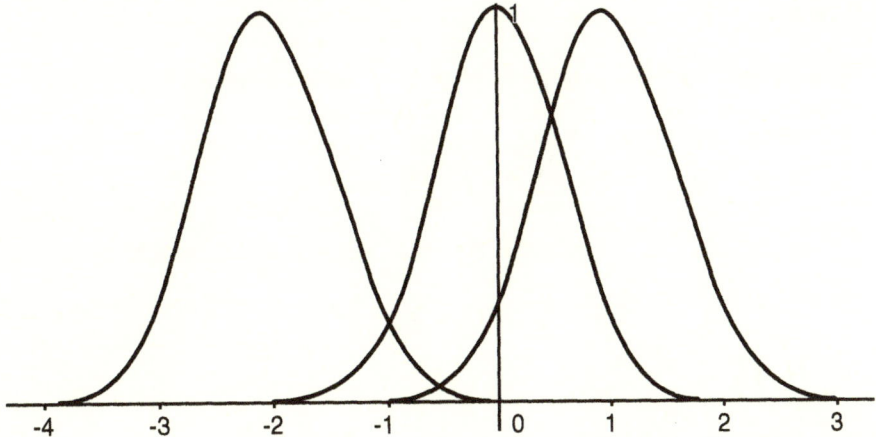

Figure 1.6. Three overlapping *B*-splines.

1.5.4 Minimization of the Strain Energy

The natural cubic spline has the following intriguing property.

Theorem 1.7 *Among all functions* $f \in C^2[x_0, x_n]$ *that interpolate a given set of points* $\{(x_i, y_i) : i = 0, 1, \ldots, n\}$, *the natural cubic spline minimizes*

$$\int_{x_0}^{x_n} [f''(x)]^2 \, dx.$$

Proof. Let $s(x)$ be the natural spline, and let $f(x)$ be some other function in $C^2[x_0, x_n]$ that interpolates the given points. Consider

$$E = \int_{x_0}^{x_n} [f''(x) - s''(x)]^2 \, dx.$$

Since the integrand is nonnegative, $E \geq 0$. We have

$$E = \int_{x_0}^{x_n} [f''(x)]^2 \, dx - 2 \int_{x_0}^{x_n} f''(x) s''(x) \, dx + \int_{x_0}^{x_n} [s''(x)]^2 \, dx,$$

which can be written as

$$E = \int_{x_0}^{x_n} [f''(x)]^2 \, dx - \int_{x_0}^{x_n} [s''(x)]^2 \, dx + 2 \int_{x_0}^{x_n} s''(x)[s''(x) - f''(x)] \, dx.$$

However,

$$\int_{x_0}^{x_n} s''(s'' - f'')\,dx = [s''(s' - f')]_{x_0}^{x_n} - \int_{x_0}^{x_n} s'''(s' - f')\,dx$$

$$= -\sum_{i=1}^{n} s_i''' \int_{x_{i-1}}^{x_i} (s' - f')\,dx$$

[since $s''(x_0) = s''(x_n) = 0$ and s_i''' is constant],

$$= -\sum_{i=1}^{n} s_i'''[s_i - f]_{x_{i-1}}^{x_i}$$

$$= 0,$$

since both s_i and f interpolate the same points. Therefore,

$$E = \int_{x_0}^{x_n} [f''(x)]^2\,dx - \int_{x_0}^{x_n} [s''(x)]^2\,dx \geq 0,$$

which gives the required result.   ∎

The integral $\int_a^b (f'')^2\,dx$ represents an approximation to the strain energy, and the above theorem is often stated by saying that the natural cubic spline minimizes the strain energy.

1.5.5 Splines under Tension

Although the B-spline representation produces a smooth curve interpolating the data, the procedure gives no direct control over the curvature or the number and position of inflection points, and there is the possibility of the spline having unwanted inflections. The representation also does not give any control over changes in curvature. We will now discuss a method which attempts to achieve this.

If we imagine an interpolating curve $f_i(x)$ between points (x_{i-1}, y_{i-1}) and (x_i, y_i) as a flexible entity and apply gradually increasing tension to the curve, then the function $f_i(x)$ will ultimately reduce to the straight line between the two points. In the absence of tension, we would like to have the ordinary cubic spline as the interpolating curve. The tension causes a perturbation of the cubic spline, with the value of the tension parameter $-\sigma^2$ indicating the amount of perturbation. When $\sigma^2 = 0$ we have the cubic spline, and when $\sigma \to \infty$ we have, in the limit, the piecewise linear curve interpolating the data. The cubic spline has a linear second derivative. Therefore, let us demand that the quantity

$$f_i''(x) - \sigma^2 f_i(x)$$

vary linearly in $[x_{i-1}, x_i]$. Using notation parallel to that in our development of cubic splines, this implies

$$f_i''(x) - \sigma^2 f_i(x)$$

$$= \frac{1}{h_i}[(m_i - \sigma^2 y_i)(x - x_{i-1}) - (m_{i-1} - \sigma^2 y_{i-1})(x - x_i)]$$

for $x \in [x_{i-1}, x_i]$. This is a second-order, linear, constant-coefficient, non-homogenous differential equation defining $f_i(x)$, so we know that the solution exists, can easily be found, and, in this case, is continuous with continuous derivatives everywhere—although we are only interested in the interval $[x_0, x_n]$. We also know that, for $\sigma = 0$, we have precisely the relation with which we started our construction of the cubic splines. Dividing by σ^2 and then letting $\sigma \to \infty$, we get

$$f_i(x) = \frac{1}{h_i}[y_i(x - x_{i-1}) - y_{i-1}(x - x_i)],$$

which is the linear interpolant.

We now proceed in a similar fashion to the development of the cubic splines, namely, solve the differential equation with boundary conditions

$$f_i(x_{i-1}) = y_{i-1} \quad \text{and} \quad f_i(x_i) = y_i,$$

then differentiate the solution and impose the condition for continuity of the first derivatives at the points $x = x_i, i = 1, 2, \ldots, n-1$. As with the cubic splines, $f_i(x)$ involves m_{i-1} and m_i, and hence our first-derivative continuity condition

$$f_{i+1}'(x_i) = f_i'(x_i), \qquad i = 1, 2, \ldots, n-1,$$

will result in a tridiagonal system for the determination of the m_i. This system is a system of $n-1$ equations in $n+1$ unknowns and (again analogously to cubic splines), we shall require two end conditions in order to obtain the final solution. The solution to this system is called a *spline under tension*. From the form of the governing differential equation when $\sigma \neq 0$, we note that these splines are definitely not piecewise cubics.

1.6 Unisolvence

We have seen that when interpolating function values or data at points, it is essential that the points be distinct. Interpolation of function values at points is referred to as *pointwise interpolation*, and we have seen that in the case of one-dimensional polynomial interpolation the system has the finite interpolation property, provided that the points are distinct. If a pointwise interpolation problem defined on some space S has the finite interpolation property for any set of distinct points in S, it is said to be *unisolvent*.

One-dimensional Lagrange polynomial interpolation is therefore unisolvent on any interval of the real line. Cubic spline interpolation, with the restriction on the end conditions cited above, is also unisolvent.

Example 1.13 *Show that the system* $\{x, x^3\}$ *is unisolvent on* (a, b) *if and only if a and b are of the same sign. Construct a normalized basis in this case.*

Let $x_1, x_2 \in (a, b)$, and take $v_1 = x$ and $v_2 = x^3$ as a basis. Then

$$\det(L_i(v_j)) = \det \begin{pmatrix} x_1 & x_1^3 \\ x_2 & x_2^3 \end{pmatrix} = x_1 x_2 (x_2 + x_1)(x_2 - x_1).$$

If a and b are of differing sign, then $0 \in (a, b)$, in which case choosing $x_1 = 0$ shows that we do not have unisolvence. In order to have unisolvence we must have $x_1 \neq 0$, $x_2 \neq 0$, and $x_1 \neq -x_2$. This is always satisfied when a and b are of the same sign.

For the normalized basis we have $v_1^*(x) = \alpha x + \beta x^3$ such that $v_1^*(x_1) = 1$ and $v_1^*(x_2) = 0$, that is,

$$\alpha x_1 + \beta x_1^3 = 1,$$
$$\alpha x_2 + \beta x_2^3 = 0.$$

Solving these equations yields

$$v_1^*(x) = \frac{x\left(x_2^2 - x^2\right)}{x_1\left(x_2^2 - x_1^2\right)}.$$

Similarly,

$$v_2^*(x) = \frac{x\left(x_1^2 - x^2\right)}{x_2\left(x_1^2 - x_2^2\right)}.$$

The question of unisolvence refers to pointwise interpolation in general and not only to the finite linear interpolation problem. Consider the following example.

Example 1.14 *Consider the unisolvence of the system*

$$\left\{ \frac{1}{x - y_i} : i = 1, 2, \dots, n \right\}$$

on the interval $[a, b]$, *where* y_i, $i = 1, 2, \dots, n$, *are distinct points that do not belong to the interval* $[a, b]$.

Since the given interpolating functions do not form a vector space with the usual addition and multiplication of real-valued functions, this problem is not a finite linear interpolation problem. This does not mean that no solution exists, but only that we cannot guarantee a unique solution for any set of linearly independent linear functionals. Nonetheless, the important matrix in this situation is

$$(a_{ij}) = \left(\frac{1}{x_i - y_j} \right).$$

The matrix does not involve the interpolating values—the function values, since this is a pointwise interpolation problem—and so, if we can show that this matrix is nonsingular, we have established unisolvence. Let

$$D_n = \det\left(\frac{1}{x_i - y_j} \right)$$

and

$$E_n(x) = \det \begin{pmatrix} \dfrac{1}{x_1 - y_1} & \dfrac{1}{x_1 - y_2} & \cdots & \cdots & \dfrac{1}{x_1 - y_n} \\[2mm] \dfrac{1}{x_2 - y_1} & \dfrac{1}{x_2 - y_2} & \cdots & \cdots & \dfrac{1}{x_2 - y_n} \\[2mm] \vdots & \vdots & & & \vdots \\[2mm] \dfrac{1}{x_{n-1} - y_1} & \dfrac{1}{x_{n-1} - y_2} & \cdots & \cdots & \dfrac{1}{x_{n-1} - y_n} \\[2mm] \dfrac{1}{x - y_1} & \dfrac{1}{x - y_2} & \cdots & \cdots & \dfrac{1}{x - y_n} \end{pmatrix}.$$

The determinant $E_n(x)$ is of the form

$$\frac{p_{n-1}(x)}{\prod_{i=1}^{n}(x - y_i)},$$

where $p_{n-1}(x) \in P_{n-1}(x)$. Furthermore, if $x = x_i$, $i = 1, 2, \ldots, n - 1$, then two rows of the determinant are equal. Hence $E_n(x_i) = 0$ for $i = 1, 2, \ldots,$ $n - 1$. Therefore,

$$E_n(x) = \frac{\alpha \prod_{i=1}^{n-1}(x - x_i)}{\prod_{i=1}^{n}(x - y_i)}.$$

Expanding the determinant $E_n(x)$ by the last row shows us that the coefficient of $1/(x - y_n)$ is D_{n-1}. Therefore,

$$E_n(x) = D_{n-1} \frac{\prod_{i=1}^{n-1}(y_n - y_i) \prod_{i=1}^{n-1}(x - x_i)}{\prod_{i=1}^{n-1}(y_n - x_i) \prod_{i=1}^{n}(x - y_i)},$$

and hence

$$D_n = D_{n-1} \frac{\prod_{i=1}^{n-1}(y_n - y_i) \prod_{i=1}^{n-1}(x_n - x_i)}{\prod_{i=1}^{n-1}(y_n - x_i) \prod_{i=1}^{n}(x_n - y_i)}.$$

Since the x_i and y_i are distinct we have $\prod_{i=1}^{n-1}(y_n - y_i) \neq 0$ and $\prod_{i=1}^{n-1}(x_n - x_i) \neq 0$. Furthermore, since $y_i \notin [a, b]$ and $x_i \in [a, b]$, we have $\prod_{i=1}^{n-1}(y_n - x_i) \neq 0$ and $\prod_{i=1}^{n}(x_n - y_i) \neq 0$. Hence, if $D_{n-1} \neq 0$, so is D_n. Now

$$D_1 = \det\left(\frac{1}{x_1 - y_1}\right) \neq 0,$$

and therefore, by induction, $D_n \neq 0$, $n \geq 1$. Hence the system is unisolvent.

1.6.1 Two-Variable Polynomial Interpolation

So far the examples of unisolvent systems have been one-dimensional. Let us consider the following simple two-dimensional pointwise interpolation problems.

Example 1.15 *Let* $V = \{1, x, y, x^2, xy, y^2\}$, *that is,* $V = P_2(x, y)$, *and consider pointwise interpolation at* $(-1, -3)$, $(1, -1)$, $(3, 1)$, $(-3, -1)$, $(-1, 1)$, *and* $(1, 3)$.

The pertinent determinant (taking the basis vectors and the linear functionals in the same order as listed) is

$$\det \begin{pmatrix} 1 & -1 & -3 & 1 & 3 & 9 \\ 1 & 1 & -1 & 1 & -1 & 1 \\ 1 & 3 & 1 & 9 & 3 & 1 \\ 1 & -3 & -1 & 9 & 3 & 1 \\ 1 & -1 & 1 & 1 & -1 & 1 \\ 1 & 1 & 3 & 1 & 3 & 9 \end{pmatrix}.$$

Performing Gaussian elimination, this determinant becomes

$$-\det \begin{pmatrix} 1 & -1 & -3 & 1 & 3 & 9 \\ 0 & 2 & 2 & 0 & -4 & -8 \\ 0 & 0 & 4 & 8 & -4 & -16 \\ 0 & 0 & 0 & 8 & 8 & 8 \\ 0 & 0 & 0 & 0 & 8 & 16 \\ 0 & 0 & 0 & 0 & 0 & 0 \end{pmatrix} = 0.$$

Hence this pointwise interpolation problem does not have the finite interpolation property, and we do not have unisolvence. From Theorem 1.2, we see that the linear functionals must be dependent.

Let us choose the same vector space but change the linear functionals (or, for pointwise interpolation, the interpolating points).

Example 1.16 *As in the previous example, let* $V = P_2(x, y)$, *with pointwise interpolation at* $\exp(ik\pi/3)$, $k = 1, 2, \ldots, 6$, *where the complex point* $r \exp(i\theta)$, $i^2 = -1$, *is the point* (a, b) *in* R^2 *such that* $r \exp(i\theta) = a + ib$. *Is this system unisolvent?*

The pertinent determinant in this case has jth row

$$\left(1 \quad \cos j\frac{\pi}{3} \quad \sin j\frac{\pi}{3} \quad \cos^2 j\frac{\pi}{3} \quad \cos j\frac{\pi}{3} \sin j\frac{\pi}{3} \quad \sin^2 j\frac{\pi}{3} \right).$$

Subtracting the fourth and sixth columns from the first gives us a column of zeros, and, once again, the linear functionals are dependent.

Lagrange two-variable polynomial interpolation is not always doomed to failure, as the following example shows.

Example 1.17 *Examine the system* $V = P_2(x, y)$, *with pointwise interpolation at* $(0, 0)$, $(1, 0)$, $(2, 0)$, $(1, 1)$, $(0, 2)$, *and* $(0, 1)$ *for unisolvence.*

The determinant here is

$$\det \begin{pmatrix} 1 & 0 & 0 & 0 & 0 & 0 \\ 1 & 1 & 0 & 1 & 0 & 0 \\ 1 & 2 & 0 & 4 & 0 & 0 \\ 1 & 1 & 1 & 1 & 1 & 1 \\ 1 & 0 & 2 & 0 & 0 & 4 \\ 1 & 0 & 1 & 0 & 0 & 1 \end{pmatrix} = -4 \neq 0.$$

The linear functionals in this case are independent, and our system therefore does have the finite interpolation property.

1.6.2 Haar's Theorem

Examples 1.15 and 1.16 clearly show that even such a simple problem as pointwise interpolation using a second-degree polynomial in two variables is not unisolvent. Is the case of quadratic two-variable interpolation an exception rather than the rule? Unfortunately, it is the one-variable case that is the exception, as the following remarkable theorem shows. This theorem bears the name of the twentieth-century Hungarian mathematician, Alfred Haar.

Theorem 1.8 *(Haar's Theorem) Let S be a point set in R^n that contains an interior point P, and let $v_i, i = 1, 2, \ldots, n$, be real-valued functions defined on S and continuous in some neighborhood of P. Then, if $n \geq 2$, the function set $\{v_i\}$ is not unisolvent on S.*

Proof. Let D be a point set contained in a neighborhood in which all the v_i are continuous, and select $\underline{x}_i, i = 1, 2, \ldots, n$, distinct points within D. The pertinent determinant is $\det(v_i(\underline{x}_j))$. Select any two of the points and interchange their positions. Since $n \geq 2$, this can be done in a continuous way such that, throughout the process, the n points are always distinct. Interchanging two points results in two rows of the determinant being interchanged, and hence its sign is changed. Since the determinant is a continuous function of the points and the points changed in a continuous fashion, there must have been some position of the n distinct points that resulted in the determinant being zero. Hence the system is not unisolvent.   ∎

1.6.3 Geometrical Equivalence

Haar's theorem serves to highlight the fact that we must be extremely careful in performing even the simplest multidimensional interpolation. This is a little disconcerting, since it is likely in curve design or approximation methods that we shall wish to use parametric or implicitly defined curves, and such procedures would often result in a two-variable interpolation problem. The construction of finite-element bases is also often a two-variable interpolation problem. We must analyze each situation, but fortunately we can, in many situations, facilitate this analysis by invoking equivalent geometrical considerations. Let us reexamine Example 1.15.

Example 1.15
Alternative solution. Let us once again consider the system with $V = P_2(x, y)$ and pointwise interpolation at $(-1, -3)$, $(1, -1)$, $(3, 1)$, $(-3, -1)$, $(-1, 1)$, and $(1, 3)$.

Using Theorem 1.4, we know that this system will have the finite interpolation property if and only if the only polynomial of degree two that is zero at these six points is the zero polynomial. Consider the locus of all zeros of a polynomial of degree two in two variables, that is, the set of points (x, y), or curve, such that

$$ax^2 + bxy + cy^2 + dx + ey + f = 0.$$

We are interested in determining whether or not the six given points belong to such a locus where not all the a, b, c, d, e, and f are zero. If this is the case, then the system does not possess the finite interpolation property. Our geometrical isomorphism is then the following: the determination of whether or not there is a polynomial in $P_2(x, y)$ that has zero value at the six given points is equivalent to determining if there is a conic section (or conic, for short) that passes through the six given points. This geometrical equivalence makes the analysis very much easier, since the first three points lie on the line $x - y - 2 = 0$ while the second three points lie on the line $y - x - 2 = 0$ and, hence, all six points lie on the conic

$$(x - y - 2)(y - x - 2) = 0,$$

that is, on

$$x^2 - 2xy + y^2 - 4 = 0.$$

This line pair, as well as the interpolated points, is shown in Figure 1.7.

If we look again at the original determinant, we now see that the sum of the fourth column and the sixth column, when subtracted from the sum of four times the first column and twice the fifth column, results in a column of zeros. ∎

The use of the equivalence theorem, Theorem 1.4, together with the geometrical isomorphism, has enabled us to avoid the determinant calculation technique

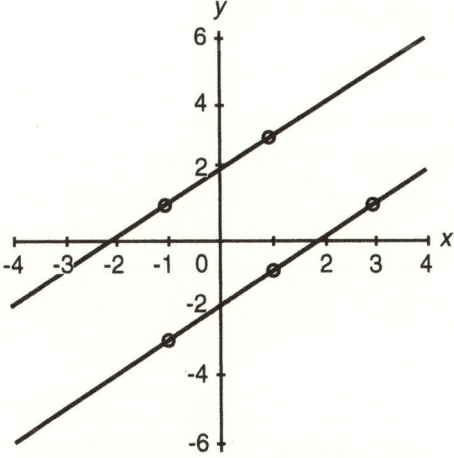

Figure 1.7. Interpolating line pair.

that we employed in our first solution to this problem. This new approach will prove quite potent in the analysis of much more sophisticated problems that we shall discuss later. For the moment, let us practice the procedure on the simple cases of Examples 1.16 and 1.17.

Example 1.16

Alternative solution. Once again, let $V = P_2(x, y)$, with pointwise interpolation at the points in R^2 equivalent to the complex points $\exp(ik\pi/3)$, $k = 1, 2, \ldots, 6$, $i^2 = -1$.

These points are on the circle $x^2 + y^2 - 1 = 0$, that is, the polynomial $x^2 + y^2 - 1$ interpolates the zero value at the six given points. Hence, by Theorem 1.4, since there exists a nontrivial vector in the given vector space that interpolates the homogenous values, the system does not have the finite interpolation property.   ∎

In our reappraisal of Example 1.17 we will introduce another geometrical concept concerning the intersections of curves. This concept, when used in conjunction with the geometrical isomorphism, simplifies the analysis of a wide class of two-dimensional polynomial interpolation problems, including Lagrange interpolation, and a variety of Hermite interpolation problems. For the moment we will introduce the idea in the framework of Lagrange interpolation with conics.

Consider the intersection of a line with a conic. A line has an equation of the form

$$\alpha x + \beta y + \gamma = 0,$$

and a conic has an equation of the form

$$ax^2 + bxy + cy^2 + dx + ey + f = 0.$$

To find where the line meets the conic, we solve these two equations simultaneously. Eliminating either x or y will lead to a quadratic equation in the other variable, which, by the fundamental theorem of algebra, has two roots, which may be identical or may be complex. Therefore, if a line has more than two intersections with a conic, the conic should be reducible to a line pair, one of which is the intersecting line. Let us use this idea in Example 1.17.

Example 1.17

Alternative solution. Let $V = P_2(x, y)$, with pointwise interpolation at $(0, 0)$, $(1, 0)$, $(2, 0)$, $(1, 1)$, $(0, 2)$, and $(0, 1)$.

Using Theorem 1.4 (our equivalence theorem), we seek a $p(x, y) \in P_2(x, y)$ such that

$$p(0, 0) = p(1, 0) = p(2, 0) = p(1, 1) = p(0, 2) = p(0, 1) = 0.$$

This system will have the finite interpolation property if and only if $p(x, y) \equiv 0$. The points $(0, 0)$, $(1, 0)$, and $(2, 0)$ lie on the line $y = 0$. Therefore, either $p(x, y) \equiv 0$ or $p(x, y) = (ax + by + c)y$. Similarly, $(0, 0)$, $(0, 2)$, and $(0, 1)$ lie on the line $x = 0$. Therefore, either $p(x, y) \equiv 0$ or $p(x, y) = axy$. However, $(1, 1)$ does not lie on either $x = 0$ or on $y = 0$. Therefore, $p(1, 1) = 0$ implies $a = 0$, and hence $p(x, y) \equiv 0$, and the system has the finite interpolation property. We cannot include any nonconstant factors apart from the x and y; for, if we did, $p(x, y) \notin P_2(x, y)$. ∎

1.6.4 Two-variable interpolation and conics

Theorem 1.9 *Pointwise interpolation in $P_2(x, y)$ at the six distinct points $(x_i, y_i), i = 1, 2, \ldots, 6$, has the finite interpolation property if and only if there is no conic that passes through all six points.*

Proof. From Theorems 1.2 and 1.3, the problem will have the finite interpolation property if and only if

$$\begin{pmatrix} x_1^2 & x_1 y_1 & y_1^2 & x_1 & y_1 & 1 \\ x_2^2 & x_2 y_2 & y_2^2 & x_2 & y_2 & 1 \\ \vdots & \vdots & \vdots & \vdots & \vdots & \vdots \\ x_6^2 & x_6 y_6 & y_6^2 & x_6 & y_6 & 1 \end{pmatrix} \neq 0,$$

that is, if and only if the columns are independent or, equivalently, if and only if $ax_i^2 + bx_i y_i + cy_i^2 + dx_i + ey_i + f = 0$ for $i = 1, 2, \ldots, 6$ implies $a = b = c = d = e = f = 0$. This implies that the problem will have the finite interpolation property if and only if there is no nontrivial set $\{a, b, c, d, e, f\}$ such that the polynomial $p_2(x, y) = ax^2 + bxy + cy^2 + dx + ey + f$ is zero at the six points. This is true if and only if there is no nontrivial set $\{a, b, c, d, e, f\}$ such that the locus of all zeros of $p_2(x, y)$ includes the six points, that is, if and only if there is no conic that passes through all six points. ∎

This theorem immediately suggests the following generality.

Theorem 1.10 *Pointwise interpolation in $P_n(x, y)$ at $\frac{1}{2}(n + 2)(n + 1)$ distinct points has the finite interpolation property if and only if there is no curve of the form $P_n(x, y) = 0$ that passes through all the given points.*

The interested reader can furnish a proof of this theorem by a simple extension of the proof of Theorem 1.9. We will develop more general methods for analyzing two-variable polynomial interpolation problems once we have discussed algebraic curves, multiple points, and the Maclaurin–Bézout Theorem in Chapter 5.

1.7 Bibliographical Notes

For an introductory view of computer-aided design and finite-element analysis in industry we recommend the book *Computer Aided Design and Manufacture* (2nd edition) by C.B. Besant, 1980, Ellis Horwood Ltd. Of a more mathematical nature and also published by Ellis Horwood is the text *Computational Geometry for Design and Manufacture* by I.D. Faux and M.J. Pratt, 1979. This text is highly recommended as supplementary reading. Also pertinent to our study is the survey article "A Review of Methods for Curve and Function Drawing" by K.W. Brodlie, in *Mathematical Methods in Computer Graphics and Design* (ed. K.W. Brodlie), 1980, Academic Press.

An outstanding text on interpolation is *Interpolation and Approximation* by P.J. Davis, 1975, Dover Publications. Davis covers much more material on classical interpolation than we have incorporated in our discussion.

Most texts on numerical analysis contain chapters on interpolation. Recommended examples are *An Introduction to Numerical Analysis* by K.E. Atkinson, 1978, Wiley, and *A First Course in Numerical Analysis* by A. Ralston and P. Rabinowitz, 1978, McGraw-Hill.

Exercises

1.1. Does the set of all polynomials with (a) integer coefficients, (b) real coefficients, (c) rational coefficients, form a field?

1.2. Let F be the set of ordered pairs of real numbers, and define addition and multiplication by

$$(\alpha, \beta) + (\gamma, \delta) = (\alpha + \gamma, \beta + \delta),$$

$$(\alpha, \beta)(\gamma, \delta) = (\alpha\gamma, \beta\delta).$$

Does F, with such addition and multiplication, form a field?

1.3. Let F be the set of ordered pairs of reals, and define addition and multiplication by

$$(\alpha, \beta) + (\gamma, \delta) = (\alpha + \gamma, \beta + \delta),$$

$$(\alpha, \beta)(\gamma, \delta) = (\alpha\gamma - \beta\delta, \alpha\delta + \beta\gamma).$$

Does F form a field under such addition and multiplication?

1.4. Show that polynomials in one variable with complex coefficients form a vector space over the field of complex numbers.

1.5. Let S be a point set on the real axis and consider the set T of all real-valued functions on S. Define addition and scalar multiplication by

$$(f + g)(x) = f(x) + g(x), \qquad (\alpha f)(x) = \alpha f(x).$$

Show that T is a vector space of real-valued functions over the reals.

1.6. Let V be the set of all linear combinations of $\sin k\theta, k = 1, 2, \ldots, n$. Show that V is a vector space over the reals.

1.7. Show that the set of functions of the form $(ax + b)/(x + c)$, with $a, b, c \in R$, does not form a vector space.

1.8. Consider the following interpolation problem: find a function $r(x) = (ax + b)/(x + c)$ such that $r(x_1) = r_1, r(x_2) = r_2$, and $r(x_3) = r_3$, with x_1, x_2, and x_3 distinct. Show that it leads to a linear system in which the matrix concerned is a function of r_1, r_2, and r_3. For such an interpolation problem we cannot expect the matrix to be nonsingular regardless of r_1, r_2, and r_3.

1.9. Show that the set of functions of the form

$$\{ax^2 + bxy + cy^2 + dx + ey + f : b^2 = 4ac\}$$

does not form a vector space. We could say that parabolas do not form a vector space. This means that, although conics do form a vector space, parabolas do not form a subspace.

1.10. Show that the six-parameter family of functions of the form

$$\left\{ \frac{\sum_{i,j=0}^{i+j=2} a_{ij} x^i y^j}{1 + bx + cy} \right\},$$

with b and c fixed, forms a vector space.

1.11. Show that the set $\{e^{ik\theta} : \theta \in [0, \pi], k = 0, \pm 1, \pm 2, \ldots, \pm n\}$, with $i^2 = -1$, forms a vector space of dimension $2n + 1$.

1.12. Show that a subset S of a vector space V will be a subspace if it is closed under addition and scalar multiplication and contains the zero vector.

1.13. If f_1, f_2, and f_3 are linearly independent, for what values of α are the vectors $g_1 = f_1 + f_2 + \alpha f_3, g_2 = f_1 + \alpha f_3$, and $g_3 = f_2 + f_3$ independent?

1.14. Prove that if the sets $\{f_i : i = 1, 2, \ldots, n\}$ and $\{g_i : i = 1, 2, \ldots, m\}$ are both bases for V, then $m = n$.

1.15. Find a basis for the vector space of 2×2 matrices.

1.16. Show that if S_1 and S_2 are both two-dimensional subspaces of R_3, then they must have a nonzero vector in common.

1.17. Show that 1, $\sin^2 x$, and $\cos 2x$ are dependent on $x \in [0, \pi]$.

1.18. Show that $y = x^2$ and $y = x^3$ are linearly independent solutions of the differential equation

$$x^2 y'' - 4xy' + 6y = 0, \qquad x \in (0, \infty).$$

1.19. Show that the set $\{\cos^i x : i = 0, 1, \ldots, n\}$ is linearly independent on $x \in [-\pi, \pi]$.

1.20. Show that the functionals L_i of Theorem 1.1 are linear functionals.

1.21. Show that the dual space V^* is a vector space.

1.22. Consider the vector space of complex numbers as a real vector space with vectors of the form $a + ib$ with the usual addition and multiplication associated with complex numbers. Which of the following functionals are linear?

(a) $f(z) = b$,

(b) $f(z) = ab$,

(c) $f(z) = a + b$,

(d) $f(z) = a^2 + b^2$.

1.23. If V is the vector space P of polynomials with real coefficients, which of the following are linear functionals?

(a) $f(p) = \int_a^b p(t)\, dt$,

(b) $f(p) = \int_a^b tp(t)\, dt$,

(c) $f(p) = \int_a^b t^2 p(t)\, dt$,

(d) $f(p) = \int_a^b p^2(t)\, dt$,

(e) $f(p) = \dfrac{d^2 p}{dt^2}$,

(f) $f(p) = \dfrac{dp}{dt}\dfrac{d^2 p}{dt^2}$.

1.24. The concept of pointwise interpolation can be extended by replacing the usual linear functionals $L_i(f) = f(x_i)$ by $L_i(f) = \lim_{x \to x_i} f(x)$. Show that, when such limits exist, the new definition defines a linear functional.

1.25. Let $V = R(a, b)$, the set of real-valued functions on the closed interval $[a, b]$. Let $\{x_i : i = 1, 2, \ldots, n\}$, be distinct points in $[a, b]$, and let $L_i(f) = f(x_i), i = 1, 2, \ldots, n$. Show that the functionals L_i are independent in V^*.

1.26. An operator L on the elements of a vector space V is *linear* over the field F if, for $f, g \in V$ and $\alpha, \beta \in F$,

$$L(\alpha f + \beta g) = \alpha L(f) + \beta L(g).$$

Show that L defined on the space of twice differentiable functions of a single real variable x by

$$L(f) = \frac{d^2 f}{dx^2} - \sigma^2 f, \qquad \sigma \in F,$$

is a linear operator.

1.27.   Show that L defined on the space of twice differentiable real-valued functions of the single real variable $x \in (0, \pi/2)$ and given by

$$L(f) = (\tan x - x \sec^2 x)\frac{d^2 f}{dx^2}$$
$$+ 2x \tan x \sec^2 x \frac{df}{dx} - 2 \tan x \sec^2 x\, f,$$

is a linear operator.

1.28.   Let

$$L(f) = \frac{d^n f}{dx^n} + a_{n-1}(x)\frac{d^{n-1} f}{dx^{n-1}} + \cdots + a_1(x)\frac{df}{dx} + a_0(x)f,$$

where $f \in C^n[a, b]$ and $a_i(x) \in C[a, b]$. Show that the set of solutions of the equation $L(f) = 0$ forms a vector space.

1.29.   Show that each of the following systems has the finite interpolation property.

(a) $V = P_n(x)$, $L_i(f) = f^{(i)}(x_i)$, $i = 0, 1, \ldots, n$, x_i distinct.

(b) The vector space V spanned by $\{\cos kx, \sin lx : k = 0, 1, \ldots, n, l = 0, 1, \ldots, n\}$, with $L_i(f) = f(x_i)$, $i = 0, 1, \ldots, 2n$, and with the x_i distinct points in $(-\pi, \pi)$.

(c) V as in (b), with

$$L_{2k}(f) = \frac{1}{\pi}\int_{-\pi}^{\pi} f(x)\cos kx\, dx, \quad k = 0, 1, \ldots, n,$$

$$L_{2k-1}(f) = \frac{1}{\pi}\int_{-\pi}^{\pi} f(x)\sin kx\, dx, \quad k = 1, 2, \ldots, n.$$

1.30.   Find an expression for the slope at (x_n, y_n) of the circle defined by the points (x_{n-1}, y_{n-1}), (x_n, y_n), and (x_{n+1}, y_{n+1}).

1.31.   Let $\{x_i : i = 0, 1, \ldots, n\}$ be a discrete set of points with $x_i > x_{i-1}$, $i = 1, 2, \ldots, n$, and let V be the set of piecewise cubics defined on $[x_0, x_n]$, i.e., if $f \in V$, then $f = f_i \in P_3(x)$ for $x \in [x_{i-1}, x_i]$. Let the cubics be such that $f \in V \Rightarrow f \in C^2[x_0, x_n]$. Then V is the set of C^2 continuous functions that are defined by piecewise cubics. Show that V, with the usual addition and multiplication, is a real vector space. What is its dimension?

1.32. Show that no cubic spline exists that is identically zero outside an interval $x \in (x_{i-1}, x_{i+2})$ but is nonzero at x_i and x_{i+1}.

1.33. Solve the equation

$$f_i''(x) - \sigma^2 f_i(x)$$
$$= \frac{1}{h_i}[(m_i - \sigma^2 y_i)(x - x_{i-1}) - (m_{i-1} - \sigma^2 y_{i-1})(x - x_i)],$$

where $h_i = x_i - x_{i-1}$ and $x \in [x_{i-1}, x_i]$, $i = 1, 2, \ldots, n$, and show that the restrictions

$$f_i(x_{i-1}) = y_{i-1}, \qquad f_i(x_i) = y_i, \qquad f_{i+1}'(x_i) = f_i'(x_i),$$

for $i = 1, 2, \ldots, n - 1$, lead to a tridiagonal system of $n - 1$ equations in the $n + 1$ unknowns m_i, $i = 0, 1, \ldots, n$.

1.34. Show that $\{1, x^2, x^4\}$ is unisolvent on $[a, b]$ if and only if a and b are of the same sign. Construct a normalized basis in this case.

1.35. Let $w(x) > 0$, for all $x \in [a, b]$. Show that $\{w(x)x^i : i = 0, 1, \ldots, n\}$ is unisolvent on $[a, b]$.

1.36. Show that $\{1, x, y\}$ is not unisolvent on any open set in R^2.

1.37. Let $h = (b - a)/n$, $x_i = a + ih$ and

$$v_i(x) = \frac{h \sin[\pi(x - x_i)/h]}{\pi(x - x_i)}, \qquad i = 0, 1, \ldots, n.$$

Using the extended definition of pointwise interpolation $L_i(f) = \lim_{x \to x_i} f(x)$, show that the $\{v_i(x)\}$ is a normalized set and, hence, that $f(x) = \sum_{i=0}^n f(x_i)v_i(x)$ is the solution to the interpolation problem of finding an $f(x) = \sum_{i=0}^n a_i v_i(x)$ such that $L_i(f) = f(x_i)$.

1.38. Show that $\sum_{i=0}^n x_i^j l_i(x) = x^j$, $j = 0, 1, \ldots, n$, where the $l_i(x)$ are defined in Example 1.11.

1.39. Let x^i, $i = 0, 1, \ldots, 2n$, be $2n + 1$ distinct points, and define $v_i(x)$ by

$$v_i(x) = \frac{\prod_{\substack{j=0 \\ j \neq i}}^{2n} \sin\left(\dfrac{x - x_j}{2}\right)}{\prod_{\substack{j=0 \\ j \neq i}}^{2n} \sin\left(\dfrac{x_i - x_j}{2}\right)}, \qquad i = 0, 1, \ldots, 2n.$$

Show that these functions form a normalized basis for the interpolation problem of Exercise 1.29(b).

1.40. Determine the interval of support of the quintic B-spline (which will have C^4 continuity). You may assume the uniqueness of the quintic B-spline when it exists.

1.41. Show, by simple algebraic manipulation, that a conic through $(0, 0)$ and $(1, 1)$ with a unit slope at the origin must be a line pair.

1.42. Prove that if $x_i^2 + y_i^2 - 1 = 0$, $i = 1, 2, \ldots, 5$, and $x_6^2 + y_6^2 - 1 \neq 0$, then the pointwise interpolation problem in $P_2(x, y)$ at the six points $\{(x_i, y_i) : i = 1, 2, \ldots, 6\}$ has the finite interpolation property.

1.43. Use algebraic manipulation to show that if $n > 2$ and n collinear points lie on a conic, then the conic is a line pair.

2

Conic Sections

2.1 Introduction

The conic sections, or conics, are the simplest example of nontrivial algebraic curves, since their loci satisfy a quadratic equation in two variables. They have proved useful in engineering drawing for many decades and remain so in many applications, from computer-aided design to simple drawing programs that are popular for artistic design and business or technical charts. They are also most appropriate as curve approximates in the finite-element method. They can be represented implicitly, or explicitly in parametric form. Along with straight lines, they form the only complete class of algebraic curves that can be parametrized by rational functions – that is, only a subset of cubics and higher-order curves have rational parametrizations (see Chapters 5 and 6). Furthermore, along with the straight lines, they are the only rational curves that do not possess singular points. In this chapter we develop the pertinent geometry and discuss a few of the properties of conics.

Such is the importance of the conics in engineering drawing that nearly every textbook upon the subject devotes a sizable section to the drawing of conics. All but the most recent of these texts, however, discuss little else with regard to the subject of curves. The cycloids, the helix, the spiral of Archimedes, and some involutes will appear because of their very important, although highly specialized, applications, but are not used as general tools for curve design.

Until the advent of computer-aided design and drafting, engineering drawings were hand-drawn by highly skilled draftsmen. The process of executing a drawing must be reproducible: two different draftsmen must be able to produce the same drawings from the same given specifications. There must, then, be a procedure that enables a curve to be produced in a simple and reliable fashion and requires the use of nothing more sophisticated than a few simple drawing instruments. The conics are ideally suited to hand drawing; thus

48

the textbooks on drawing discuss conics to the almost total exclusion of anything else. In computer-aided design, the conics fall from their former state of preeminence, although they remain important design tools. The spline curves, including parametric splines, are just as simple to plot by computer. There is, however, a change of emphasis from the kind of skills required of the draftsman to those necessary for a skillful practitioner of computer-aided design. The computer programmer operates in a more abstract setting. Just consider how difficult it would be (and how voluminous) to express in writing every detail of information that passes from eye to mind to hand during the process of a single, careful sketch, and you begin to appreciate just how much detailed mathematical knowledge is necessary to specify, within the limits of a programming language, precisely what curve you need and show *a priori* that such a curve exists.

Any given computer graphics language is unlikely to provide more than one or two ways to specify a conic. One language may, perhaps, require the coefficients of the polynomial that defines the conic in an algebraic way, while another may use separate specifications for the different kinds of conics, the ellipse, for example, being specified by its center together with the major and minor axes. On the other hand, the designer may wish to be able to use a wide variety of specifications, selecting the most appropriate one in each situation. It is absolutely crucial, therefore, that we know a variety of specifications and can relate each one to the particular definition required by the computer language being used. In different applications we may wish to find conics that pass through four points and touch a given line, pass through five given points, touch four lines, have a given focus and pass through three points, pass through three points with prescribed slopes at two of the points, and so on. Do all these different sets of conditions actually specify a unique conic? We will discuss several different ways of approaching the elementary study of conics, for there is much to be gained by being aware of more than one approach. We will progress fairly quickly to the concept of a conic as the locus of points whose coordinates satisfy a quadratic equation in two variables, for it is this approach that sets the stage for our discussion of algebraic curves, which in their turn lead to rational curves and their applications to parametric splines and certain aspects of curved finite elements. Although we will discuss only a few of the more fundamental properties of conics, we will emphasize ways of defining them and will illustrate this aspect with numerous examples.

Euclid wrote four books on conics, which were to a great extent a compendium and discussion of the state of the art in his time. Aristaeus's five books on the *Elements of Conic Sections* are lost to us. Archimedes was the next major writer on the subject. What remained the definitive treatise on the subject for centuries was written by Apollonius, who was one of Archimedes's contemporaries, although a score or more years his junior. Prior to the work of Apollonius the conics had been produced by cutting a cone with a plane that was at right angles to one of the cone's generating lines. Thus the three different conic sections came from cutting three different cones, a right-angled one producing the parabola, an acute-angled cone the ellipse, and an obtuse-angled one the hyperbola. Apollonius produced all three conics from a single cone by varying the angle of the cutting plane, and it is to him, also, that we owe the current names of these conic sections. We illustrate his concept by describing the ellipse and deducing one of its most fundamental properties. Although the concept of the conic as a section of a cone is of Greek origin, the arguments of deduction based upon the introduction of spheres within the cone are much more recent, dating from the first part of the nineteenth century, and are ascribed to the Belgian mathematicians Germinal Dandelin and Adolphe Quételet.

2.2 The Ellipse

2.2.1 The Focal-Distance Property

Consider Figure 2.1, where a part of the lower half of a right circular cone is depicted. The vertex is at O, and points D and B lie on the circle formed by cutting the cone by a plane that is at right angles to the axis of the cone. We now cut the cone with a different plane, which is not parallel to the first plane but which does cut OD and OB in points that lie on the same side of O The locus of intersection of this latter plane and the cone is called an *ellipse*.

We introduce the *Dandelin spheres* within the cone, one on each side of the plane of the ellipse and of such size that each sphere touches both the plane of the ellipse and the cone, this latter contact being round the entire cone and hence producing a circle of contact. The planes of these two circles must be perpendicular to the axis of the cone. Let the smaller circle be denoted by cir_1 and the larger one by cir_2. Let the contact between the smaller sphere and the plane of the ellipse be at the point F_1, and let the contact between the larger sphere and the plane of the ellipse be at the point F_2. Let P be any point on the ellipse, and let OP meet cir_1 in Q_1 and cir_2 in Q_2.

The lines PF_1 and PQ_1 are tangents from P to the same sphere. Since $PF_1 = PQ_1$ and $PF_2 = PQ_2$, we have

$$PF_1 + PF_2 = PQ_1 + PQ_2 = Q_1 Q_2,$$

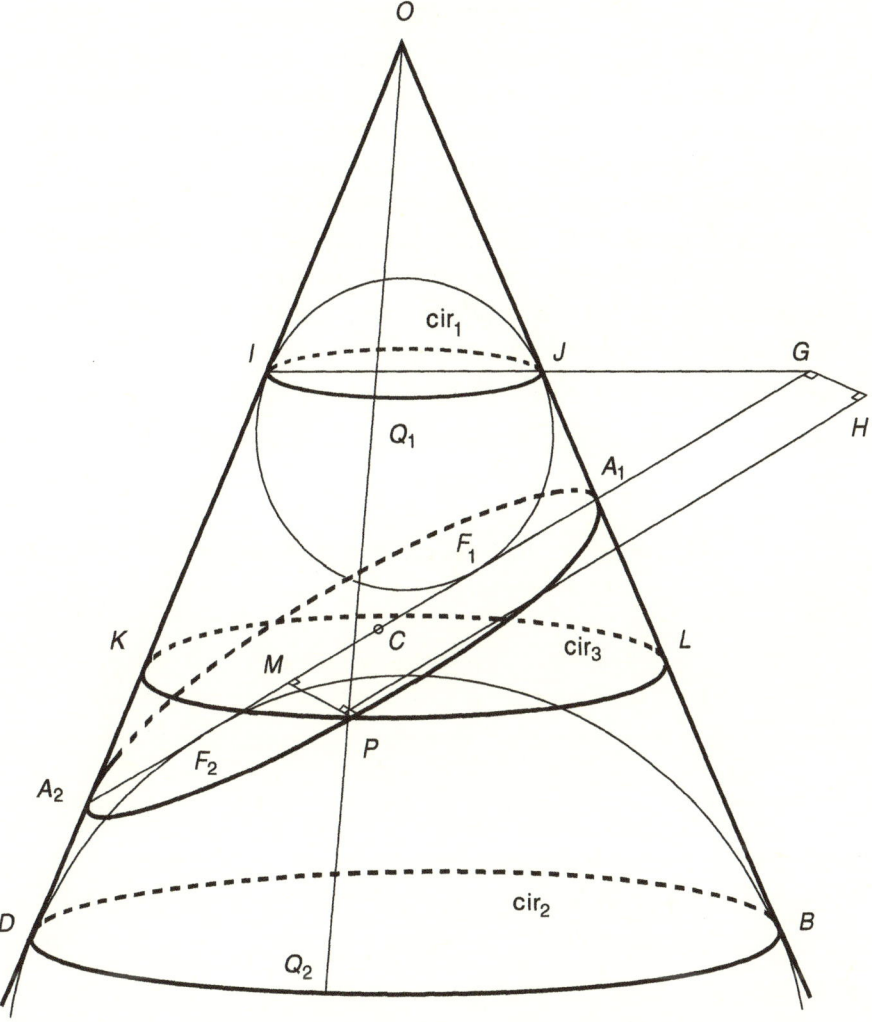

Figure 2.1. The ellipse as a section of the cone.

which is the constant distance along a *generator* (a line on the cone through
its vertex O) between the parallel circles cir_1 and cir_2. Hence, for any point P
on the ellipse the sum of the distances from P to F_1 and from P to F_2 is a
constant.

The points F_1 and F_2 are called the *foci* of the ellipse. The points A_1 and A_2
where the line $F_1 F_2$ extended meets the ellipse give us the ends of the *major
axis*. The points where the perpendicular bisector of $A_1 A_2$ meets the ellipse
are the ends of the *minor axis*. The point of intersection of the major and minor
axes is the *center* of the ellipse.

2.2.2 The Focus–Directrix Property

In Figure 2.1 let the plane through the circle cir_1 meet the plane of the ellipse in the line defined by G and H, where H has been chosen so that PH is perpendicular to GH, and G has been chosen to lie in the plane OA_1A_2, thus making GH perpendicular to A_1G. Introducing the notation $\pi(R, S, T)$ for the plane defined by three noncollinear points R, S, T, let $\pi(O, A_1, A_2)$ meet cir_1 in the points I and J. Let the plane through P perpendicular to the axis of the cone meet the cone in the circle cir_3, and let $\pi(O, A_1, A_2)$ meet cir_3 in the points K and L. Let $\pi(K, L, P)$ meet the major axis A_1A_2 in the point M. The distance PM is called the *ordinate* of P, and we have $PM \perp A_1A_2$. Therefore, $PHGM$ is a rectangle. However,

$$PF_1 = PQ_1 \quad \text{(tangents to the same sphere)},$$

and

$$PQ_1 = LJ \quad \text{(cir_1 and cir_3 both $\perp$ to cone axis)}.$$

Also,

$$PH = MG \quad \text{(\textit{PHGM} is a rectangle)};$$

therefore,

$$\frac{PF_1}{PH} = \frac{LJ}{MG}.$$

However,

$$\frac{LJ}{MG} = \frac{A_1J}{A_1G} \quad \text{(A_1ML cut by a line $JG \parallel ML$)},$$

and

$$A_1J = A_1F \quad \text{(tangents to the same sphere)};$$

therefore,

$$\frac{PF_1}{PH} = \frac{A_1F}{A_1G},$$

which is a constant independent of P.

This constant ratio is called the *eccentricity* and is usually denoted by e. It is left as an exercise (Exercise 2.2) to show that, for the ellipse, $e < 1$. The line GH is the *directrix*. For the hyperbola $e > 1$, and for the parabola $e = 1$. Many textbooks consider the focus–directrix properties, first published by Pappus of Alexandria in the late third century A.D., as defining the various conics.

2.2.3 Another Ratio Property

Returning to Figure 2.1, we illustrate another important property, also true for the hyperbola, namely that $(PM)^2/(A_1M \times MA_2)$ is a constant.

The angle LPK is a right angle (angle in a semicircle). Therefore, $\triangle LMP$ is similar to $\triangle PMK$ (three angles in common); hence

$$\frac{PM}{KM} = \frac{ML}{MP}, \qquad \text{or} \quad (PM)^2 = KM \times ML.$$

Now $\triangle A_1ML$ is similar to $\triangle A_1GJ$ ($ML \parallel JG$, and M, L, A_1, G, and J are in the same plane). Therefore,

$$\frac{ML}{A_1M} = \frac{JG}{A_1G} = \text{constant}, \qquad \text{independent of } M.$$

Also, $\triangle A_2KM$ is similar to $\triangle A_2IG$ ($KM \parallel IG$, and K, M, A_2, G, and I are in the same plane), which implies

$$\frac{KM}{MA_2} = \frac{IG}{A_2G} = \text{constant}, \qquad \text{independent of } M.$$

Hence,

$$\frac{ML}{A_1M} \times \frac{KM}{MA_2} = \frac{JG}{A_1G} \times \frac{IG}{A_2G} = \text{constant}, \qquad \text{independent of } M.$$

On substituting $(PM)^2$ for $KM \times ML$, we have the final result.

If we take P to be an extremity of the minor axis, then PM is of length equal to the length b of the semiminor axis, and $A_1M = MA_2$, both having length equal to the length a of the semimajor axis. Thus the constant ratio must be b^2/a^2. That is, for any point P on the ellipse,

$$\frac{(PM)^2}{A_1M \times MA_2} = \frac{b^2}{a^2}.$$

If we denote the center of the ellipse by C, then $A_1M = A_1C + CM$ and $MA_2 = A_1C - CM$, hence

$$\frac{(PM)^2}{(A_1C)^2 - (CM)^2} = \frac{b^2}{a^2}.$$

The following example illustrates an application of this last result. A rod of fixed length is constrained at its ends to slide along two fixed lines that are at right angles to each other. Show that any point on the sliding rod describes an ellipse.

In Figure 2.2, let the rod be of length XY and let P be an arbitrary point on the rod. We can assume, without loss of generality, that P divides XY in the ratio $k : l$, with $k > l$. Drop a perpendicular from P to the fixed line through

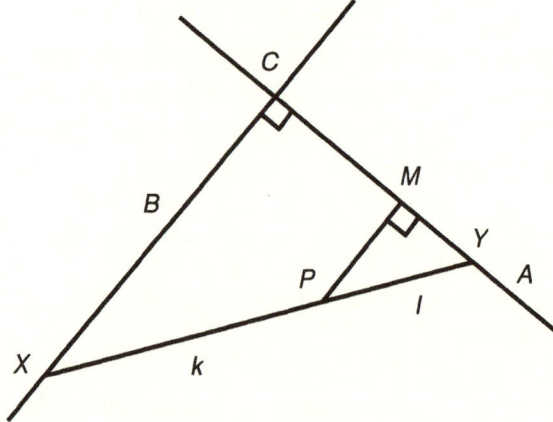

Figure 2.2. Constrained rod.

Y, and let the perpendicular meet this line at M. Let the fixed line through X meet the perpendicular fixed line through Y at C. From the symmetry of the situation, C will be the center of the locus of P. Since $PM \| XC$, we have

$$\frac{CM}{MY} = \frac{k}{l}.$$

Since $\triangle PMY$ is right-angled, $(PM)^2 + (MY)^2 = (PY)^2$. Let A be the point where the locus of P meets the fixed line through Y, and let B be that point where the locus of P meets the fixed line through X. Then $AC = ck$ and $BC = cl$, where c is some constant. Hence,

$$(AC)^2 - (CM)^2 = c^2 k^2 - \frac{k^2}{l^2}(MY)^2$$

$$= c^2 k^2 - \frac{k^2}{l^2}[(PY)^2 - (PM)^2]$$

$$= \frac{k^2}{l^2}(PM)^2 \qquad \text{(since } PY = cl).$$

Therefore,

$$\frac{(PM)^2}{(AC)^2 - (CM)^2} = \frac{l^2}{k^2} = \frac{(BC)^2}{(AC)^2},$$

showing us that the locus of P is an ellipse, PM being the ordinate and A and B being one end of the major and minor axes, respectively.

When the cutting plane is parallel to one of the generators of the cone, the conic so produced is a *parabola*, in which case we have a single focus F. If the line through F perpendicular to the major axis meets the parabola at B_1 and B_2, then $B_1 B_2$ is referred to as the *latus rectum*. The point A where the major axis meets the curve is called the *vertex*. It can be shown that, if PM is an ordinate of a parabola, then

$$(PM)^2 = 4AF \times AM \quad \text{and} \quad B_1 B_2 = 4AF.$$

2.3 Conics with Three Common Points and Common Focus

In Chapter 1 we stressed both the power and the limitations of the basic theory of the finite linear interpolation problem. Here we stress the importance of deciding whether the problem is indeed a finite linear interpolation problem or something more sophisticated. Imposing the same number of conditions as we have degrees of freedom may result in the nonexistence of a solution, whereas imposing fewer conditions, rather than providing an infinity of solutions, may give a finite number of possibilities. The following example, which was first solved by Edmund Halley, is not equivalent to a finite linear interpolation problem.

Example 2.1 *Show that, in general, there are four conics that pass through three given points and have a fourth given point as a common focus.*

In Figure 2.3 let A, B, and C be the given points on the conics, with F as common focus. For any particular value of the eccentricity e the directrix must be a distance AF/e from A and a distance BF/e from B. The directrix must, therefore, be tangent to the circles centered at A and B with radii AF/e and

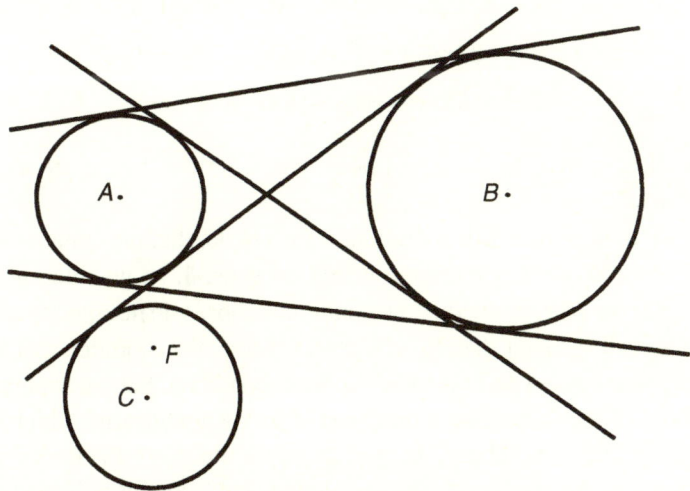

Figure 2.3. Conics with a common focus.

BF/e respectively. There are, in general, four such tangents. As e increases from zero, these tangents sweep out the plane in a continuous fashion, and hence the circle centered at C with radius CF/e (which is also changing in a continuous fashion) will become tangent to each of these four lines in turn. For these four situations we have the focus, directrix, and eccentricity, and hence a conic is specified. There are therefore, in general, four conics which satisfy the given conditions.

Let us test this conjecture by considering an algebraic example of this problem. We desire to construct conics through the three points $(1, 2)$, $(-1, \frac{3}{2})$, and $(0, 0)$, with common focus at $(1, 0)$. Since the conics pass through the origin, the directrix cannot, and we can write the equation of the directrix in the general form $ax + by + 1 = 0$. The points on the conics must, therefore, satisfy the equation

$$(x - 1)^2 + y^2 = e^2 \frac{(ax + by + 1)^2}{a^2 + b^2}. \tag{2.1}$$

We can substitute the given point coordinates in (2.1), solve one of the three resulting equations for e^2 in terms of a and b, substitute this expression into the remaining two equations, and end up with two quadratic equations in a and b. Two quadratic equations will, in general, have four solutions. Each of the four solution sets for e^2, a, and b gives us a conic with the required properties. Their equations are

$$88x^2 + 45xy - 20y^2 - 56x - 21y = 0,$$
$$20x^2 + 15xy + 12y^2 - 28x - 35y = 0,$$
$$120x^2 - 85xy - 12y^2 + 168x - 35y = 0,$$
$$12x^2 + 65xy - 60y^2 - 84x + 91y = 0.$$

These conics, as well as the given points and common focus, are shown in Figure 2.4.

Were we to impose a fifth condition, for example, that the conic must pass through one additional point, there would, in general, be no solution to this problem. On the other hand, the four conditions chosen do not specify a unique conic. We shall see later that, in an algebraic sense, the condition that a conic passes through a given point is indeed a linear condition, but the stipulation of the position of the focus is nonlinear and it is this nonlinearity which results in the multiplicity of solutions. In most practical situations it is undesirable to have either no solution or a nonunique solution, and from this point of view we say that such problems are ill posed.

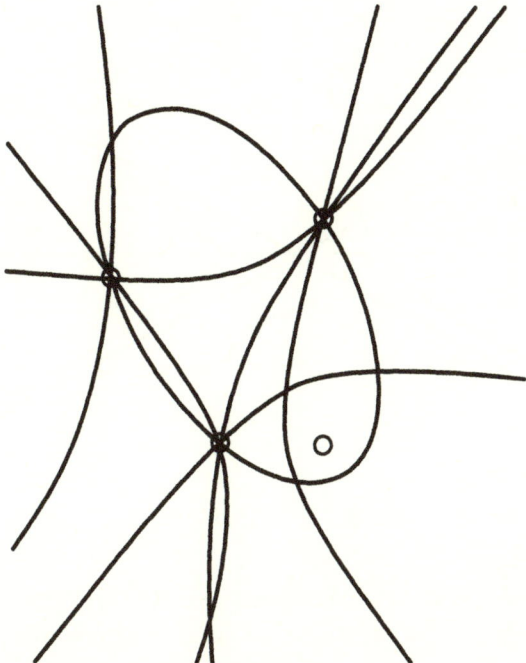

Figure 2.4. Example: conics with three common points and common focus.

2.4 The Equation of a Conic

We have stressed the importance of the ability to approach a problem in more than one way, and our examples and exercises give practice to this end. To highlight the radically different interpretations of the same equations viewed as the Greeks would have and as rendered by the analytical geometer, we use as example the equation $a^2 + b^2 = c^2$. The Greek interpretation would say that this equation stated that there exist three lengths, namely a, b, and c, such that the sum of the areas of the squares erected on a and b, respectively, equals the area of the square erected on c. In terms of analytical geometry, this equation says that the point with coordinates (a, b) lies on a circle, centred at the origin, with radius c. These are two very different interpretations. The Euclidean concepts such as area, length, angle, and order are unnecessary in the purely algebraic approach that we shall develop, but we must always remember that we are interested in real-world applications with real properties.

Example 2.2 *Defining the hyperbola as the locus of points the difference of whose distances from two fixed points is constant, obtain an equation for the curve in rectangular (Cartesian) coordinates.*

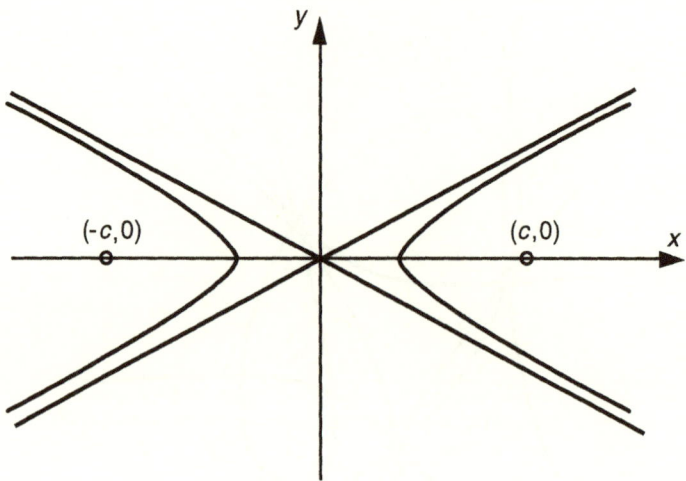

Figure 2.5. The hyperbola.

Let the two fixed points have coordinates $(-c, 0)$ and $(c, 0)$ (see Figure 2.5). Then, for any point P with coordinates (x, y) on the hyperbola, we must have

$$\sqrt{(x + c)^2 + y^2} - \sqrt{(x - c)^2 + y^2} = 2a,$$

where a is a constant. Rearranging and then squaring both sides, we get

$$cx - a^2 = a\sqrt{(x - c)^2 + y^2}.$$

Squaring again and rearranging, we finally get

$$x^2(c^2 - a^2) - y^2 a^2 = a^2(c^2 - a^2). \tag{2.2}$$

The point $(a, 0)$ clearly satisfies the required conditions provided that $|a| < |c|$. Hence, $c^2 - a^2$ is positive. Let $b^2 = c^2 - a^2$. Then the coordinates of P must satisfy the equation

$$\frac{x^2}{a^2} - \frac{y^2}{b^2} = 1.$$

This equation is usually referred to as the *standard form* of the equation for the hyperbola.

The next example is of a more general nature and is of paramount importance, as it links together three different ways of expressing the conic.

Example 2.3 *Using the focus–directrix property of a conic, deduce an equation that must be satisfied by the Cartesian coordinates of a point on the curve.*

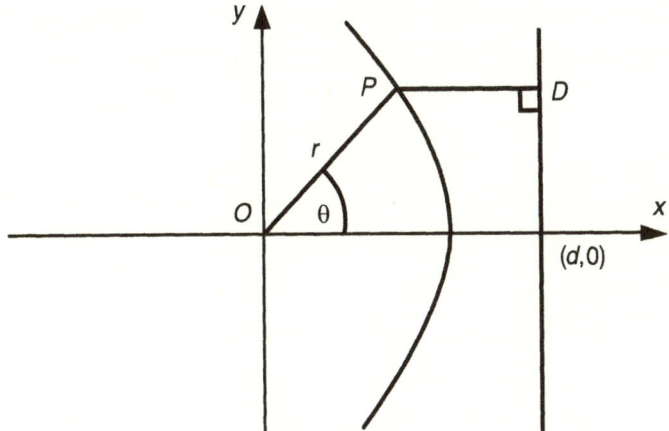

Figure 2.6. The conic in polar coordinates.

A simple algebraic manipulation will result if we initially use polar coordinates r and θ and place a focus at the origin O and the directrix at $x = d$ (see Figure 2.6). Let e denote the eccentricity. If P is on the curve,

$$PO = r \quad \text{and} \quad PD = d - r\cos\theta.$$

Therefore, $r = e(d - r\cos\theta)$, or

$$r = \frac{ed}{1 + e\cos\theta} \tag{2.3}$$

This equation is referred to as the *polar equation* of the conic. The relationship between Cartesian and polar coordinates is given by $x = r\cos\theta$, $y = r\sin\theta$. Therefore,

$$x = \frac{ed\cos\theta}{1 + e\cos\theta}, \qquad y = \frac{ed\sin\theta}{1 + e\cos\theta}. \tag{2.4}$$

These equations provide a parametric definition of the conic. This parametrization is not rational. However, upon making the substitution $t = \tan(\theta/2)$, these equations become

$$x = \frac{ed(1 - t^2)}{1 + e + t^2(1 - e)}, \qquad y = \frac{2edt}{1 + e + t^2(1 - e)}, \tag{2.5}$$

and hence all conics permit a rational parametrization. We shall return to this aspect of conics in Chapter 5.

Using the fact that $\cos\theta = x/\sqrt{x^2 + y^2}$ and eliminating $\cos\theta$ in (2.3), we obtain a Cartesian equation in the form

$$(1 - e^2)x^2 + y^2 = e^2d^2 - 2e^2dx. \tag{2.6}$$

The Cartesian equation (2.2) is more symmetric than that of (2.6), but the procedure used in Example 2.4 produced both an alternative equation in polar coordinates and an explicit parametric expression defining the curve. The algebraic approach need not be any better than the purely abstract geometrical one with which we introduced this chapter. It is an advantage to be able to see the relationships among various viewpoints in order to be able to select the most appropriate procedure for the task in hand.

2.5 The General Equation of the Second Degree

In (2.6), if $e < 1$ (the case of the ellipse), the coefficients of the x^2 and y^2 terms are of the same sign. When the curve is a hyperbola, $e > 1$ and these coefficients have different signs. For the parabola, $e = 1$ and the x^2 term vanishes. We also note that any linear transformation of the coordinate axes would still result in a quadratic equation. These facts prompt us to inquire if the locus of points whose coordinates satisfy an equation of the second degree is necessarily a conic.

The most general equation of the second degree can be written in the form

$$ax^2 + 2hxy + by^2 + 2gx + 2fy + c = 0, \tag{2.7}$$

or, alternatively, as

$$\begin{pmatrix} x & y & 1 \end{pmatrix} \begin{pmatrix} a & h & g \\ h & b & f \\ g & f & c \end{pmatrix} \begin{pmatrix} x \\ y \\ 1 \end{pmatrix} = 0. \tag{2.8}$$

This equation represents an ellipse if $h^2 - ab < 0$, a parabola if $h^2 - ab = 0$, and a hyperbola if $h^2 - ab > 0$. The expression $h^2 - ab$ is called the *discriminant* of the quadratic equation. The circle is a special case of the ellipse. Relative to rectangular axes, the circle corresponds to $a = b$ and $h = 0$. The product of two first-degree polynomials, with each factor representing a line, will be a second-degree polynomial. Hence a pair of lines (a line pair) belongs to the loci of points whose coordinates satisfy a quadratic equation. The condition for the general equation to represent a line pair is that

$$\det \begin{pmatrix} a & h & g \\ h & b & f \\ g & f & c \end{pmatrix} = 0.$$

We will consider a line pair, including the case of a repeated line (the two lines of the pair are identical), as a special case of a conic and refer to it as a *degenerate* or *reducible* conic. Hence, the general equation of the second

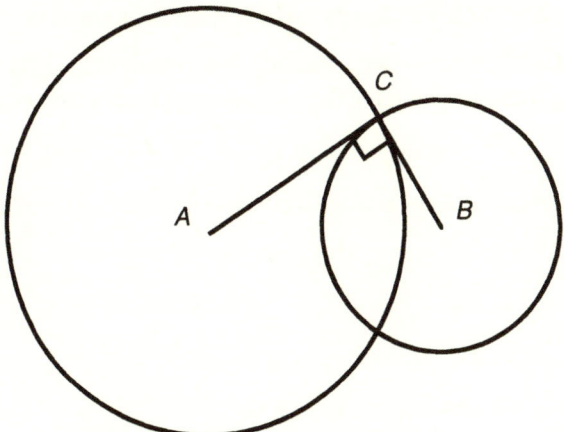

Figure 2.7. Circles meeting at right angles.

degree will always represent a conic. This equation is referred to as the *point equation* of the conic.

Example 2.4 *Find the algebraic condition for two circles to cut each other at right angles.*

The situation is depicted in Figure 2.7, where A and B are the centers of the two circles which meet at C, with the angle ACB a right angle. Then $(AB)^2 = (AC)^2 + (CB)^2$. If the circles have equations

$$x^2 + y^2 + 2gx + 2fy + c = 0 \quad \text{and} \quad x^2 + y^2 + 2g'x + 2f'y + c' = 0,$$

the centers are $(-g, -f)$ and $(-g', -f')$. Writing the equations in the form

$$(x + g)^2 + (y + f)^2 = g^2 + f^2 - c \quad \text{and}$$
$$(x + g')^2 + (y + f')^2 = g'^2 + f'^2 - c',$$

we see that the squares of the two radii are $g^2 + f^2 - c$ and $g'^2 + f'^2 - c'$, respectively. Hence, the algebraic condition required must be

$$(g - g')^2 + (f - f')^2 = g^2 + f^2 - c + g'^2 + f'^2 - c',$$

which reduces to

$$2gg' + 2ff' = c + c'.$$

2.6 Parallel Chords and Conjugate Diameters

The change of coordinate system obtained by a translation of the origin to the point (x_0, y_0) gives an equation of the same general form with new coefficients,

$$a'x^2 + 2h'xy + b'y^2 + 2g'x + 2f'y + c' = 0,$$

where the (x, y) coordinates refer to the new coordinate system, and

$$a' = a, \qquad b' = b, \qquad h' = h,$$

$$f' = hx_0 + by_0 + f, \qquad g' = ax_0 + hy_0 + g, \qquad (2.9)$$

$$c' = ax_0^2 + 2hx_0y_0 + by_0^2 + 2gx_0 + 2fy_0 + c.$$

If this is written in polar coordinates, we get

$$(a' \cos^2 \theta + 2h' \cos \theta \sin \theta + b' \sin^2 \theta)r^2$$

$$+ 2(g' \cos \theta + f' \sin \theta)r + c' = 0. \qquad (2.10)$$

For any point (x_0, y_0), if a line through this point has slope defined by

$$g' \cos \theta + f' \sin \theta = 0,$$

the coefficient of r vanishes and the two solutions of (2.10) are equal and of opposite sign. Therefore, this line meets the conic in two points symmetrically placed with respect to (x_0, y_0). Hence, through any point there is a chord of the conic which is bisected at that point. By a similar argument, we see that if $g' = f' = 0$ then, in general, *every* chord through (x_0, y_0) would be bisected there. Solving these equations for x_0 and y_0, we obtain

$$x = \frac{hf - bg}{ab - h^2}, \qquad y = \frac{hg - af}{ab - h^2}, \qquad (2.11)$$

which must then (see Exercise 2.15) be the coordinates of the *center* of the conic. Neither x_0 nor y_0 is defined when the discriminant $h^2 - ab = 0$ (i.e., in the case of the parabola), and we deduce that a *parabola has no center*. We will remove this restriction in Chapter 4 when we develop homogenous coordinates for projective, rather than Euclidean, two-space.

   If we think of a new origin as a variable point (x, y) but hold θ fixed, considering parallel chords through the variable point (x, y), we see that these chords will be bisected at (x, y) if $g' \cos \theta + f' \sin \theta = 0$, that is, if

$$(ax + hy + g) \cos \theta + (hx + by + f) \sin \theta = 0,$$

which must be the equation of the locus of midpoints of the chords with slope $\tan \theta$. Regardless of the value of θ, all these loci pass through the point given by

$$ax + hy + g = 0, \qquad hx + by + f = 0,$$

which is, as we have seen, the center of the conic. Hence, the locus of midpoints of parallel chords passes through the center. The locus is called a *diameter*, and the lines it bisects are called *ordinate lines*. The *ordinate* of a point on the conic with respect to a diameter is the distance along the ordinate line from the point to the diameter. For the parabola, since $h^2 - ab = 0$, the lines $ax + hy + g = 0$ and $hx + by + f = 0$ are parallel and hence all diameters are parallel, with slopes given by $\tan \phi = -a/h = -h/b$.

If $\tan \phi$ is the slope of a diameter bisecting chords with slope $\tan \theta$, we see that

$$\tan \phi = -\frac{a + h \tan \theta}{h + b \tan \theta},$$

or

$$b \tan \theta \tan \phi + h (\tan \theta + \tan \phi) + a = 0. \qquad (2.12)$$

From the symmetry of this equation, we see that in this case the diameter with slope $\tan \theta$ will also bisect chords with slopes $\tan \phi$. Two diameters whose slopes satisfy the above equation, and hence for which each diameter bisects chords parallel to the other, are referred to as *conjugate diameters*. The following example gives some additional practice and, via the analytical approach, generalizes the ratio result given in Section 2.2.3.

Example 2.5 *Let O be any point, and let two chords AB and CD of a conic be drawn through O (Figure 2.8). Show that*

$$\frac{OA \times OB}{OC \times OD},$$

is independent of the point O, provided that the directions of the chords remain fixed. Use this result to show that, if N is any other point and the chord EF

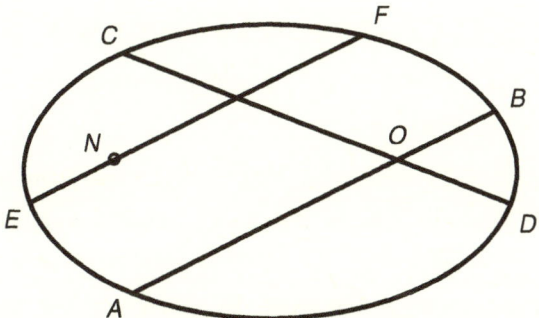

Figure 2.8. Intersecting chords of a conic.

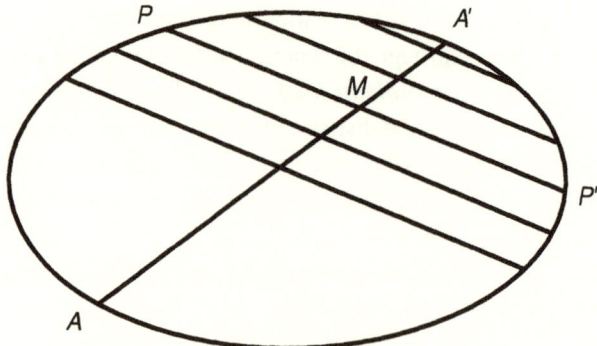

Figure 2.9. Intersecting chords of a conic.

parallel to AB is drawn through N then the ratio

$$\frac{OA \times OB}{NE \times NF},$$

is independent of the direction of the chord, provided that the points N and O remain fixed. Finally, show that if P is any point of the conic and PM is an ordinate to some diameter AA' (Figure 2.9), then

$$\frac{(PM)^2}{AM \times MA'}$$

is independent of P.

Consider the general equation of the conic written in polar form with O as the origin of coordinates. Then, as in (2.10), we have an equation of the form

$$(a\cos^2\theta + 2h\cos\theta\sin\theta + b\sin^2\theta)r^2 + 2(g\cos\theta + f\sin\theta)r + c = 0,$$

from which we see that, if r_1 and r_2 are the roots, then their product is

$$r_1 r_2 = \frac{c}{a\cos^2\theta + 2h\cos\theta\sin\theta + b\sin^2\theta}.$$

Hence, if AB and CD make angles ϕ and ψ, respectively, with the x-axis through O, then

$$\frac{OA \times OB}{OC \times OD} = \frac{a\cos^2\psi + 2h\cos\psi\sin\psi + b\sin^2\psi}{a\cos^2\phi + 2h\cos\phi\sin\phi + b\sin^2\phi}.$$

If we move to parallel axes anywhere else, then a, h, and b remain fixed, and hence, since ϕ and ψ are given as fixed, the ratio is constant. Since $EF \| AB$,

$$NE \times NF = \frac{c'}{a\cos^2\theta + 2h\cos\theta\sin\theta + b\sin^2\theta}.$$

Therefore,

$$\frac{OA \times OB}{NE \times NF} = \frac{c}{c'}.$$

Finally, draw a diameter AA' (Figure 2.9), and let PM be the ordinate. Extend PM along the ordinate line to P'. Then

$$\frac{PM \times MP'}{AM \times MA'}$$

is constant. Since $PM = MP'$, this gives us that

$$\frac{(PM)^2}{AM \times MA'},$$

is a constant, with that particular constant depending on ϕ and ψ. Holding the diameter AA' fixed, we clearly hold one of the angles, for example ϕ fixed. Also, as P moves on the conic, the angle between PM and AA' remains fixed. Therefore, ψ is also fixed and we have the desired result. This extends the result of Section 2.2.3 to the case of any diameters, not just perpendicular ones.

Example 2.6 *In Figure 2.10 a semicircle is drawn with AB as diameter, and a line LA making an angle α with AB is drawn such that LB is perpendicular to AB. X is a variable point on AB, and XP is drawn perpendicular to AB, meeting LA in M and the semicircle in Y and such that $PM = XY$ Show that P lies on a conic.*

Since triangle AYB is right-angled (angle in a semicircle), $\triangle AXY$ and $\triangle YXB$ are similar. This implies that

$$\frac{XY}{BX} = \frac{AX}{XY},$$

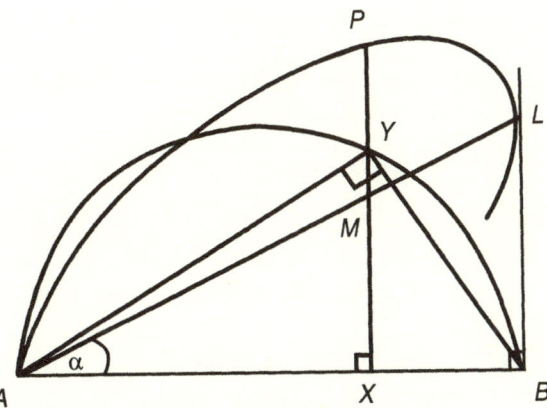

Figure 2.10. A conic construction.

or

$$(XY)^2 = AX \times BX.$$

However,

$$PM = XY;$$

therefore,

$$(PM)^2 = AX \times BX.$$

Also,

$$AX = AM \cos \alpha \quad \text{and} \quad BX = ML \cos \alpha.$$

Therefore,

$$\frac{(PM)^2}{AM \times ML} = \cos^2 \alpha.$$

Therefore, P lies on a conic with LA as a diameter, and conjugate diameter perpendicular to AB.

2.7 Central Conics

From the algebraic method of defining the center we see that, upon moving the origin of the axes to the center of the conic, we obtain the equation of the conic in the form

$$ax^2 + 2hxy + by^2 + c' = 0. \tag{2.13}$$

A conic written in this form is called a *central conic*. The c', as in (2.10), is given by

$$c' = ax_0^2 + 2hx_0y_0 + by_0^2 + 2gx_0 + 2fy_0 + c$$
$$= (ax_0 + hy_0 + g)x_0 + (hx_0 + by_0 + f)y_0 + gx_0 + fy_0 + c,$$

where (x_0, y_0) are the original coordinates of the center. Substituting the values for the coordinates of the center, we find that

$$(ab - h^2)c' = g(hf - bg) + f(hg - af) + c(ab - h^2)$$
$$= \det \begin{pmatrix} a & h & g \\ h & b & f \\ g & f & c \end{pmatrix}.$$

Example 2.7 *Find the condition for the equation $Ax^2 + 2Hxy + By^2 = 0$ to represent conjugate diameters to a central conic.*

Let the diameters have equations $y = mx$ and $y = m'x$. From (2.12) we know that, if the conjugate diameters have slopes $\tan\theta$ and $\tan\phi$, then

$$b\tan\theta\tan\phi + h(\tan\theta + \tan\phi) + a = 0,$$

which, in our present notation, is

$$bmm' + h(m + m') + a = 0,$$

where the equation of the conic is $ax^2 + 2hxy + by^2 + c = 0$. The product of the conjugate diameters is

$$(y - mx)(y - m'x) = y^2 - (m + m')xy + mm'x^2 = 0.$$

Hence,

$$mm' = \frac{A}{B} \quad \text{and} \quad m + m' = -\frac{2H}{B}.$$

Therefore,

$$b\frac{A}{B} - h\frac{2H}{B} + a = 0, \quad \text{or} \quad bA + aB - 2hH = 0.$$

Example 2.8 *Find the diameter conjugate to $y = 0$ with respect to the central conic $x^2 + 2xy + 4y^2 - 1 = 0$.*

Let the required diameter have equation $y = mx$. Then the pair of conjugate diameters can be written as

$$y^2 - mxy = 0.$$

In the notation of Example 2.7 we have $a = 1$, $b = 4$, $h = 1$, $A = 0$, $B = 1$, and $H = -m/2$. Then $1 - 2(-m/2) = 0$, giving $x + y = 0$ as the diameter conjugate to $y = 0$.

If the axes are conjugate diameters then, since their equation is $xy = 0$, we have $A = B = 0$. Hence, $h = 0$ and the conic must be of the form

$$ax^2 + by^2 + c = 0.$$

Hence, rewriting the equation of the conic with reference to any pair of conjugate diameters as the coordinate axes will result in an equation which does not possess an xy term. The angles θ and ϕ made with an axis by two conjugate diameters are connected by the relation (2.12), namely

$$b\tan\theta\tan\phi + h(\tan\theta + \tan\phi) + a = 0.$$

If these diameters are perpendicular to each other, then $\tan\phi\tan\theta = -1$, so that

$$h\tan^2\theta + (a - b)\tan\theta - h = 0.$$

Using the fact that $\tan\phi = y/x$, this equation can be written in Cartesian form as

$$hx^2 - (a - b)xy - hy^2 = 0,$$

which represents a real line pair through the origin. Hence, there is always precisely one pair of conjugate diameters that cut each other at right angles. These diameters are called the *axes* of the central conic, and the points where they meet the conic are called the *vertices*. Recall that the conjugate diameters pass through the center. Hence, the axes must pass through the center. Since the center of a parabola is not defined in the Euclidean plane, the parabola cannot have a pair of axes. Its axis is defined to be that diameter (recall that all the diameters of a parabola are parallel) for which the corresponding bisected chords (ordinate lines) are perpendicular to the diameter.

Example 2.9 *Find the axes of the conic* $2x^2 + 4xy - y^2 + 1 = 0$.

Here we have $a = 2$, $b = -1$, and $h = 2$, and hence the axes are given by $2x^2 - 3xy - 2y^2 = 0$, that is,

$$2x + y = 0 \quad \text{and} \quad x - 2y = 0.$$

The conic can be written in the form

$$3(2x + y)^2 - 2(x - 2y)^2 + 5 = 0.$$

2.8 The Number of Degrees of Freedom

In Chapter 1 we drew a connection between the interpolation problem in $P_2(x, y)$ and the geometrical problem of defining conics that satisfy certain conditions. It is important to be able to determine if the desired conditions determine a unique conic and, in the cases where they do, to produce some equation for the curve.

The general quadratic equation in two variables has six terms and, since any multiple of the equation represents the same conic, we conclude that the general conic has five degrees of freedom. The equation is (2.7), namely,

$$ax^2 + 2hxy + by^2 + 2gx + 2fy + c = 0.$$

Since at least one of the coefficients must be nonzero, we can normalize that coefficient to unity and be left with five coefficients. Since we do not know in advance which of the coefficients are nonzero, it is a better policy, in general, to solve for five of the coefficients in terms of the sixth. Imposing five conditions in order to determine the coefficients will, in general, only have a chance of success if the conditions imposed are linearly independent and result in linear conditions on the coefficients. The parabola, with $h^2 - ab = 0$, has only four degrees of freedom, and we might think that there is a unique parabola through four points. However, the condition for a conic to be a parabola is nonlinear in the coefficients, and there are, in general, two parabolas through four points (see Exercise 28).

If we demand that a conic pass through the point (x_1, y_1), we have to satisfy the linear condition

$$ax_1^2 + 2hx_1y_1 + by_1^2 + 2gx_1 + 2fy_1 + c = 0,$$

which leads us to the following example.

Example 2.10 *Construct the conic through five given points* (x_i, y_i) *for* $i = 1, 2, \ldots, 5$.

We have the five equations

$$ax_i^2 + 2hx_iy_i + by_i^2 + 2gx_i + 2fy_i + c = 0, \qquad i = 1, 2, \ldots, 5,$$

or, in matrix form,

$$\begin{pmatrix} x_1^2 & 2x_1y_1 & y_1^2 & 2x_1 & 2y_1 \\ x_2^2 & 2x_2y_2 & y_2^2 & 2x_2 & 2y_2 \\ \vdots & \vdots & \vdots & \vdots & \vdots \\ \vdots & \vdots & \vdots & \vdots & \vdots \\ x_5^2 & 2x_5y_5 & y_5^2 & 2x_5 & 2y_5 \end{pmatrix} \begin{pmatrix} a \\ h \\ b \\ g \\ f \end{pmatrix} = -c \begin{pmatrix} 1 \\ 1 \\ 1 \\ 1 \\ 1 \end{pmatrix}.$$

At first sight it may seem appropriate to consider the solution of this system, but further inspection shows the fallacy of this conjecture. Even supposing that the five given points do indeed uniquely determine a conic, there is no guarantee that the matrix on the left of the above equation is nonsingular. Certainly there must be one of the coefficients in the equation of the conic that will be nonzero, but we do not know *a priori* which one. This difficulty can be overcome by posing the problem in the strict framework of the finite linear interpolation problem. In this approach we seek to determine a polynomial [a vector in $P_2(x, y)$] rather than a conic. The polynomial has six degrees of freedom, and hence we must impose a sixth linear condition. We saw in Example 1.18 and Exercise 1.43 that a conic with three points in common with a straight line must have that line as a factor. Therefore, the conic defined by the five given points cannot pass through any point collinear with any pair from these five points. Let us take a sixth point distinct from the given five and collinear with two of them. Let us consider the finite linear interpolation problem of determining a polynomial of degree two that assumes zero value at the given five points and any nonzero value, say unity, at the sixth point. Theorem 1.9 guarantees a unique solution to this problem, and the locus of all zeros of the polynomial so determined is the conic sought. We have thus circumvented any possibility of having to deal with a singular matrix. The system of equations in this case is

$$\begin{pmatrix} x_1^2 & 2x_1y_1 & y_1^2 & 2x_1 & 2y_1 & 1 \\ x_2^2 & 2x_2y_2 & y_2^2 & 2x_2 & 2y_2 & 1 \\ \vdots & \vdots & \vdots & \vdots & \vdots & \vdots \\ \vdots & \vdots & \vdots & \vdots & \vdots & \vdots \\ x_5^2 & 2x_5y_5 & y_5^2 & 2x_5 & 2y_5 & 1 \\ x_6^2 & 2x_6y_6 & y_6^2 & 2x_6 & 2y_6 & 1 \end{pmatrix} \begin{pmatrix} a \\ h \\ b \\ g \\ f \\ c \end{pmatrix} = \begin{pmatrix} 0 \\ 0 \\ 0 \\ 0 \\ 0 \\ 1 \end{pmatrix}.$$

2.8.1 Pencils. The Conic through Five Points

Although the procedure described above is straightforward, it is cumbersome and gives little geometrical insight into the problem at hand. We now develop a more efficient procedure for solving this problem.

Assume that, in general, two conics have four points in common. Let $C_1 = 0$ and $C_2 = 0$ be the equations of two conics. We will from now on often refer to C_1 and C_2 as being conics, where a more precise terminology would demand the statements that $C_1 = 0$ and $C_2 = 0$ are the equations of conics. There are then four points common to C_1 and C_2. The equation

$$C_1 + \lambda C_2 = 0,$$

is quadratic and hence must represent a conic. For any value of the parameter λ this equation will be satisfied by the coordinates of each of the four points common to C_1 and C_2 and hence must represent a one-parameter family of conics through the four common points of C_1 and C_2. We saw that demanding that a conic pass through a point imposes one linear condition on the coefficients defining the equation of the conic. Hence, forcing the conic to pass through four points will, in general, impose four linear conditions. However, the conic has only five degrees of freedom, so there can only be a one-parameter family of conics through four points. That is, any conic through the four points common to C_1 and C_2 can be written in the form $C_1 + \lambda C_2 = 0$. A one-parameter family of curves is called a *pencil*. We now return to Example 2.10.

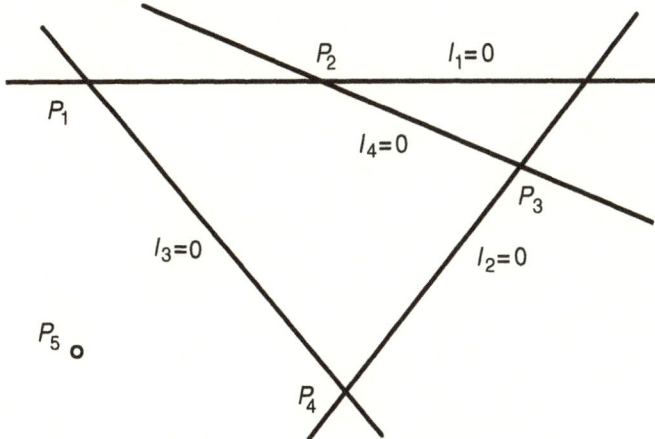

Figure 2.11. A pencil of conics through four points: construction.

Example 2.10

Alternative solution. Construct the conic through the five points $(x_i, y_i), i = 1, 2, \ldots, 5$.

Let P_i be the point with coordinates $(x_i, y_i), i = 1, 2, \ldots, 5$, and let $l_i = 0, i = 1, 2, 3, 4$, be the lines defined by the points $P_1 P_2$, $P_3 P_4$, $P_1 P_4$, and $P_2 P_3$, respectively (see Figure 2.11). Then $l_1 l_2 + \lambda l_3 l_4 = 0$ is a pencil of conics through the points P_1, P_2, P_3, and P_4. If we now require that this conic pass through P_5, then

$$\lambda = -\frac{l_1(x_5, y_5) l_2(x_5, y_5)}{l_3(x_5, y_5) l_4(x_5, y_5)},$$

and we have the equation of the conic without any cumbersome elimination on 5×5 or 6×6 matrices.

Example 2.11 *Construct the pencil of conics through the four points $(-1, -1)$, $(0, -2)$, $(1, 2)$, and $(-2, 5)$.*

Let $l_1 = 0$ be the line through $(-1, -1)$ and $(0, -2)$, $l_2 = 0$ the line through $(0, -2)$ and $(1, 2)$, $l_3 = 0$ the line through $(1, 2)$ and $(-2, 5)$, and $l_4 = 0$ the line through $(-1, -1)$ and $(-2, 5)$. We use the line pairs $l_1 l_3$ and $l_2 l_4$ in the construction, and the pencil is given by $l_1 l_3 + \lambda l_2 l_4 = 0$ or, in expanded form, by

$$(x + y + 2)(x + y - 3) + \lambda(4x - y - 2)(6x + y + 7) = 0.$$

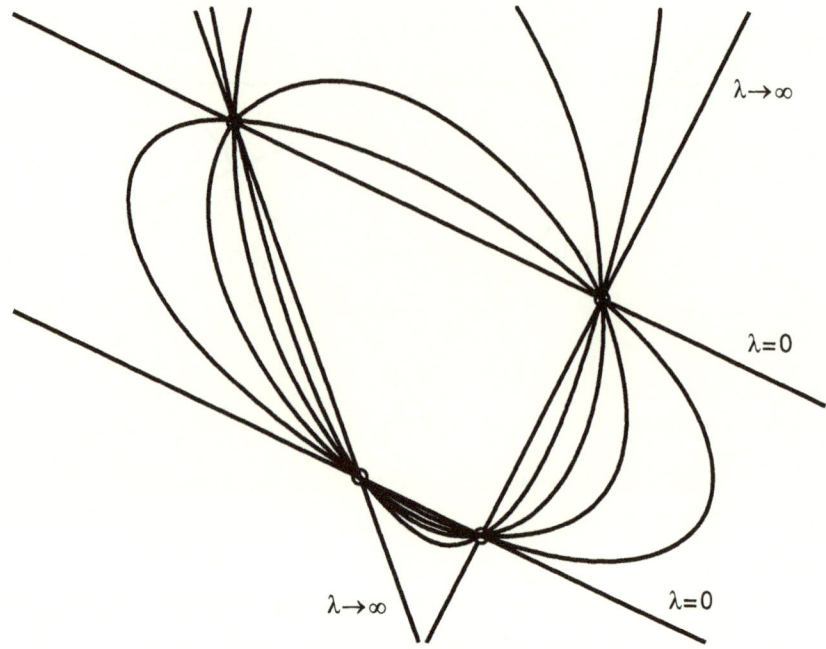

$\lambda \to \infty$

$\lambda = 0$

$\lambda \to \infty$ $\lambda = 0$

Figure 2.12. A pencil of conics through four points: example.

The four common points of the pencil, the line pairs (i.e., the conics for the parameter values $\lambda = 0$ and $\lambda \to \infty$), and other members of the pencil are shown in Figure 2.12.

Example 2.12 *Show that, in general, there are three values of λ which result in the conic $C_1 + \lambda C_2 = 0$ becoming a line pair.*

Since the pencil represents the entire system of conics through the four points common to the two conics, say the points P_1, P_2, P_3, and P_4, and since the three line pairs formed by $(P_1 P_2)(P_3 P_4)$, $(P_1 P_3)(P_2 P_4)$, and $(P_1 P_4)(P_2 P_3)$ represent conics through these four points, they must belong to the pencil, each one corresponding to some value of λ.

Alternatively, the condition for a conic to be a line pair was given as

$$\det \begin{pmatrix} a & h & g \\ h & b & f \\ g & f & c \end{pmatrix} = 0.$$

This equation results, in general, in a cubic equation in λ, giving three roots if we include complex roots and the multiplicities of repeated roots.

Example 2.13 *Find the equation of the locus of centers of conics through four points.*

Construct the pencil of conics through the four points, with an equation of the form

$$(a + \lambda a')x^2 + 2(h + \lambda h')xy + (b + \lambda b')y^2 + 2(g + \lambda g')x$$
$$+ 2(f + \lambda f')y + (c + \lambda c') = 0.$$

The coordinates of the center of this conic must satisfy the equations

$$(a + \lambda a')x + (h + \lambda h')y + (g + \lambda g') = 0,$$
$$(h + \lambda h')x + (b + \lambda b')y + (f + \lambda f') = 0.$$

Eliminating λ, we get

$$(ax + hy + g)(h'x + b'y + f') - (hx + by + f)(a'x + h'y + g') = 0,$$

which is itself the equation of a conic.

As a numerical illustration, we consider the pencil of conics through the points $(0, 0)$, $(1, 0)$, $(1, 1)$, and $(-1, 2)$ (Figure 2.13) with equation

$$(x^2 - y^2 - x + y) + \lambda(2x^2 + xy - 2x - y) = 0.$$

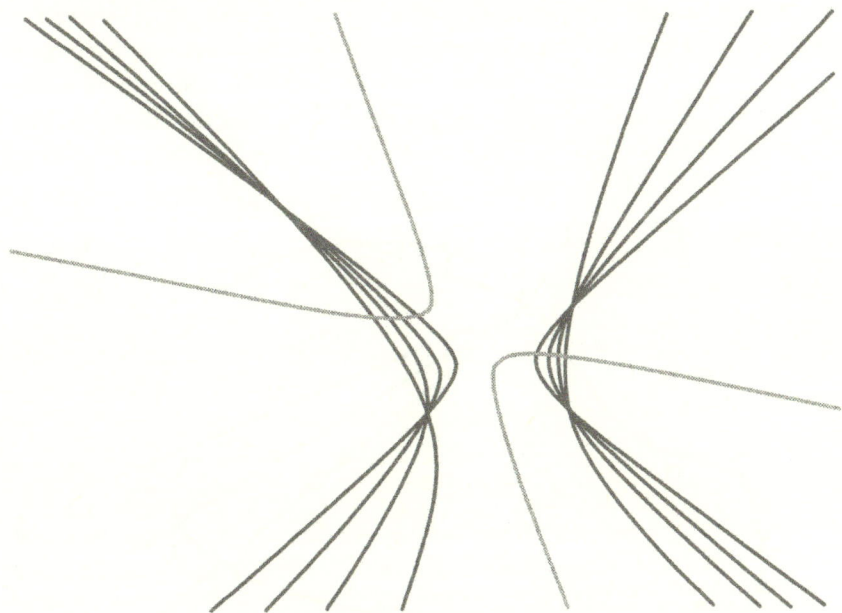

Figure 2.13. The locus of centers of a conic pencil.

The coordinates of the center must satisfy the equations

$$2x - 1 + \lambda(4x + y - 2) = 0,$$

$$-2y + 1 + \lambda(x - 1) = 0,$$

and elimination of the parameter λ results in the locus of the centers, namely

$$2x^2 + 8xy + 2y^2 - 7x - 5y + 3 = 0.$$

This hyperbola is also shown in Figure 2.13. The problem of obtaining parametric equations for the center is left for Exercise 2.30.

2.9 The Tangent to a Curve

The usual way of thinking of a tangent to a curve is as the line whose slope is the same as the slope of the curve at some point common to both. This approach would require the calculation of the slope of the curve before the equation of the tangent could be found. An alternative way of defining a tangent might be as a line that meets the curve in at least two coincident points. This latter approach, in a modified and more precise form, is the way we will ultimately define tangents to algebraic curves in Chapter 5.

Consider Figure 2.14. Let A and B_1 be two points on a curve, and draw the line AB_1. Consider a sequence of points $B_2, B_3, \dots$, each on the curve and each one closer to A than the previous point (in the sense of arc length along the curve). For each B_i there is a unique line AB_i, which, by construction, meets the curve in the points A and B_i (and perhaps other points in addition to these

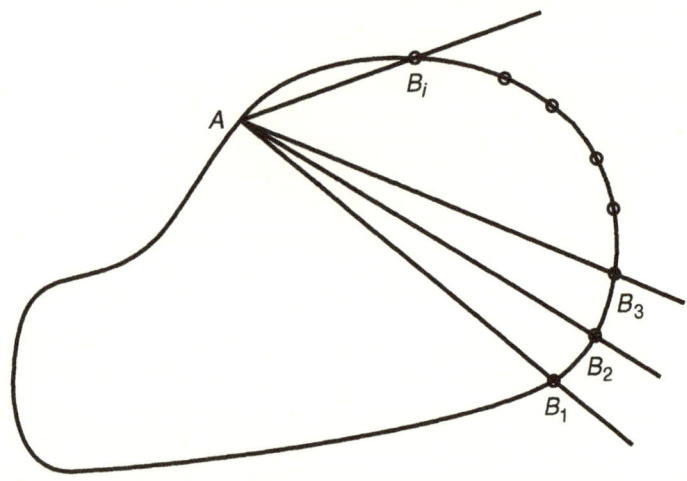

Figure 2.14. The tangent as a limiting chord.

two). If, in the limit as B_i tends to A, the chord AB_i is still well defined, then we shall have a line that meets the curve twice at the point A.

Example 2.14 *By considering the tangent as the limiting case of a chord as two points of contact coalesce, find the equation of the tangent to the curve $y = x^2$ at the point* $(1, 1)$.

Let (x_i, y_i) be a point on the curve other than the point $(1, 1)$. Then the chord has equation

$$y - 1 = \frac{y_i - 1}{x_i - 1}(x - 1).$$

Since (x_i, y_i) lies on the curve, we know that $y_i = x_i^2$. Therefore, the equation of the chord can be written as

$$y - 1 = (x_i + 1)(x - 1),$$

or

$$y - 1 - (x_i + 1)(x - 1) = 0.$$

Then

$$\lim_{x_i \to 1} [y - 1 - (x_i + 1)(x - 1)] = y - 2x + 1.$$

Therefore, the equation of the tangent is $y - 2x + 1 = 0$.

This solution paves the way for the following purely algebraic solution.

Example 2.14
Alternative solution. An arbitrary line through the point $(1, 1)$ has equation $y - 1 = m(x - 1)$, and will meet the curve $y = x^2$ where

$$x^2 - mx + m - 1 = 0.$$

This quadratic will have coincident roots if and only if

$$m^2 - 4(m - 1) = 0,$$

that is,

$$(m - 2)^2 = 0,$$

which implies that the slope of the tangent has the value $m = 2$, and the required tangent has the equation $y - 1 = 2(x - 1)$.

It is left as an exercise (Exercise 2.35) to show that the equation of the tangent at the point (x_1, y_1) to the general conic (2.7) is

$$axx_1 + h(xy_1 + yx_1) + byy_1 + g(x + x_1) + f(y + y_1) + c = 0, \quad (2.14)$$

or, alternatively,

$$(x_1 \quad y_1 \quad 1) \begin{pmatrix} a & h & g \\ h & b & f \\ g & f & c \end{pmatrix} \begin{pmatrix} x \\ y \\ 1 \end{pmatrix} = 0.$$

2.10 Conics That Touch Given Lines

It is common in curve design to require that the curve must touch one or more given lines. We examine, therefore, what kind of condition is imposed on the equation of the conic by such a restriction.

2.10.1 Conics through Four Points Touching a Given Line

Since, in general, there is a one-parameter family of conics through four points, it seems reasonable that imposing the further condition that the curve touches a particular line would then define a unique conic.

The equation of the tangent to the general conic at a point (x_1, y_1) is given by (2.14), namely

$$axx_1 + h(xy_1 + yx_1) + byy_1 + g(x + x_1) + f(y + y_1) + c = 0.$$

Since the conic is required to be tangent to the line $\alpha x + \beta y + \gamma = 0$, it must touch it somewhere, say at (x_1, y_1). Then the two equations must represent the same line, and hence,

$$\frac{\alpha}{ax_1 + hy_1 + g} = \frac{\beta}{hx_1 + by_1 + f} = \frac{\gamma}{gx_1 + fy_1 + c}.$$

Furthermore, we know that the point (x_1, y_1), wherever it is, must lie on the line $\alpha x + \beta y + \gamma = 0$. Therefore,

$$\alpha x_1 + \beta y_1 + \gamma = 0.$$

We have deduced three equations involving the coordinates x_1 and y_1 under the assumption only that the conic touched a given line. We now eliminate x_1 and y_1 among these three equations. After a great deal of algebraic manipulation, we find that the final condition for the general conic to touch the line $\alpha x + \beta y + \gamma = 0$ is

$$A\alpha^2 + B\beta^2 + C\gamma^2 + 2F\beta\gamma + 2G\alpha\gamma + 2H\alpha\beta = 0, \qquad (2.15)$$

where $A, B, C, \ldots$ are the cofactors of $a, b, c, \ldots$ in the coefficient matrix

$$\begin{pmatrix} a & h & g \\ h & b & f \\ g & f & c \end{pmatrix}.$$

The equation (2.15) is referred to both as the *tangential equation* and as the *line equation* of the conic. The cofactors are quadratic in the coefficients that determine the general conic, and hence, without further restriction, the problem of defining a conic touching a certain given line is a nonlinear interpolation problem. Since there is a pencil of conics through four points, there will be a quadratic equation imposed on the single remaining parameter in order that a member of the pencil shall touch a given line. There will be, in general, two solutions to this equation, and hence we conclude that there are, in general, two conics through four points and touching a given line.

Example 2.15 *Find all conics through the points* $(-1, -1)$, $(-1, 1)$, $(1, -1)$, *and* $(1, 1)$ *and touching the line* $y = x + \sqrt{6}$.

The pencil of conics through the four points can be written in the form

$$x^2 - 1 + \lambda(y^2 - 1) = 0.$$

The condition that $y = x + \sqrt{6}$ is a tangent reduces to

$$\lambda^2 - 4\lambda + 1 = 0, \qquad \text{whence} \quad \lambda = 2 \pm \sqrt{3}.$$

Hence, the conics

$$x^2 - 1 + (2 + \sqrt{3})(y^2 - 1) = 0 \quad \text{and}$$
$$x^2 - 1 + (2 - \sqrt{3})(y^2 - 1) = 0,$$

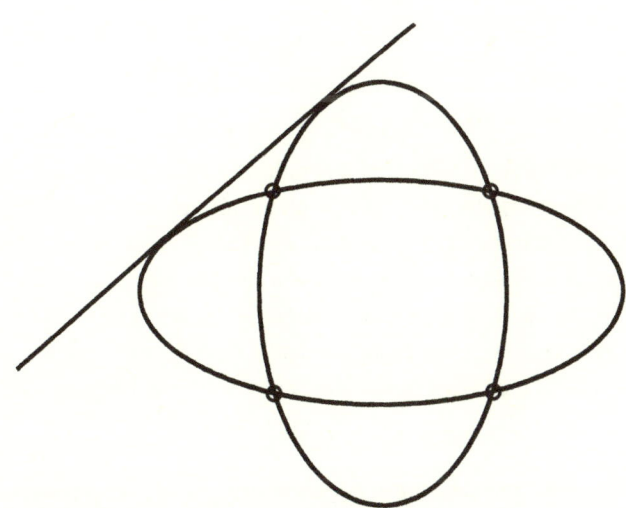

Figure 2.15. Conics through four points tangent to a given line.

are both tangent to $y = x + \sqrt{6}$ and pass through the required four points. The calculated conics, as well as the four given points and the tangent line, are shown in Figure 2.15.

The condition for the line $\alpha x + \beta y + \gamma = 0$ to be tangent to the general conic is a homogenous quadratic equation in α, β, and γ. Assuming that the line does not pass through the origin, we can normalize γ to unity, and hence we have a quadratic equation in α and β that must be satisfied. Demanding that the line be tangent to a second conic will give a second quadratic equation that must be satisfied by the parameters α and β. In the (α, β) parameter plane, these two equations represent conics with, in general, four intersections. Therefore, if we seek a line tangent to two given conics, we must then find (α, β) values that satisfy both quadratic equations in the parameter plane, so that, in general, there are four tangents to two conics. We assumed this fact in the solution of Example 2.1.

2.10.2 The Conic Touching Five Lines

The condition for a given line to be a tangent to a conic, although quadratic in the coefficients $a, b, c, \ldots$ that determine the conic, is linear in the coefficients $A, B, C, \ldots$. Therefore, the tangential equation (2.15) of the conic will, in general, be uniquely specified when five tangents are given. Furthermore, if

$$\Delta = \det \begin{pmatrix} a & h & g \\ h & b & f \\ g & f & c \end{pmatrix}$$

and

$$M = \begin{pmatrix} A & H & G \\ H & B & F \\ G & F & C \end{pmatrix},$$

then the cofactor of A in M is $a\Delta$, the cofactor of B in M is $b\Delta$, and so on. Hence, the ordinary equation (2.7) of the conic, also called the *point equation* of the conic, is derivable from the tangential equation in exactly the same manner as the tangential equation is derivable from the point equation. This, in turn, means that a conic will, in general, be uniquely determined when five tangents are given. This duality between points and lines is something we shall refer to again, though somewhat briefly, a little later in this chapter. The general concept of duality will be discussed in detail in Chapters 3 and 4.

Example 2.16 *Find the tangential equation of the conic with point equation*

$$x^2 + y^2 - 2x - 4y + 4 = 0,$$

and the point equation of the conic with tangential equation

$$9\alpha^2 + 4\beta^2 - 12\alpha\beta - 2\beta\gamma + 2\alpha\gamma = 0.$$

The matrix of coefficients of the point equation is

$$\begin{pmatrix} 1 & 0 & -1 \\ 0 & 1 & -2 \\ -1 & -2 & 4 \end{pmatrix}.$$

Hence $A = 0, B = 3, C = 1, F = 2, G = 1, H = 2$, and the tangential equation is

$$3\beta^2 + \gamma^2 + 4\alpha\beta + 2\alpha\gamma + 4\beta\gamma = 0.$$

This means that, if α, β, and γ satisfy this equation, then the line $\alpha x + \beta y + \gamma = 0$ will be tangent to $x^2 + y^2 - 2x - 4y + 4 = 0$. For example, we see that (α, β, γ) equal to $(1, 0, -2), (1, 0, 0), (0, 1, -1), (0, 1, -3)$, or $(1, 1, -3+\sqrt{2})$ satisfies the equation and, indeed, $x - 2 = 0, x = 0, y - 1 = 0, y - 3 = 0$, and $x + y - 3 + \sqrt{2} = 0$ are all tangents to the given conic.

For the second part of the problem, the matrix of coefficients is

$$\begin{pmatrix} 9 & -6 & 1 \\ -6 & 4 & -1 \\ 1 & -1 & 0 \end{pmatrix}.$$

Hence $a\Delta = -1, b\Delta = -1, c\Delta = 0, f\Delta = 3, g\Delta = 2$, and $h\Delta = -1$, giving the point equation of the conic

$$x^2 + 2xy + y^2 - 4x - 6y = 0.$$

Example 2.17 *Find the equation of the conic that touches the five lines*

$$x = 0, \qquad y = 0, \qquad x - 2y + 1 = 0,$$
$$2x - y - 1 = 0, \quad and \quad x + y - 3 = 0.$$

Using, as above, the equation of a line in the form $\alpha x + \beta y + \gamma = 0$, we note that

$x = 0$	corresponds to	$\alpha = 1,$	$\beta = 0,$	$\gamma = 0,$
$y = 0$	corresponds to	$\alpha = 0,$	$\beta = 1,$	$\gamma = 0,$
$x - 2y + 1 = 0$	corresponds to	$\alpha = 1,$	$\beta = -2,$	$\gamma = 1,$
$2x - y - 1 = 0$	corresponds to	$\alpha = 2,$	$\beta = -1,$	$\gamma = -1,$
$x + y - 3 = 0$	corresponds to	$\alpha = 1,$	$\beta = 1,$	$\gamma = -3.$

The condition for the line to be tangential to the conic is [Eq. (2.15)]

$$A\alpha^2 + B\beta^2 + C\gamma^2 + 2F\beta\gamma + 2G\alpha\gamma + 2H\alpha\beta = 0.$$

Hence,

$$A = 0,$$

$$B = 0,$$

$$A + 4B + C - 4F + 2G - 4H = 0,$$

$$4A + B + C + 2F - 4G - 4H = 0,$$

$$A + B + 9C - 6F - 6G + 2H = 0.$$

Therefore, $A = B = 0$, $19C = 26F$, $G = F$, and $19H = -3F$, giving the tangential equation of the conic as

$$13\gamma^2 + 19\beta\gamma + 19\alpha\gamma - 3\alpha\beta = 0.$$

Then the coefficient matrix of the point equation is

$$M = \begin{pmatrix} 0 & -\frac{3}{2} & \frac{19}{2} \\ -\frac{3}{2} & 0 & \frac{19}{2} \\ \frac{19}{2} & \frac{19}{2} & 13 \end{pmatrix},$$

so that the point equation of the conic is

$$361(x^2 + y^2) - 878xy + 114(x + y) + 9 = 0.$$

The five tangent lines and the calculated hyperbola are shown in Figure 2.16.

Let us now shift the last of the five lines a little closer to the origin, to become $2x + 2y - 3 = 0$, so that the lines enclose a convex region in the first quadrant. The tangential equation of the conic now becomes

$$28\gamma^2 + 22\beta\gamma + 22\alpha\gamma + 3\alpha\beta = 0,$$

and the corresponding point equation is

$$484(x^2 + y^2) - 632xy - 132(x + y) + 9 = 0,$$

which is the ellipse shown in Figure 2.17.

2.10.3 *More Conics through Points and Touching Lines*

So far we have considered the unique conic touching five lines, and the conic pair through four points and touching one line.

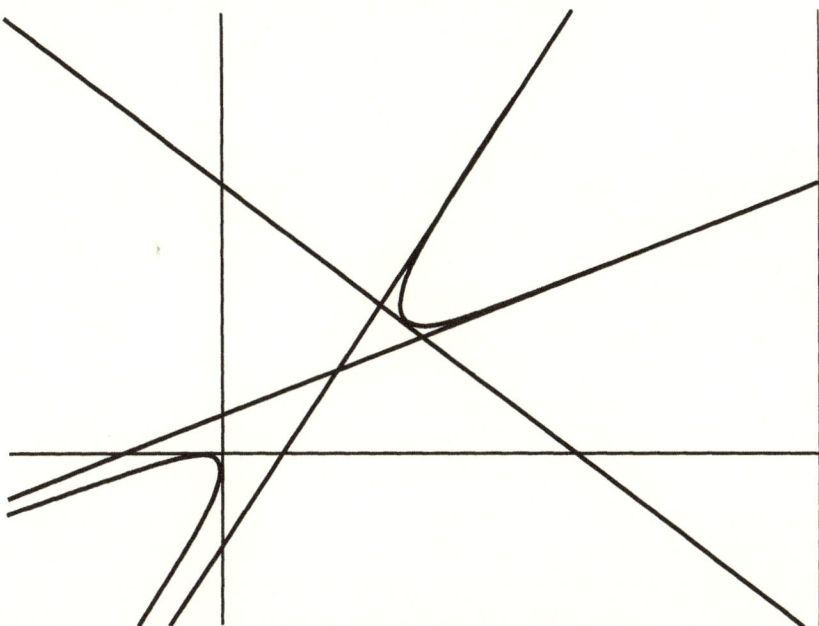

Figure 2.16. Hyperbola tangent to five lines.

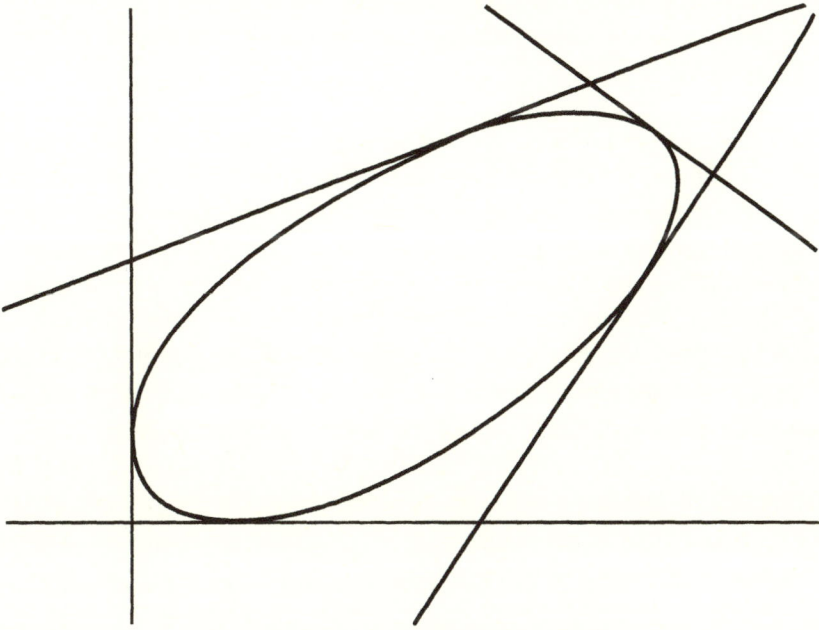

Figure 2.17. Ellipse tangent to five lines.

Since passing a conic through a point imposes a linear condition on the coefficients in the point equation, there is a two-parameter family or *net* of conics through three points. The condition of tangency imposes a quadratic condition on the coefficients in the point equation, and hence demanding that the conic be tangent to one of the lines would result in the two parameters having to satisfy a quadratic equation. There would then be infinitely many solutions to this problem, each one corresponding to a pair of parameters which lie on a conic in the parameter plane. However, forcing the conic to be tangent to another line will give us a second quadratic equation in the two parameters, that is, a second conic in the parameter plane, on which the parameters must lie. Since the conic must be tangent to both lines, the parameter pairs must lie on both conics in the parameter space. We assume that two conics meet in four points, and hence there are four pairs of parameters that satisfy both quadratic equations. There are therefore, in general, four conics that satisfy the five given conditions.

The condition of tangency is linear in terms of the coefficients of the tangential equation. We see that in this form there will be a two-parameter family or net of conics tangent to three lines. The condition for a point to lie on the conic is quadratic in these coefficients, and hence, demanding that the conic pass through two points will give us two quadratic conditions in the coefficients of the tangential equation. There are therefore, in general, four conics through two points and touching three lines.

A similar argument shows that there are two conics through a single point and tangent to four lines.

An appreciation of the style of argument we used here is essential for a complete understanding of our future development of algebraic curves, rational curves, and the applications to parametric splines, curved local domains, and the concept of closeness. The arguments we have used give more understanding of the problem than algebraic calculations and are good examples of synthetic arguments (Chapter 3). We will use more elegant methods to solve the following problems when we have studied the principle of the duality of lines and points (Chapters 3 and 4). For the moment we will use algebraic methods. Since we can write all the conditions referred to in our arguments in explicit form, the steps described can be used to find the solutions explicitly.

Example 2.18 *Produce a sequence of conics through the points $(0, 0)$, $(1, 0)$, and $(0, 1)$ and touching the line $2x = 3$. Which of these conics are also tangent to the line $2y = 3$?*

There are three lines $l_1 = 0$, $l_2 = 0$, and $l_3 = 0$ defined by three noncollinear points, and hence there are three distinct line pairs $l_1 l_2 = 0$, $l_2 l_3 = 0$, and

$l_3 l_1 = 0$. Therefore, the equation

$$\alpha l_1 l_2 + \beta l_2 l_3 + \gamma l_3 l_1 = 0$$

must represent a conic through the three points. This represents a two-parameter (since one of α, β, and γ can be chosen as unity) family or net of conics, and hence every conic through the three points can be written in this form. In our specific example the net of conics through the three points can be written as

$$xy + \mu x(1 - x - y) + \nu y(1 - x - y) = 0,$$

or

$$\mu x^2 + xy(\mu + \nu - 1) + \nu y^2 - \mu x - \nu y = 0.$$

The matrix of coefficients of this conic is

$$\begin{pmatrix} \mu & \dfrac{\mu + \nu - 1}{2} & -\dfrac{\mu}{2} \\[2ex] \dfrac{\mu + \nu - 1}{2} & \nu & -\dfrac{\nu}{2} \\[2ex] -\dfrac{\mu}{2} & -\dfrac{\nu}{2} & 0 \end{pmatrix},$$

and hence the $A, B, C, \ldots$, of Section 2.9 are given by

$$A = -\tfrac{1}{4}\nu^2, \qquad H = \tfrac{1}{4}\mu\nu, \qquad B = -\tfrac{1}{4}\mu^2,$$

$$C = \tfrac{1}{4}[4\mu\nu - (\mu + \nu - 1)^2],$$

$$F = \tfrac{1}{4}[2\mu\nu - \mu(\mu + \nu - 1)], \qquad G = \tfrac{1}{4}[2\mu\nu - \nu(\mu + \nu - 1)].$$

The line $2x = 3$ corresponds to $\alpha = 2$, $\beta = 0$, and $\gamma = -3$, and hence the condition for the conic to be tangent to the line $2x = 3$ is $4A + 9C - 12G = 0$. Upon substitution this condition becomes

$$9\mu^2 - 6\mu\nu + \nu^2 - 18\mu - 6\nu + 9 = 0.$$

We can produce infinitely many conics satisfying the required conditions by selecting some value of μ (or ν) and solving the quadratic equation for the corresponding ν (or μ). We select a sequence of values μ_i, $i = 1, 2, \ldots$, and let ν_i, $i = 1, 2, \ldots$, be one of the two corresponding ν-values. Then each member of the sequence of conics $C_i(x, y)$, $i = 1, 2, \ldots$, defined by

$$C_i(x, y) \equiv xy + \mu_i x(1 - x - y) + \nu_i y(1 - x - y) = 0,$$

$$i = 1, 2, \ldots,$$

passes through the three given points and is tangent to the line $2x = 3$.

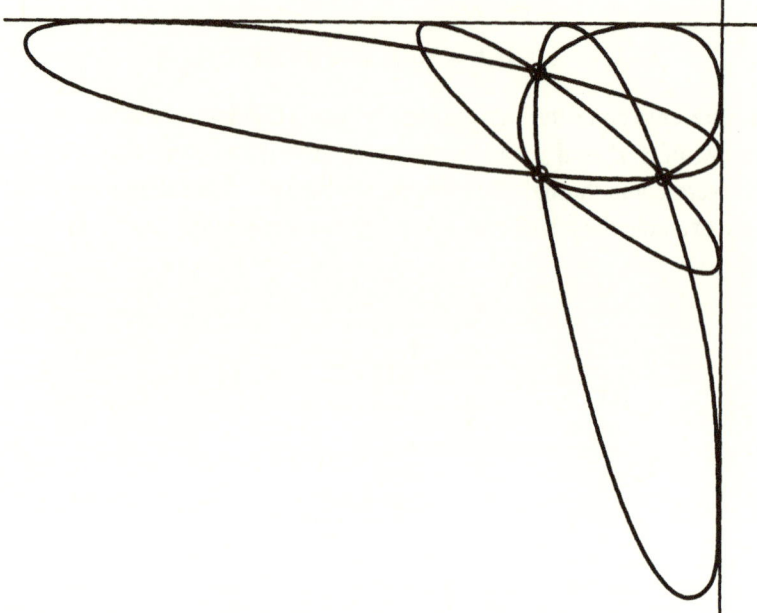

Figure 2.18. Four conics through three points tangent to two lines.

If we seek the conics from this class that are also tangent to the line $2y = 3$, we must impose the condition

$$\mu^2 - 6\mu\nu + 9\nu^2 - 6\mu - 18\nu + 9 = 0.$$

Subtracting this equation from the previous tangency condition, we have

$$(\mu - \nu)(2\mu + 2\nu - 3) = 0,$$

which leads to the two possibilities $\mu = \nu$ and $\mu = \frac{1}{2}(3 - 2\nu)$. From these two possibilities we get the four pairs of (μ, ν)-values

$$\tfrac{1}{2}(6 + 3\sqrt{3}, 6 + 3\sqrt{3}), \qquad \tfrac{1}{2}(6 - 3\sqrt{3}, 6 - 3\sqrt{3}),$$
$$\tfrac{1}{8}(6 + 3\sqrt{3}, 6 - 3\sqrt{3}), \qquad \tfrac{1}{8}(6 - 3\sqrt{3}, 6 + 3\sqrt{3}),$$

each pair giving one of the four conics passing through the given points and tangent to both the given lines. The three points, the two tangent lines, and the four conics are shown in Figure 2.18.

Example 2.19 *Indicate explicitly how to calculate the conics tangent to the four lines $x + y - 1 = 0$, $x - y - 1 = 0$, $x - y + 1 = 0$, and $x + y + 1 = 0$, and through the point $(1/\sqrt{6}, 1/\sqrt{6})$.*

The condition for a line to be tangent to a conic now has to be satisfied by the line coordinates of the four given lines, namely $(1, 1, -1)$, $(1, -1, -1)$, $(1, -1, 1)$, and $(1, 1, 1)$. We solve the corresponding equations to find that

$$F = G = H = 0, \qquad A + B + C = 0,$$

so that the tangential equation of the pencil of conics touching the four lines is given by

$$A\alpha^2 + B\beta^2 - (A + B)\gamma^2 = 0,$$

with the corresponding pencil of point equations

$$B(A + B)x^2 + A(A + B)y^2 - AB = 0.$$

We select those members of the pencil that pass through the point $(1/\sqrt{6}, 1/\sqrt{6})$. This means that

$$A^2 - 4AB + B^2 = 0, \qquad \text{or} \quad A = (2 \pm \sqrt{3})B.$$

The two conics are obtained by substituting these ratios in the point equation. The resulting ellipses, together with the lines of tangency and the common point, are shown in Figure 2.19.

If we shift the point through which the conics are required to pass to $(3, 1)$, we find that A and B have to satisfy the equation

$$A^2 + 9AB + 9B^2 = 0, \qquad \text{or} \quad 2A = (-9 \pm 3\sqrt{5})B.$$

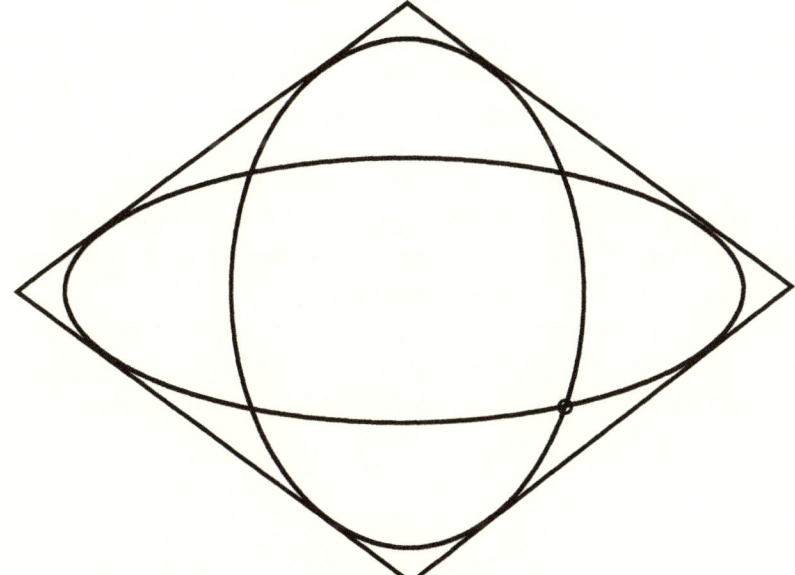

Figure 2.19. Two ellipses through a point and touching four lines.

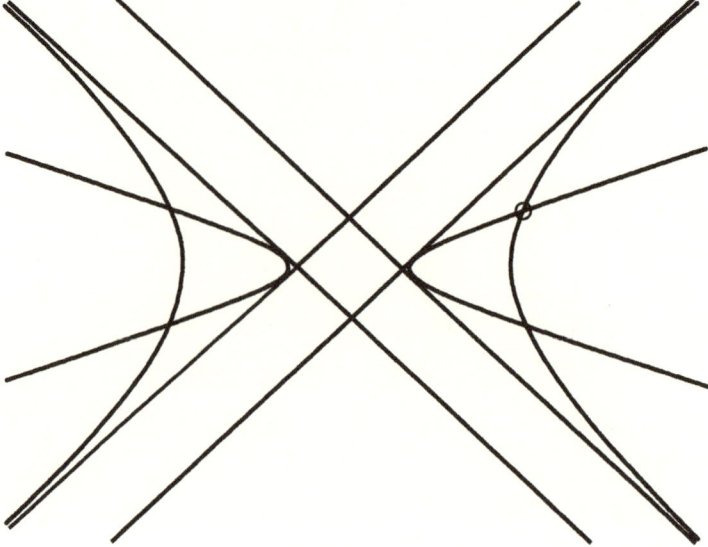

Figure 2.20. Two hyperbolas through a point touching four lines.

The two conics that satisfy all these conditions are hyperbolas, and are shown in Figure 2.20.

2.11 Tangency to a Line at a Prescribed Point

We have analyzed all the possibilities of conics passing through arbitrary points and being tangent to arbitrary lines. The situation is a little disconcerting in that we found only two cases where there is a unique solution, namely, where the conic passes through five points and where it is tangent to five lines. However, by making specific choices of points and lines in other cases, we can ensure uniqueness. If we require that a conic pass through a point and *at that point* be tangent to a given line, we shall have imposed two linear constraints on the conic rather than a linear and a quadratic one.

Let (x_1, y_1) be the point through which the conic must pass, and let the line to which the conic must be tangent at the point (x_1, y_1) have equation $\alpha x + \beta y + \gamma = 0$. Then, from (2.14), the tangent at (x_1, y_1) is

$$axx_1 + h(xy_1 + yx_1) + byy_1 + g(x + x_1) + f(y + y_1) + c = 0,$$

or

$$x(ax_1 + hy_1 + g) + y(hx_1 + by_1 + f) + gx_1 + fy_1 + c = 0.$$

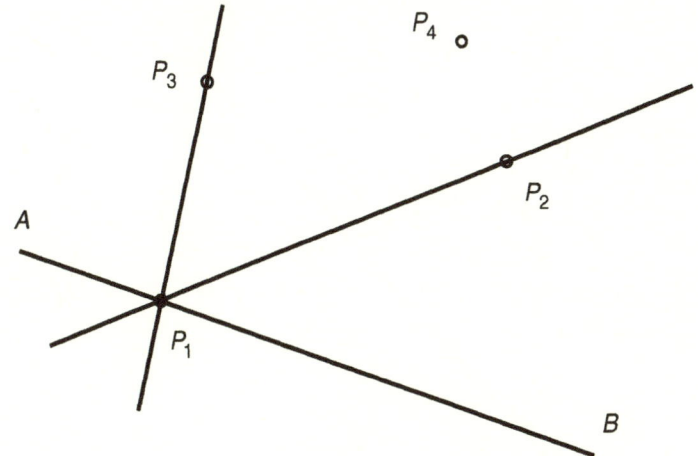

Figure 2.21. A pencil through three points, tangent to a line.

Since this must be the line $\alpha x + \beta y + \gamma = 0$ we have the two linear conditions

$$\frac{ax_1 + hy_1 + g}{\alpha} = \frac{hx_1 + by_1 + f}{\beta} = \frac{gx_1 + fy_1 + c}{\gamma}.$$

Example 2.20 *Show that there is a unique conic through four points and tangent to a line at one of the four points.   In particular, take the points* $(0, 0)$, $(1, 0)$, $(0, 1)$, *and* $(2, 3)$ *and the tangent line* $y = 2x$.

With five degrees of freedom and five linear conditions we can, in general, construct a unique solution. We have seen, however, that in some cases the use of the idea of pencils of curves produced the desired solution in a more efficient manner. We will use this idea here. In Figure 2.21, let AB be the given tangent line with equation $t = 0$, and let the given points be P_1, P_2, P_3 and P_4, with the equation of line $P_1 P_2$ being $l_1 = 0$, that of line $P_1 P_3$ being $l_2 = 0$, and that of line $P_2 P_3$ being $l_3 = 0$. The pencil of conics

$$l_1 l_2 + \alpha t l_3 = 0,$$

passes through points P_1, P_2, and P_3 and is tangent to the line AB at point P_1. The tangency condition can be seen by putting $t = 0$ in the equation of the pencil; for then, since both l_1 and l_2 pass through point P_1, the coordinates of this point will appear as a squared term. We demand that the conic pass through point P_4. This implies

$$l_1(x_4, y_4)\, l_2(x_4, y_4) + \alpha t (x_4, y_4)\, l_3(x_4, y_4) = 0,$$

where (x_4, y_4) are the coordinates of P_4 to find the appropriate value of α.

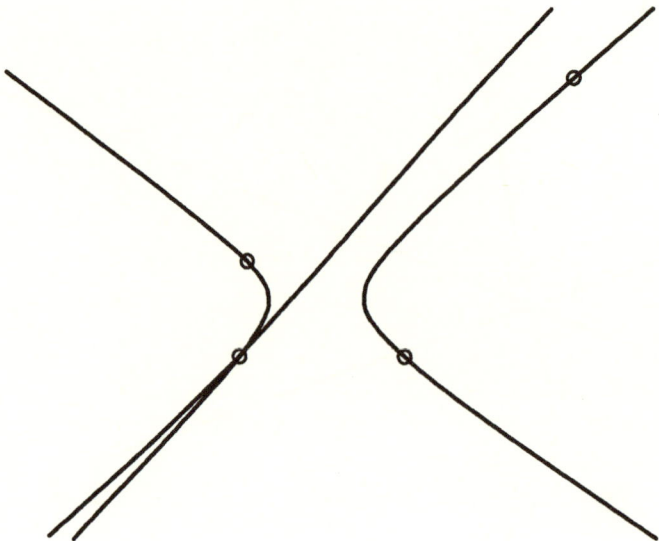

Figure 2.22. Conic through four points, tangent to a line at one point.

In particular, let the coordinates of points P_1, P_2, P_3, and P_4 be $(0, 0)$, $(1, 0)$, $(0, 1)$, and $(2, 3)$, respectively, and let the equation of the tangent be $2x - y = 0$. Then the pencil becomes

$$xy + \alpha(2x - y)(1 - x - y) = 0.$$

Requiring that this conic pass through $(2, 3)$, we obtain $4\alpha = 6$, and hence the conic (Figure 2.22)

$$2xy + 3(2x - y)(1 - x - y) = 0,$$

satisfies all the required conditions.

We can use arguments similar to those in the previous subsection to see that there are, in general, two conics through two points with given slopes at these points and touching a third line. Let us consider an example that serves as a warning to be careful when using the terminology "in general."

Example 2.21 *Determine all conics through the points* $(-3, -1)$ *and* $(-1, 1)$, *and tangent to the lines* $y + 1 = 0$ *and* $x + 1 = 0$ *at these points, that are also tangent to the line* $2x + 3y + 1 = 0$.

The pencil of conics through $(-3, -1)$ and $(-1, 1)$ and tangent to $y + 1 = 0$ and $x + 1 = 0$ is given by

$$(x - y + 2)^2 + \alpha(x + 1)(y + 1) = 0,$$

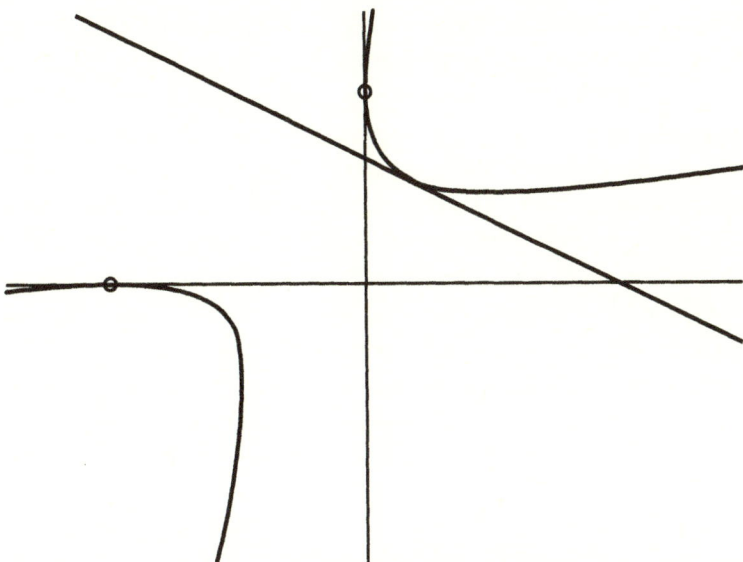

Figure 2.23. A unique conic through two points, tangent to three lines.

or, in expanded form,

$$x^2 + (\alpha - 2)xy + y^2 + (\alpha + 4)x + (\alpha - 4)y + \alpha + 4 = 0.$$

We are looking for those members of this pencil that are also tangent to the line $2x + 3y + 1 = 0$, that is, the line and the selected conics must have double intersections. On substitution of $y = -(2x+1)/3$ in the conic equation, we find that the tangency condition (the equation must be a complete square) reduces to $\alpha(\alpha + 4) = 0$. There are two solutions, but the case $\alpha = 0$ corresponds to the repeated line $(x - y + 2)^2 = 0$, which is a degenerate conic. The only solution to the problem corresponds to $\alpha = -4$, namely

$$x^2 - 6xy + y^2 - 8y = 0.$$

This hyperbola, the given points, and the three tangents are shown in Figure 2.23.

2.12 Inflection Points and Curvature

We now give an intuitive description of an inflection point of a curve in a similar vein to our discussion in Section 2.9, in which we introduced the idea of a tangent as a line that meets a curve at two coincident points. Continuing with this idea, we consider an inflection point of a curve as a point where the tangent must meet the curve in *at least* three points.

The line $y = 0$ meets the curve $y = x(x - x_i)^2$ in the three distinct points $x = 0, \pm x_i$. Since this line is defined by the points $(0, 0)$ and $(x_i, 0)$, we know that, as $x_i \to 0$, the line will become the tangent at $(0, 0)$ to the curve $y = x^3$. For this particular curve the line $y = 0$ meets the curve in the three coincident points $(0, 0)$, $(0, 0)$, and $(0, 0)$. In general, this will not happen, and hence an arbitrary point of a curve is not likely to be an inflection point. For curves that do not possess loops or cusps it is reasonable to consider an inflection point as a point of the curve where the tangent meets the curve in three coincident points. This in turn leads us to the assertion that, since any line meets a conic in at most two points, a conic cannot have an inflection point.

Since a circle is uniquely determined by three distinct noncollinear points, there is, in general, a unique circle through three points of a curve. Will this circle remain well defined if the three points coalesce? Consider the following example.

Example 2.22 *Take the three points $(0, 0)$, (x_i, x_i^2), and $(-x_i, x_i^2)$ on the curve $y = x^2$, and consider what happens to the circle defined by these points as they coalesce to the point $(0, 0)$. Show that the circle is well defined in the limit and that it has four-point contact with the given curve at the origin.*

First we must construct the circle through the three points. This gives us some further practice in the use of linear systems of conics. The three points define the lines $y = \pm x_i x$ and $y = x_i^2$, and hence the net of conics given by

$$\left(y^2 - x_i^2 x^2\right) + \alpha(y - x_i x)\left(y - x_i^2\right) + \beta(y + x_i x)\left(y - x_i^2\right) = 0$$

is the two-parameter family of conics passing through the three specified points. This can be rewritten as

$$-x_i^2 x^2 + x_i(\beta - \alpha)xy + (1 + \alpha + \beta)y^2 + x_i^3(\alpha - \beta)x - x_i^2(\alpha + \beta)y = 0.$$

This equation represents a circle, provided that $\beta - \alpha = 0$ and $1 + \alpha + \beta = -x_i^2$. The required circle is then

$$x^2 + y^2 - \left(x_i^2 + 1\right)y = 0.$$

Upon taking the limit as $x_i \to 0$, we obtain

$$x^2 + y^2 - y = 0.$$

The circle is therefore well defined in the limit, and meets $y = x^2$ where $x^4 = 0$, which we interpret as four coincident points.

If the circle defined by three points of a curve is well defined in the limit as the points coalesce, the circle is called the *circle of curvature*. The *radius of curvature* is the radius of the circle of curvature, the *center of curvature* is the center of this circle, and the *curvature* is the reciprocal of the radius of curvature.

2.13 Conics and Curvature

2.13.1 Conics with Specified Curvature at a Point

The imposition of a specific circle of curvature at a point on a conic must be equivalent to three linear conditions on the coefficients of the conic. More generally, the requirement of three-point contact of a conic with a given conic must impose three linear conditions. Hence, there must be a net of conics that meet a given conic in three coincident points.

Let $C = 0$ be the equation of the given conic through a point P. Let $t = 0$ be the equation of the tangent to the given conic at P, and let $l_\alpha = 0$ be the equation of any line through P, where by using the subscript α we denote the single degree of freedom of the line. Consider the conic

$$C + \beta t l_\alpha = 0.$$

This conic meets the conic $C = 0$ where the lines $t = 0$ and $l_\alpha = 0$ meet $C = 0$. The line $t = 0$ meets $C = 0$ twice at the point P, and $l_\alpha = 0$ meets $C = 0$ once at the point P. Therefore, the conic $C + \beta t l_\alpha = 0$ has three-point contact with $C = 0$ at the point P.

Example 2.23 *Describe the net of conics that have three-point contact with the conic $2x^2 - 2xy + 3y^2 - 3x - 4y + 1 = 0$ at the point $(0, 1)$.*

We verify that the given point lies on the given conic. The tangent to the given conic at $(0, 1)$ is $5x - 2y + 2 = 0$, and an arbitrary line through $(0, 1)$ has equation $mx - y + 1 = 0$. The net of conics satisfying the required conditions is therefore

$$2x^2 - 2xy + 3y^2 - 3x - 4y + 1 + \beta(5x - 2y + 2)(mx - y + 1) = 0,$$

or

$$2x^2 - 2xy + 3y^2 - 3x - 4y + 1$$
$$+ \beta(5x - 2y + 2)(1 - y) + \gamma x(5x - 2y + 2) = 0,$$

where we have written $\gamma = \beta m$ in order to remove the inessential nonlinearity of the coefficients. Any member of the net meets the given conic three times at

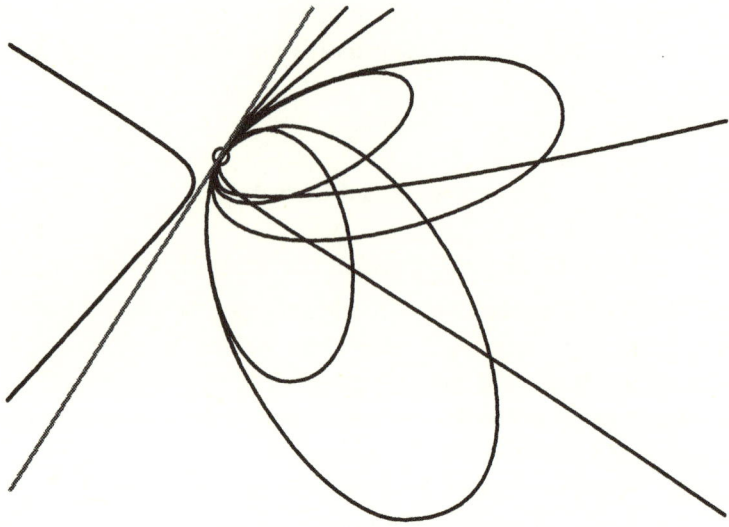

Figure 2.24. Conics with three-point contact.

(0, 1) and once at

$$\left(\frac{5 - 2m}{3m^2 - 2m + 2}, \frac{m^2 + 3m + 2}{3m^2 - 2m + 2} \right).$$

The given conic, the common tangent, and five members of the net, corresponding to the parameter value pairs (β, γ) equal to $(1, 0)$, $(\frac{1}{2}, \frac{1}{2})$, $(-1, \frac{1}{2})$, $(2, -3)$ and $(-1, 1)$, respectively, are shown in Figure 2.24.

Construct the net of conics whose center of curvature at the point $(1, 2)$ is at $(1, 0)$. The radius of curvature is 2, and the circle of curvature has equation

$$(x - 1)^2 + y^2 - 4 = 0.$$

The tangent to the circle of curvature at the point $(1, 2)$ has equation $y = 2$, and an arbitrary line through the point $(1, 2)$ has equation

$$y - 2 = m(x - 1).$$

The net of conics can then be written in the form

$$(x - 1)^2 + y^2 - 4 + \beta(y - 2)(mx - y + 2 - m) = 0,$$

or, equivalently,

$$(x - 1)^2 + y^2 - 4 + \beta(y - 2)^2 + \gamma(y - 2)(x - 1) = 0.$$

Since the center of curvature is specified, all the members of the net lie to the same side of the tangent at the given point. The circle and center of curvature,

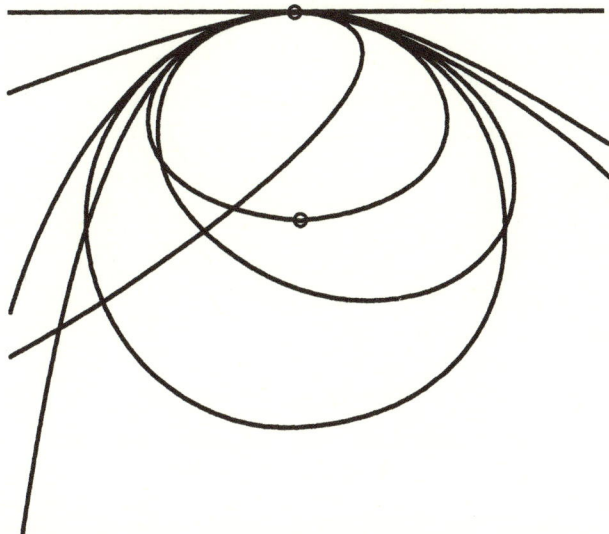

Figure 2.25. Conics with prescribed curvature at a point.

the tangent, and the five members of the net corresponding to the same parameter values as in Example 2.23 are shown in Figure 2.25.

2.13.2 Osculating Conics

Continuing the above argument, we could demand that two conics meet in four coincident points. When this happens the conics are said to *osculate*. There will be a pencil of conics that have four-point contact with a given conic at a prescribed point. The pencil is easily constructed, as follows. Let $C = 0$ be the equation of the given conic, let P be the point at which osculation is required, and let $t = 0$ be the equation of the tangent to the given conic at P. Then $C + \alpha t^2 = 0$ is the equation of the pencil of conics osculating at P.

Example 2.24 *Construct the pencil of conics with four-point contact with the circle* $(x - 1)^2 + y^2 - 4 = 0$ *at the point* $(1, 2)$.

The given circle is the circle of curvature in the second part of Example 2.23. In that example, our constructed conics were required to have three-point contact with this circle. We now want only those members of the net that will have four-point contact with the circle. This pencil of conics is given by

$$(x - 1)^2 + y^2 - 4 + \alpha(y - 2)^2 = 0.$$

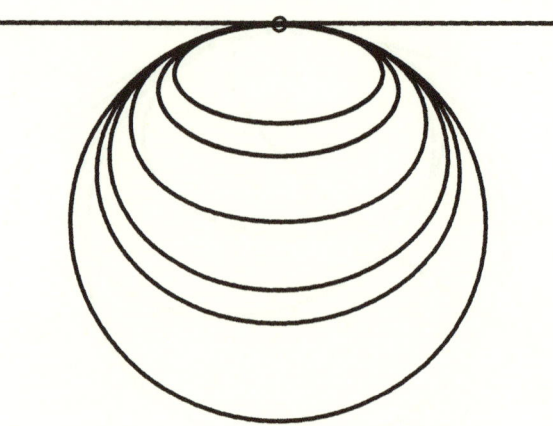

Figure 2.26. Conics with four-point contact.

The circle, the tangent line, and the five members of the pencil corresponding to parameter values $\alpha = 1, 2, 3, \frac{1}{2}$, and $\frac{1}{3}$ are shown in Figure 2.26.

What is referred to as the osculating circle is the circle that has three-point contact with the conic, and happens to meet the conic once more in the same point. This enables us to find the circle of curvature at any point of a conic in a very straightforward fashion, as follows.

Let $C = 0$ be the equation of the given conic, and let P be the point on the conic at which we wish to determine the circle of curvature. Let $t = 0$ denote the tangent to the conic at P, and let $l_\alpha = 0$ be any line through P. Then $C + \beta t l_\alpha = 0$ represents the net of conics having three-point contact with the conic at P. This is a circle if the coefficient of the xy term is zero and if the coefficients of x^2, and y^2 are the same. These conditions give two linear equations in the two unknowns and hence determine a unique member of the net, which must then be the circle of curvature of the given conic at P.

Example 2.25 *Find the curvature and center of curvature of the conic*

$$x^2 - 2xy + y^2 - x + 3y - 4 = 0$$

at the point $(0, 1)$.

The tangent to the given conic at $(0, 1)$ is $3x - 5y + 5 = 0$, and an arbitrary line through $(0, 1)$ has equation $mx - y + 1 = 0$. The net of conics having three-point contact with the given conic at $(0, 1)$ is then given by

$$x^2 - 2xy + y^2 - x + 3y - 4 + \beta(3x - 5y + 5)(mx - y + 1) = 0,$$

or

$$x^2(1 + 3\gamma) - xy(2 + 3\beta + 5\gamma) + y^2(1 + 5\beta)$$
$$+ x(3\beta + 5\gamma - 1) + y(3 - 10\beta) - 4 + 5\beta = 0,$$

where $\gamma = m\beta$. This will be a circle, provided that

$$2 + 3\beta + 5\gamma = 0 \quad \text{and} \quad 1 + 3\gamma = 1 + 5\beta,$$

implying $\beta = -\frac{3}{17}, \gamma = -\frac{5}{17}$, and $m = \frac{5}{3}$. The equation of the circle of curvature is

$$2x^2 + 2y^2 - 51x + 81y - 83 = 0.$$

The center of curvature is at $(\frac{51}{4}, -\frac{81}{4})$, and the curvature is $2\sqrt{2}/(17\sqrt{17})$.

2.14 Algebraic Dependence of Interpolatory Conditions

We have described several of the more common ways of defining conics. We have seen that some of the interpolation conditions impose nonlinear constraints upon the coefficients that define the conic. All the constraints were algebraic, that is, there were some polynomial equations that had to be satisfied by the coefficients in the equation of the conic. Therefore, we call these interpolation problems *algebraic*.

With our finite linear interpolation problem, we considered a vector in some vector space. Having solved the problem, we then wrote an equation representing a curve that satisfied the interpolation conditions, moving easily between the vector and a curve, and between the function and an equation. By contrast, in the present chapter we have restricted our attention to describing the curve rather than the polynomial that is associated with it. It is a simple matter to formulate any conic interpolation problem in a framework similar to that of the finite linear interpolation problem by adding one additional linear condition to the set of conditions already being considered. For example, the problem of seeking the conic satisfying the five conditions of passing through three points with prescribed slopes at two of these points can be identified with the problem of finding a polynomial $f(x, y) \in P_2(x, y)$ that satisfies the six conditions of evaluating to zero at the three points, having $-(\partial f/\partial x)/(\partial f/\partial y)$ prescribed at two of the points, and, in addition, satisfying $f(x_0, y_0) \neq 0$ at some sixth point with coordinates (x_0, y_0). This specifies a polynomial, but if we now set that polynomial equal to zero, we shall obtain an equation of exactly the same conic as we would have obtained using the five conditions to specify the curve.

In the final chapter, when we consider the construction of bases for local interpolation, the polynomial and the curve will be of equal importance, and we shall find it more appropriate to consider the construction of the polynomial. In the meantime, it remains important for us to appreciate the equivalence between the two problems. Haar's theorem highlights the linear dependence of the interpolating conditions that can arise in multivariable interpolation. Since our more general technique of conic definition is equivalent to an algebraic interpolation problem that includes the linear case, we have all the potential difficulties implied by Haar's theorem, as well as difficulties arising from the dependence of nonlinear constraints. There is, in general, a unique conic through five points, but if four of the points are collinear, the solution will no longer be unique. There are, in general, four conics through three points and tangent to two lines, but if these lines are parallel, there will be infinitely many solutions. There is, in general, a pencil of conics through two points with prescribed slopes at the two points. If one of the slopes is parallel to the chord defined by the two points, then the pencil of conics satisfying these conditions is always a line pair. There is, in general, a unique conic having a prescribed circle of curvature at a point and passing through two other points, but there is no solution if one of these points lies on the tangent to the circle of curvature at the point of contact.

We include both linear and nonlinear (though still polynomial) conditions in the term *algebraic conditions* and will refer to dependency among these conditions as *algebraic dependence*.

2.15 Qualitative Properties

In our introduction to Chapter 1, we drew a distinction between design and approximation and made some remarks concerning a suitable meaning of "good" approximation when using implicit curves. In the design context it is likely that the criteria determining how good our curve is will vary substantially from problem to problem, and it is therefore impossible to give any hard and fast rules. These criteria will depend both on the particular type of curve being used and on the particular application. Although applications vary widely, we make some comments about qualitative properties of curves likely to affect the determination of a satisfactory curve for design purposes. With practice one may perceive attributes of a curve that would not be perceptible to the untrained eye. Some things, however, are fairly readily appreciated. It is difficult for the untrained eye to distinguish changes in third derivatives, but inflection points are easily noticed, and fair guesses can be made at asymptotes. It is easy to see whether or not a drawn curve is of one or more continuous pieces and whether it contains a cusp or loop. The eye can tell if the curve passes through certain points, and it can distinguish differences in slope and curvature. The existence of cusps and loops is particularly important in the study of third- and higher-order rational curves, for example parametric splines, and we will discuss these and other qualitative properties in Chapter 5.

We have already discussed the construction of conics through points, touching lines, and with specified curvature, but have made no comments about the fact that certain conics, the hyperbolas have two pieces or branches. The conic with its center as the origin of the coordinate axes has its equation in the form (2.13), namely

$$ax^2 + 2hxy + by^2 + c' = 0.$$

This is a hyperbola if $h^2 - ab > 0$. In this situation the quadratic part $ax^2 + 2hxy + by^2$ factors into two real linear factors. For example, if $a \neq 0$,

$$ax^2 + 2hxy + by^2 = \frac{1}{a}[ax + (h + \mu)y][ax + (h - \mu)y],$$

where $\mu^2 = h^2 - ab$. (If $a = 0$, $b \neq 0$, use b instead of a. If both a and b are zero, the hyperbola is already in the desired form.)

The lines

$$ax + (h + \mu)y = 0 \quad \text{and} \quad ax + (h - \mu)y = 0,$$

are called the *asymptotes* of the hyperbola. If $h^2 - ab \leq 0$, it is customary to say that there are no asymptotes. Hence, a hyperbola has two asymptotes; the parabola and the ellipse have none. If we refer the equation of the hyperbola to its asymptotes as axes, then the equation becomes $xy = -ac'$, and the asymptotes separate two disjoint pieces of the curve. In any curve design or approximation problem we shall only be using a piece of the curve. The use of hyperbolas is likely to be perfectly satisfactory if the piece of the curve being used lies on a single branch, but totally unsuitable if it includes pieces from both branches.

Sometimes we can construct a technique of curve specification that, although not directly connected with qualitative properties, manages to give us an intuitive feel for and control over these properties. In the following sections, we give examples related to conics.

2.16 Parametrizations

In Section 2.4 we showed two common parametrizations (2.4) and (2.5) for the conics. We could produce other parametrizations by, for example, substituting some other expression for t in (2.5). There are therefore infinitely many different parametrizations of the same curve. Usually the most suitable one is the one that is the easiest to compute. It is an added advantage if some physical meaning can be attached to the parameter and/or its coefficients. As regards the parameter itself, its meaning will be clear if we know how the parametrization was produced. The meaning of the angle θ and of $t = \tan(\theta/2)$ in (2.4) and (2.5) are obvious from the constructions that produced them. We shall see in Section 5.5 that, given a rational parametrization of a curve, there is a straightforward procedure for producing the algebraic form. However, even when we know that an algebraic curve is rational, it is not always easy to produce a parametrization. For the conics a parametrization can be produced from the following very simple argument.

A line meets a conic twice, and hence a line through some fixed point of a conic meets the conic again in exactly one other point. If we seek the points where the arbitrary line through the fixed point meets the conic, we will have a quadratic equation which, by construction, must factor. One linear factor will correspond to the fixed point, and the other will correspond to a variable point. This latter factor will give us a parametrization of the conic, as in the following example.

Example 2.26 *Produce a parametrization of the conic*

$$2x^2 + xy - 3y^2 - x + 2y = 0$$

by taking arbitrary lines through the origin.

Since the origin lies on the conic, the procedure is guaranteed to work. An arbitrary line through the origin has equation $y = mx$. This line meets the conic when

$$x[(2 + 3m)(1 - m)x - 1 + 2m] = 0.$$

The factor x corresponds to the origin. The other point where the line meets the curve is given by

$$x = \frac{1 - 2m}{(2 + 3m)(1 - m)}, \qquad y = \frac{m(1 - 2m)}{(2 + 3m)(1 - m)}.$$

This is therefore a parametrization of the conic.

We can think of a curve as a one-dimensional entity – like a piece of string that has taken up a particular orientation in the plane. The single parameter can be thought of as a point moving along a straight line while the corresponding point on the curve traces out the curve. In this sense a planar parametric curve is a mapping from a line to points in the plane. In most of our applications we shall be considering only a curve segment between two given points on the curve. These points determine a chord. It often gives geometrical insight if the parameter of our parametrization is associated with a point on this chord in such a way that the parameter changes from zero to unity as the point on the curve moves from one endpoint of the curve segment to the other. Let us clarify these ideas by considering the construction of such a parametrization for a conic.

Example 2.27 *Produce a parametrization of the conic*

$$2x^2 + xy + 3y^2 - x - 2y = 0$$

such that the arc of the conic from $(0, \frac{2}{3})$ *clockwise to* $(\frac{1}{2}, 0)$ *corresponds to parameter values from zero increasing to one and such that a point on the arc can be associated with a point on the chord.*

The chord defined by the points $(0, \frac{2}{3})$ and $(\frac{1}{2}, 0)$ can be expressed in parametric form as

$$x = at + b, \qquad y = ct + d.$$

We want $t = 0$ to correspond to $x = 0$, $y = \frac{2}{3}$, and we want $t = 1$ to correspond to $x = \frac{1}{2}$, $y = 0$. Therefore,

$$x = \tfrac{1}{2}t, \qquad y = \tfrac{2}{3}(1 - t).$$

We take any point on the conic, for example $(0, 0)$. The line through $(0, 0)$ and $(\frac{1}{2}t, \frac{2}{3}(1 - t))$ has equation $\frac{1}{2}ty = \frac{2}{3}(1 - t)x$ and meets the conic when

$$x[(18t^2 - 28t + 16)x + 5t^2 - 8t] = 0.$$

Hence, a suitable parametrization is

$$x = \frac{(8 - 5t)t}{2(9t^2 - 14t + 8)}, \qquad y = \frac{2(8 - 5t)(1 - t)}{3(9t^2 - 14t + 8)}.$$

Notice that the denominator does not become zero for any real parameter value.

A point $(\bar{x}, \bar{y})$ on the conic is identified with the point on the chord where the line $\bar{x}y = \bar{y}x$ meets the chord $4x + 3y - 2 = 0$, namely

$$\left(\frac{2\bar{x}}{4\bar{x} + 3\bar{y}}, \frac{2\bar{y}}{4\bar{x} + 3\bar{y}} \right).$$

2.16.1 A Parametrization of the Parabola

The general conic is a parabola if the discriminant is zero. The quadratic terms then form a perfect square, and the parabola can be written in the form

$$(\beta x - \alpha y)^2 + 2gx + 2fy + c = 0.$$

From this equation we see that any line of the form $\beta x - \alpha y + \gamma = 0$ meets the parabola only once, and hence the axis of the parabola has slope β/α. Consider the parabola through two points and with a specified slope for the axis. For ease

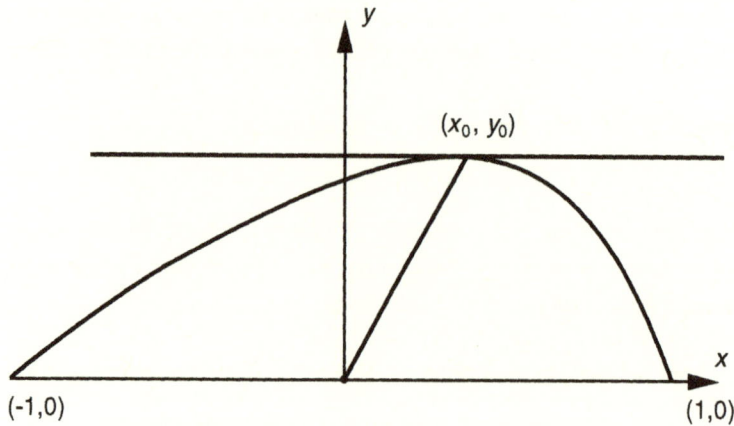

Figure 2.27. Parametric equation of the parabola.

of algebraic manipulation and without loss of generality, we take the points as $(-1, 0)$ and $(1, 0)$ and the slope as β/α (Figure 2.27). The parabola then has the equation

$$(\beta x - \alpha y)^2 + 2fy - \beta^2 = 0.$$

The origin is the midpoint of the chord defined by $(-1, 0)$ and $(1, 0)$. The line through the origin with slope β/α meets the parabola where

$$x = \frac{\alpha\beta}{2f}, \qquad y = \frac{\beta^2}{2f}.$$

If this point is (x_0, y_0), then our parabola will be uniquely determined and the equation can be written

$$(y_0 x - x_0 y)^2 + y_0(y - y_0) = 0,$$

where $\beta/\alpha = y_0/x_0$.

Let us produce a parametrization by projecting through points on the chord from $x = -1$ to $x = 1$. If we take a projection along some arbitrary line through a point of the chord, then we shall obtain a quadratic, neither of whose zeros we know. If we select some arbitrary point on the parabola and project from this point through a point of the chord, we shall certainly, as in Example 2.27, produce a parametrization. We choose, however, to project along lines parallel to the axis of the parabola.

Take $x = t$, $y = 0$, as a parametrization of the chord. The line through $(t, 0)$ parallel to the axis of the parabola has equation

$$y = \frac{y_0}{x_0}(x - t),$$

and this line meets the parabola in the unique point given parametrically by

$$x = t + x_0(1 - t^2), \qquad y = y_0(1 - t^2).$$

The point (x_0, x_0) corresponds, by construction, to $t = 0$, and the tangent at this point is parallel to the chord used for the parametrization, that is, $dy/dx = 0$ at $t = 0$, in agreement with the properties of conjugate diameters. Therefore, the point (x_0, y_0) is the furthest extent of the curve from the chord. This point then determines precisely how much the curve "bows out" (or in), and also determines the degree of asymmetry about the perpendicular bisector of the chord. This particular parametrization is therefore easy to use, even if we do not have a detailed knowledge of the underlying geometry.

2.16.2 A Parametrization of a Circle

We discuss one other example of a parametrization. Consider Figure 2.28, where a circle is drawn through the points $(-1, 0)$ and $(1, 0)$. We wish to produce a parametrization of the arc from $(-1, 0)$ through $(0, y_0)$ to $(1, 0)$, where y_0 represents the depth of the curve. The equation of the circle is

$$x^2 + y^2 + \frac{1 - y_0^2}{y_0} y - 1 = 0.$$

The point diametrically opposite $(0, y_0)$ is $(0, -1/y_0)$. Take the chord from $(-1, 0)$ to $(1, 0)$ as being parametrized by $x = t$, $y = 0$. Then the line through

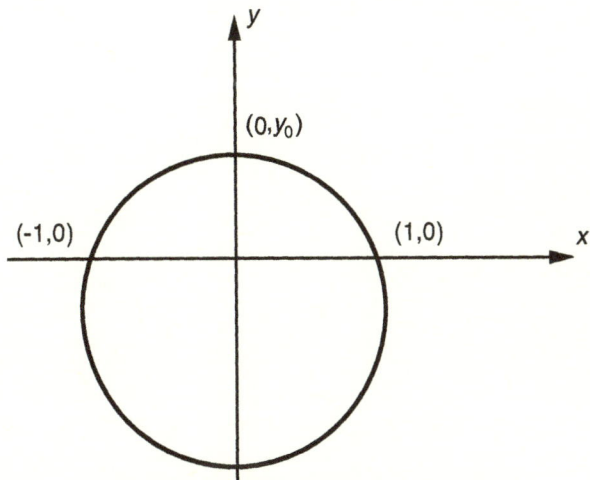

Figure 2.28. Parametric equation of the circle.

a point of this chord and the point $(0, -1/y_0)$ has equation

$$y = \frac{1}{y_0 t}(x - t),$$

and, by construction, will meet the circle in one other point, which is given by

$$x = \frac{(1 + y_0^2)t}{1 + y_0^2 t^2}, \qquad y = \frac{y_0(1 - t^2)}{1 + y_0^2 t^2}.$$

Again both the parameter and the coefficients have, by construction, a very easily understood meaning.

2.17 Piecewise Conic Curves

In most applications, the curves used are defined in a piecewise fashion. We saw in the first chapter that the cubic splines were actually a collection of cubics, each one being defined on its own interval. Let us look at a few of the possible ways of defining piecewise conic curves.

2.17.1 Piecewise Conic C^0 Curves

Suppose we are given an ordered set of points as in Figure 2.29 and are required to pass a curve through these points such that, for any given interval, the curve is the arc of a conic. The most obvious procedure would be to take the points five at a time, defining the unique conic through points 1 to 5 as the

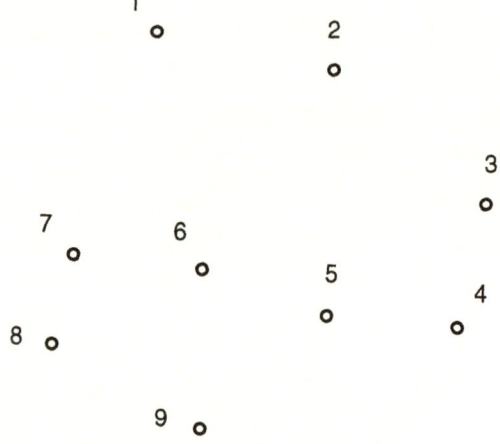

Figure 2.29. An ordered set of points.

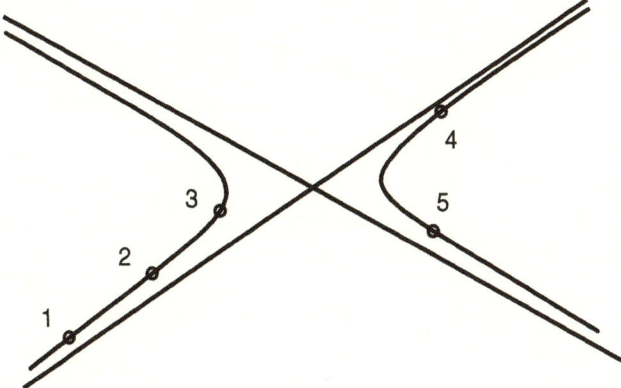

Figure 2.30. A hyperbola through five points.

first arc, the unique conic through points 5 to 9 as the second arc, and so on. This procedure could run into trouble, since even a very reasonable collection of five points might imply that the curve was a hyperbola with some of the points on one branch and the remainder on the other. Consider Figure 2.30, where a hyperbola is sketched. This hyperbola is uniquely determined by any five of its points. In particular, the five marked points uniquely determine the hyperbola, and the resulting conic arc is not continuous from point 1 through to point 5.

An alternative would be to take only four points at a time. We would then have a pencil of conics defined on each set of four points and would require some way of selecting a particular conic from each pencil. The parameter defining the pencil could be used as a control parameter, with conics being plotted for a range of values of this parameter and the most suitable one being selected by eye. This procedure need not be too tedious if suitable graphics hardware and software are avaliable.

The easiest procedure would be to take the points three at a time and to restrict ourselves to piecewise parabolic curves. If we identify the middle point of three consecutive points as the image under the parametrization of the midpoint of the chord defined by the other two (Figure 2.31), we can use the parametrization of the parabola given in the Section 2.16.1. In the figure, the three consecutive points are P_i, P_{i+1}, and P_{i+2}, and the midpoint of the chord $P_i P_{i+1}$ is denoted by M_{i+1}. The parabola will pass through P_i, P_{i+1}, and P_{i+2}, and the axis of the parabola will be parallel to $P_{i+1}M_{i+1}$. This technique of piecewise parabolic curve definition provides little flexibility in either a design or approximation environment. It is, however, an extremely

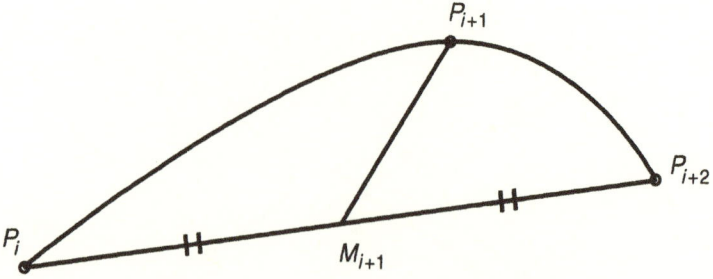

Figure 2.31. A parabola through three points.

common, perhaps the most common, method of producing piecewise conic curves.

2.17.2 Piecewise Conic C^1 Curves

If the curve is to be C^1 continuous, we must have continuity of the first derivative at the endpoints of each conic arc. Two ways of achieving this come to mind. We could interpolate point and slope at the endpoints of each piece, or we could simply demand continuity of the first derivatives at these endpoints. The first procedure is the more restrictive and requires some estimate of the endpoint slopes. If we are in an approximation situation, then there is no difficulty in estimating a slope, but if we are in a design mode or simply seeking a C^1 curve that passes through given points, then the selection of appropriate slopes is by no means obvious. This was the very situation in which we were placed in our discussion of cubic splines (Section 1.5). In that case we sought a piecewise cubic curve that was C^2 continuous and had only data points, no slopes or second derivatives, as input information. We achieved the desired end by interpolating to the given points and then imposing the conditions of continuity of first and second derivatives, which then resulted in a set of linear equations, which, when solved, produced the spline curve. It would be reasonable to attempt the same approach here. Let us implement such an argument.

Let

$$C_i(x, y) \equiv a_i x^2 + 2h_i xy + b_i y^2 + 2g_i x + 2f_i y + c_i = 0$$

denote the conic to be constructed between points P_{i-1} and P_i. Let the coordinates of P_i be (x_i, y_i). The condition that C_i passes through P_{i-1} and P_i gives us the two linear equations

$$a_i x_{i-1}^2 + 2h_i x_{i-1} y_{i-1} + b_i y_{i-1}^2 + 2g_i x_{i-1} + 2f_i y_{i-1} + c_i = 0,$$
$$a_i x_i^2 + 2h_i x_i y_i + b_i y_i^2 + 2g_i x_i + 2f_i y_i + c_i = 0.$$

Let $r_i(x, y)$ denote the slope of the conic C_i. Then

$$r_i(x, y) = -\frac{a_i x + h_i y + g_i}{h_i x + b_i y + f_i}.$$

Continuity of first derivative at the point P_i then implies

$$r_i(x_i, y_i) = r_{i+1}(x_i, y_i),$$

that is,

$$\frac{a_i x_i + h_i y_i + g_i}{h_i x_i + b_i y_i + f_i} = \frac{a_{i+1} x_i + h_{i+1} y_i + g_{i+1}}{h_{i+1} x_i + b_{i+1} y_i + f_{i+1}},$$

which is a nonlinear equation in the coefficients determining the conics. Once more we realize that, simple though this interpolation problem appears, it is not a finite linear interpolation problem. This nonlinearity can be avoided by choosing some slope at each endpoint. In the design mode, or when we are only given a set of points, we can think of the slopes as control parameters that can be varied until a suitable curve is produced.

Suppose that we have a set of points together with a set of slopes, one for each point, and we wish to construct a piecewise conic C^1 curve satisfying these interpolatory conditions. Let us construct a conic between each consecutive pair of points. Is this a linear or a nonlinear problem? The condition for a conic to be tangent to a line is quadratic in the coefficients of the point equation (Section 2.10), but the condition for it to be tangent to a line at a specific point is equivalent to two linear conditions (Section 2.11). Therefore, there is a pencil of conics through two points with prescribed slopes at these points. Consider Figure 2.32, where P_{i-1} and P_i are the consecutive points; $t_{i-1}(x, y) = 0$ and $t_i(x, y) = 0$ are the equations of the lines through P_{i-1} and P_i, respectively, having the required slopes at these points; and $l_i(x, y) = 0$ is the equation of the chord $P_{i-1} P_i$. Then

$$t_{i-1}(x, y)t_i(x, y) + \alpha l_i^2(x, y) = 0$$

represents a pencil of conics tangent to $t_{i-1} = 0$ at P_{i-1} and tangent to $t_i = 0$ at P_i. We must now select a value of α. Let us do this by forcing the conic to pass through some additional point R_i with coordinates (μ_i, ν_i). The unique conic specified in this way has the equation

$$l_i^2(\mu_i, \nu_i)t_{i-1}(x, y)t_i(x, y) - t_{i-1}(\mu_i, \nu_i)t_i(\mu_i, \nu_i)l_i^2(x, y) = 0.$$

If Q_i is the point where the lines $t_{i-1} = 0$ and $t_i = 0$ meet, then the conic arc is continuous from P_{i-1} through R_i to P_i if the point R_i lies within the triangle

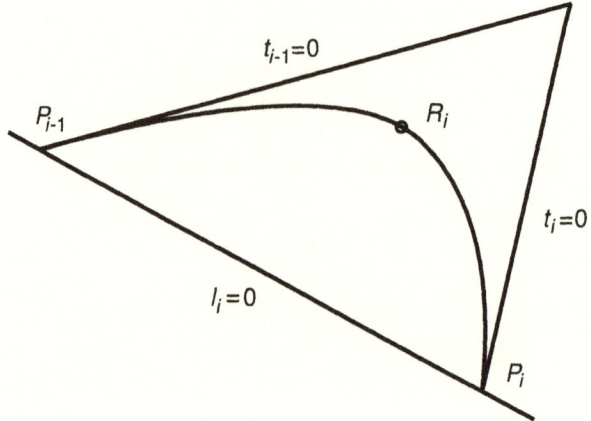

Figure 2.32. A conic through three points tangent to two lines.

$P_{i-1} Q_i P_i$. This is easily understood by the following argument. Assume that the arc is not continuous. Then the conic must be a hyperbola with one branch passing through P_{i-1} and the other branch passing through P_i. Each branch of a hyperbola is infinite in extent. Neither of the branches can meet the lines $t_{i-1} = 0$ and $t_i = 0$ at any points other than P_{i-1} and P_i; for if they did, that line would meet the conic in at least three points, the point of tangency being a point of double contact. Without loss of generality, we may assume that the branch through R_i also passes through P_{i-1}. Therefore, this branch, if it does not also pass through P_i, must cut the line $l_i = 0$ at some point other than P_{i-1}. This gives us a contradiction, since the line $l_i = 0$ would then meet the conic in three points. Therefore, the branch through P_{i-1} and R_i must also pass through P_i, and the arc is therefore continuous.

Any potential problems with disjoint arcs are hence avoided by making the conic pass through one additional point within the triangle $P_{i-1} Q_i P_i$. The particular choice of point R_i will determine whether the conic is a parabola, hyperbola or ellipse. Let us normalize our line equations by arranging that

$$t_{i-1}(x_i, y_i) = t_i(x_{i-1}, y_{i-1}) = 1.$$

Normalization of $l_i(x, y)$ to have unit value at Q_i results in

$$l_i(x, y) = 1 - t_{i-1}(x, y) - t_i(x, y).$$

We can now consider $t_{i-1} = 0$ and $t_i = 0$ as the coordinate axes of a local co-ordinate system. If (μ_i', ν_i') are the coordinates of R_i in this coordinate system, then the equation of the conic is

$$(1 - \mu_i' - \nu_i')^2 t_{i-1} t_i - \mu_i' \nu_i' (1 - t_{i-1} - t_i)^2 = 0.$$

Rearranging this equation, we obtain, in terms of the new coordinate system,

$$\mu_i' \nu_i' t_{i-1}^2 + [2\mu_i' \nu_i' - (1 - \mu_i' - \nu_i')^2] t_{i-1} t_i + \mu_i' \nu_i' t_i^2$$
$$- 2\mu_i' \nu_i' (t_{i-1} + t_i) + \mu_i' \nu_i' = 0,$$

which is an ellipse, parabola, or hyperbola depending on whether the discriminant

$$(1 - \mu_i' - \nu_i')^2 [(\mu_i' - \nu_i')^2 - 2(\mu_i' + \nu_i') + 1],$$

is less than zero, equal to zero, or greater than zero. If the first factor is zero, the conic is degenerate. Therefore, the curve

$$(t_{i-1} - t_i)^2 - 2(t_{i-1} + t_i) + 1 = 0, \tag{2.16}$$

which is a parabola, divides the triangle $P_{i-1} Q_i P_i$ into two regions. The left-hand side of (2.16) is positive at Q_i, that is, when $t_{i-1} = t_i = 0$. Therefore, if the point R_i is chosen on the same side of the dividing parabola as Q_i, the resulting conic is a hyperbola. If R_i is on the opposite side of the parabola, then the conic is an ellipse, and if R_i lies anywhere on the parabola (2.16), the resulting conic is a parabola, which is, indeed, that very same parabola.

Were this technique being used to approximate a given curve, then the appropriate values of the slopes would be known. It would usually be suitable to calculate these slopes at alternate points, using the intermediate point as our point R_i that fixes the particular conic of the pencil. The situation is then as depicted in Figure 2.33, where we assume that some curve is given. Since no conic possesses an inflection point, we must select any inflection points of the given curve as endpoints of conic arcs. We calculate (or are given) the tangents at every second point P_i, P_{i+2}, and P_{i+4} and use the technique described above to produce a piecewise conic approximation, where the conic C_{i+1} is defined between points P_i and P_{i+2}, being tangent to the prescribed tangents at these points and also passing through the point P_{i+1}. If Q_{i+1} is the point where the tangents at P_i and P_{i+2} meet, then C_{i+1} is continuous from P_i to P_{i+2} if P_{i+1} lies within the triangle $P_i Q_{i+1} P_{i+2}$. This condition would be trivial to check during the construction process. The procedure then gives a very simple way of producing piecewise conic C^1 approximations to curves.

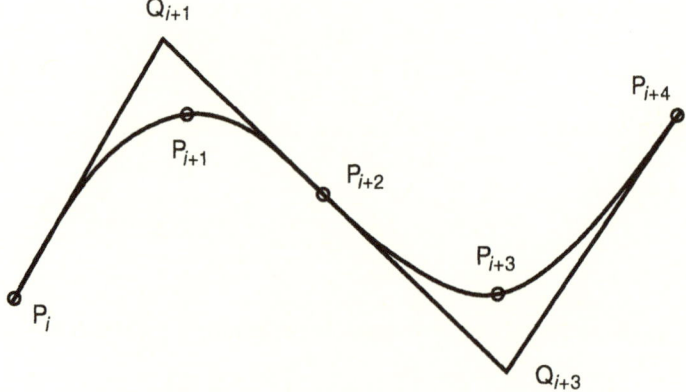

Figure 2.33. Piecewise conic C^1 curves.

2.18 Bibliographical Notes

There are many texts on engineering drawing that provide the interested reader with more detailed information on procedures for actually drawing (as opposed to mathematically defining) conics and on specific applications of such curves in various aspects of engineering. *The Fundamentals of Engineering Drawing and Graphic Technology* by Thomas E. French and Charles J. Vierk (3rd edition), McGraw-Hill (1972), and *Drafting for Industry* by Walter C. Brown, Goodheart-Willcox (1978), are two recommended texts on this subject.

Current texts on calculus and analytical geometry usually contain chapters on conics, but it is unlikely that they will be of much use to the serious student. We can do no better than to refer the reader to George Salmon's *A Treatise on Conic Sections*, which appeared in 1847 and has remained a standard text since then. The sixth edition was published in 1879 and is still available, having been reprinted by the Chelsea Publishing Company, New York. The text *Mathematical Elements for Computer Graphics* by D.F. Rogers and J.A. Adams, McGraw-Hill (1976), includes a chapter on plane curves, although restricting itself almost entirely to conics.

Exercises

2.1. Show, using the constructions of Figure 2.1, that the length of the major axis of an ellipse is equal to the distance between the points of contact of the two spheres with a generating line of the cone.

2.2. Show, using the constructions of Figure 2.1, that the eccentricity of an ellipse is less than unity.

2.3. By cutting the right circular cone with a plane that meets the cone in two disjoint curves, one on either side of the vertex, show, by an argument analogous to that given in the text for the ellipse, that the difference between the distances from any point on a hyperbola to its foci is a constant.

2.4. Show, using graphical construction, that the eccentricity of a hyperbola is greater than unity.

2.5. Given the focus, the eccentricity, and two points on a conic, how many possible positions of the directrix are there?

2.6. Let PM be an ordinate of an ellipse, A one end of the major axis, B one end of the minor axis, and C the center. Let PM be extended to P', where $(BC)(P'M) = (AC)(PM)$. Without using coordinate geometry, show that the locus of P' is a circle.

2.7. A line PP' perpendicular to the axis of a parabola with vertex A meets the parabola at P and P' and the axis at M, The circle defined by the points A, P, and P' meets the axis again at N. Without using coordinate geometry, show that the length of the latus rectum is MN.

2.8. Without using coordinate geometry, show that any chord through the center of an ellipse or a hyperbola is bisected at the center.

2.9. Let AB and CD be, respectively, the major and minor axes of an ellipse. Let X be on the minor axis, Y on the major axis, and Z on XY extended, such that $2(XZ) = AB$ and $2(YZ) = CD$. Without using coordinate geometry, show that Z lies on the ellipse.

2.10. Let A, B, and C be the vertices of a triangle, D a variable point on BC, and DM parallel to AB, with M on AC. If DP is drawn perpendicular to AC so that $(BC)(DC) = (PM)^2$, show, without the use of coordinate geometry, that the locus of P is an ellipse.

2.11. Let AB be a line containing the points $P_1, P_2, \ldots, P_i, \ldots$ such that $AP_i > AP_{i-1}$ and such that $AP_1, AP_2 \ldots, AP_i, \ldots$ are the diameters of semicircles drawn through A. Let C be on AB such that $AC < AP_1$, and let the perpendicular to AB at C meet the semicircle through P_i at the point $Q_i, i = 1, 2, \ldots$. Erect perpendiculars to AB at $P_i, i = 1, 2, \ldots$, and let $\{R_i, i = 1, 2, \ldots, \}$ be the set of points such that $P_i C Q_i R_i$ is a rectangle. Without using coordinate geometry, show that the set of points $\{R_i, i = 1, 2, \ldots\}$ lies on a parabola.

2.12. In a similar fashion to the solution displayed in Example 2.3, derive Cartesian equations for the ellipse and the parabola.

2.13. Determine polar and parametric equations for the following conics:
(a) $x^2 + y^2 = 2x$
(b) $3x^2 - y^2 + 32x + 64 = 0,$

(c) $x^2 + 2y^2 + xy - y = 0$,

(d) $y^2 - x - 1 = 0$,

(e) the parabola with focus at $(1, 1)$ and directrix $r \cos \theta = -4$,

(f) the hyperbola of eccentricity $\frac{3}{2}$ that has a focus at the origin with corresponding directrix $r \cos \theta = 9$,

(g) the ellipse of eccentricity $\frac{1}{8}$ with a focus at $(2, 1)$ and corresponding directrix $r \cos \theta = 16$.

2.14.   Determine the type of each of the following conics. Test for line pairs and repeated lines. Find the coordinates of the center in the cases when it exists:

(a) $16x^2 - 8xy + y^2 + 8x - 2y + 1 = 0$,

(b) $x^2 + 3xy - y^2 + x - y + 1 = 0$,

(c) $2x^2 + 4xy + 5y^2 - 2x + y - 1 = 0$,

(d) $4x^2 - 4xy + y^2 - x - y - 3 = 0$,

(e) $2x^2 - xy - 3y^2 - 3x + 2y + 1 = 0$.

2.15.   Taking the center of an ellipse or a hyperbola as the point of intersection of the axes, show algebraically that any chord through the center of an ellipse or hyperbola is bisected at the center.

2.16.   Find an equation for the locus of centers of circles that touch two given circles.

2.17.   Find an equation for the diameter conjugate, with respect to the conic $x^2 + 4y^2 - 2x - 16y + 13 = 0$, to the diameter that has unit slope.

2.18.   Rewrite the conic $x^2 - 2xy + 7y^2 + 14x - 26y + 30 = 0$ with its center as origin. Find the diameter conjugate to $y = 2x$, and find the axes.

2.19.   Show, using vector notation, that four points form the vertices of a parallelogram if and only if the diagonals of the quadrilateral formed by the four points bisect each other.

2.20.   Show, using vectors, that the distance form the point (x_1, y_1) to the line $ax + by + c = 0$ is

$$\frac{|ax_1 + by_1 + c|}{\sqrt{a^2 + b^2}}.$$

2.21.   Find the coordinates of the centers of the circles that touch the lines

$$y + x\sqrt{3} - \sqrt{3} = 0, \qquad y - x\sqrt{3} - \sqrt{3} = 0, \quad \text{and} \quad y = 0.$$

2.22.   Considering the conic that is represented by the general equation, show that $ax + hy + g = 0$ is the equation of the diameter that bisects chords parallel to the x-axis.

2.23. Find the equation of a circle that cuts the circles

$$x^2 + y^2 - 1 = 0, x^2 + y^2 - 4x + 6y - 12 = 0 \quad \text{and}$$

$$x^2 + y^2 - 2x + 2y + 1 = 0$$

orthogonally. Does this circle cut any other circles orthogonally?

2.24. Find the angle of intersection of two circles.

2.25. Show that when a line pair is transformed to its center, the constant term vanishes.

2.26. Find the axes of $14x^2 - 4xy + 11y^2 = 0$, and write the equation referred to these axes.

2.27. Show that two central conics have, in general, one and only one pair of common conjugate diameters.

2.28. Show that, in general, there are two parabolas through four points.

2.29. What is the equation of a conic passing through the points where a given conic meets the axes?

2.30. Find the parametric equation for the locus of centers of conics through four points.

2.31. Find the equation of the chord connecting the intersection points of two circles.

2.32. Find the equation of the conic passing through the five points

$$(2, 1), \qquad (1, 0), \qquad (3, -1), \qquad (-1, 0), \quad \text{and} \quad (3, 3).$$

2.33. Find an equation representing all conics through the points

$$(-1, -1), \qquad (2, 2), \qquad (10, 10), \qquad (1000, 1000), \quad \text{and} \quad (0, 1).$$

2.34. By (a) considering a tangent as a limiting case of a chord, and (b) considering a tangent as a line that meets a conic in two coincident points, find the tangent at the point $(0, 1)$ to the conic

$$x^2 - xy + 2y^2 - 3x + y - 3 = 0.$$

2.35. Show that the equation of the tangent to the general conic at the point (x_1, y_1) is

$$axx_1 + h(xy_1 + yx_1) + byy_1 + g(x + x_1) + f(y + y_1) + c = 0.$$

2.36. Find the conics through $(0, 0)$, $(1, 0)$, $(0, 1)$, and $(2, 0)$ that touch the line $x = 3$.

2.37. Find the conics through the points $(\pm 1, 0)$ and $(0, \pm 1)$ that touch the line $4x + 5 = 0$.

2.38. Find the conic that is tangent to the five lines $x - 2y + 2 = 0$, $2x - y + 2 = 0$, $x - 2y - 2 = 0$, $2x - y - 2 = 0$, and $x + y - 2 = 0$.

2.39. Find the conic which is tangent to the lines $x = \pm 1$, $y = \pm 1$, and $x + y - \sqrt{2} = 0$.

2.40. Find the tangential equation of the conic whose point equation is

$$4x^2 - 6xy + y^2 + 2x - 8y + 11 = 0,$$

and the point equation of the conic whose tangential equation is

$$\alpha^2 + 2\gamma^2 - \alpha\beta - 3\beta\gamma + 2\gamma = 0.$$

2.41. Find the conditions that must be satisfied by conics that pass through $(0, 0)$, $(1, 1)$, and $(-1, 2)$ and are tangent to the lines $x - y - 4 = 0$ and $x + 2y + 2 = 0$.

2.42. Find the conditions that must be satisfied by conics that are tangent to $x = 0$, $y = 0$, and $1 - x - y = 0$ and pass through the points $(2, 1)$ and $(1, 2)$.

2.43. Find the condition that must be satisfied by conics that are tangent to $x = 0$, $y = 0$, $x = 4$, and $1 - x - y = 0$ and pass through the point $(2, 4)$.

2.44. Find all conics through $(1, -1)$, $(2, 0)$, $(3, 1)$, and $(-1, -5)$ and tangent to $2x + y - 1 = 0$.

2.45. Find all conics through $(1, 0)$, $(0, 1)$, and $(4, -1)$ and tangent to $x - 2y - 1 = 0$ and $4x - 3y + 3 = 0$.

2.46. Find the conics through $(1, 0)$ and $(0, 1)$ that are tangent to $x - y + 1 = 0$, $x - y - 1 = 0$, and $x + y - 4 = 0$.

2.47. Using the concept of a net of conics tangent to three lines, show that there are four circles touching three lines. Give two arguments, one based upon the tangential equation of the net and the other based upon the point equation of the net. What is the point equation of the net?

2.48. By considering the circle of curvature at a point as the limiting case of the circle defined by the point and two neighboring points on the curve as the neighboring points coalesce to the given point, find the equation of the circle of curvature to the conic $x^2 - y^2 + 2x - y - 6 = 0$ at the point $(2, 1)$.

2.49. Find the net of all conics that have three-point contact with the conic $x^2 - y^2 + 2x - y - 6 = 0$ at the point $(2, 1)$.

2.50. Find the equation of the conic that has three-point contact with the conic $4x^2 + xy - 3y = 0$ at the origin and passes through the points $(1, 1)$ and $(0, 2)$.

2.51. Find the equation of the net of conics whose center of curvature at the point $(0, 0)$ is at $(\frac{1}{2}, \frac{1}{2})$.

2.52. Find the equation of the osculating circle, at the origin, to the conic $4x^2 + xy - 2y^2 + 3x = 0$.

2.53. Find the equation of the parabola that has four-point contact with the conic $x^2 + 2y^2 - x + 2y = 0$ at the origin.

2.54. Show that, if four of five given points are collinear, there is a pencil of conics through the five points.

2.55. Construct more than four conics through the points $(0, 0)$, $(1, 0)$, and $(0, 1)$ and tangent to the lines $x + y - 1 = 0$ and $2x + 2y - 5 = 0$.

2.56. Show that there is no conic whose circle of curvature at the point $(0, 1)$ is $x^2 + y^2 - 1 = 0$ and that passes through the point $(1, 1)$.

2.57. Construct a parametrization of the conic

$$x^2 - 3xy + 4y^2 - 8x + 5y + 1 = 0$$

by taking arbitrary lines through the point with coordinates $(1, 1)$.

2.58. Produce a parametrization of the conic

$$2x^2 + 4y^2 - 3x - 5y + 1 = 0$$

such that the arc of the conic from $(0, 1)$ clockwise to $(1, 0)$ corresponds to parameter values from zero increasing to one.

3

Synthetic Geometry

3.1 Introduction

We will now have to become familiar with homogeneous coordinates, the notion of a complex point of a curve, and the ramifications of working in projective, rather than Euclidean, space. The techniques we will develop will provide the tools necessary to analyze some very important "real" problems. However, before introducing projective space, let us continue our discussion of geometry.

3.2 Properties Invariant under Transformations

If we think for a moment of the usual high-school geometry that concerns itself with conditions under which two triangles are congruent, we realize that these conditions take no account of where in the plane the triangles are actually placed. When we studied conics, we often moved the conic to a particular position in the plane, for example, such that its center was at the origin, and hence any properties we deduced were independent of the particular position of the conic. This idea characterizes *Euclidean geometry*. These movements, or displacements, from one position to another are, simply, Euclidean transformations. We can think of the displacement that actually leaves things unchanged as an identity displacement or *identity transformation*. Finally, we notice that for any displacement there is always the "opposite" or inverse displacement, which returns things to their original position.

If a set of lines is concurrent, it will remain so under a displacement, and hence we can think of concurrency of lines as being a Euclidean property. Likewise, collinearity of points can be thought of as a Euclidean property. However, these properties are invariant under a more general transformation than a mere displacement. What we see with our eyes is a *projection* of the physical world. Under such a projection, which is also a transformation and which, because of its connection to the way we see, is a very real concept indeed, a circle is no longer invariant. A circle, when viewed from a point not on the line through the center of the circle and perpendicular to the plane of the circle, will appear as an ellipse. However, concurrent lines and collinear points will remain so even under this more general transformation. The study of those properties of objects that are invariant under this more general class of transformations is called *projective geometry*. Most modern texts on computer graphics devote substantial space to a discussion of these crucially important transformations. It is usual to refer to a particular property as being associated with the largest class of transformations under which it is an invariant. Thus we would say that concurrency of lines is a projective property rather than saying that it is both a projective and a Euclidean property.

If a beam of light spreads out from a point source and casts a shadow of some object on a flat wall, then the shadow is a *central projection* of the object. Since our eyes are fairly close together, what we see when looking at objects is, essentially, a central projection of the scene. In many applications, one wishes to look at views of objects as if the eye were so far away from the object that all rays of light from the object to the eye are parallel. In both computer graphics and geometry, this is referred to as a *parallel projection* and can be thought of as a special case of central projection, where the center of projection is "infinitely far" from the object. The geometry that concerns itself with parallel projections is called *affine geometry*.

The projective, affine, and Euclidean geometries are the three most important geometrical structures in computer graphics and computer-aided design.

3.3 Synthetic and Analytical Approaches

There are two main approaches to the study of geometry: the *synthetic* and the *analytical*. We attempt no definition of these words, but do attempt to give an idea of the circumstances under which each would be more appropriately used. In a synthetic approach, we argue directly about the geometrical entities involved, namely the lines, planes, spheres, curves, and so on. In the analytical approach, the geometrical entities are related, via the use of coordinates, to equations, and the argument then proceeds by the use of algebra. Most of the remainder of this book will concern itself with analytical or, to give it its more grandiose name, algebraic geometry. Often, however, great insight into a problem can be gained by considering it first from a synthetic viewpoint, and we proceed, in the present chapter, to give some examples of synthetic geometry.

Let us point out that the common algebraic operations of addition and multiplication can be handled in a purely synthetic manner. We illustrate the Euclidean addition of planar points in the following example.

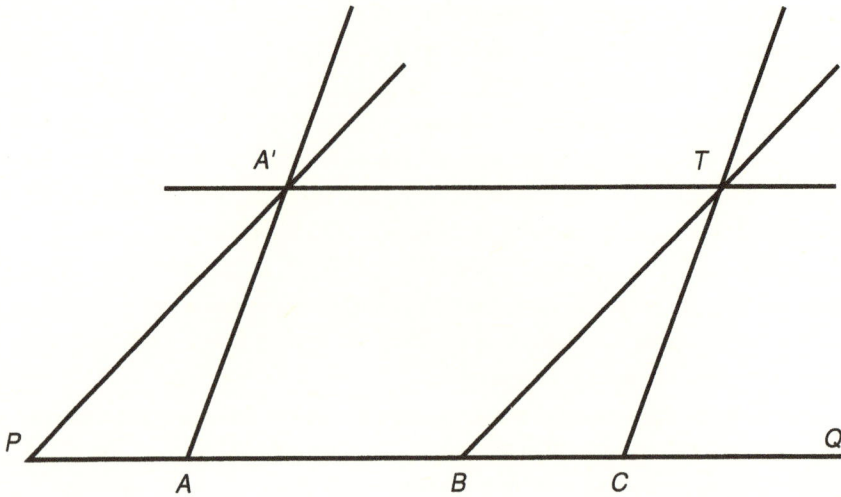

Figure 3.1. The synthetic addition of points.

Example 3.1 *Let A and B be points in the Euclidean plane, and let PQ be a line through A and B (Figure 3.1). If A and B are distinct, then the line PQ will be unique. We will define the sum of the two points A and B, written as A + B, by the following construction.*

Choose a point A' not on the line PQ, and draw the distinct lines PA' and AA', both different from PQ. Through B, draw a line parallel to PA', and through A', draw a line parallel to PQ. Since these two lines cannot be parallel (PA' and PQ are not, since they intersect), let them meet in T, and through T draw a line parallel to AA' to meet PQ in C. We now define $A + B$ by

$$A + B = C.$$

We can relate this definition to numbers in the following manner. Assume that the point P represents the origin and that the directed line segments $\overrightarrow{PA}$ and $\overrightarrow{PB}$ represent the numbers a and b, respectively. Then, since the triangles PAA' and BCT are congruent by construction, the directed line segment $\overrightarrow{BC}$ also represents the number a, so that $\overrightarrow{PC}$ represents the number $a + b$.

3.4 The Projective and Euclidean Planes

We will study algebraic curves in the projective plane and, from the results in that plane, deduce results in the Euclidean plane. We will find that a single type of curve in the projective plane may correspond to more than one type of curve in the Euclidean plane. It is clear, therefore, that there is a very profound difference between these planes. Three ideas are particularly important. These are: the idea of points lying on a line (being collinear) or lines passing through a point (being concurrent), the idea of parallelism, and the idea of an ordering of points on a line.

The Euclidean system as studied by the Greeks had certain flaws in connection with undefined terms, existence, congruence, and order, and these problems were addressed in the late nineteenth and early twentieth centuries. Many different sets of axioms and different lists of undefined terms were proposed around that time, but it is David Hilbert's that has become the standard. His treatise *Foundations of Geometry* was published in 1899. In this work he set Euclidean geometry on a rigorous axiomatic foundation. He took the words "point" and "line" to be undefined terms. The idea of a point lying on a line or a line passing through a point is referred to as an *incidence* relation. Hilbert's first set of axioms concerns existence and incidence. The axioms are:

Axiom E1 There exists at least one line.

Axiom E2 On each line there exist at least two points.

Axiom E3 Not all points lie on the same line.

Axiom E4 There is one and only one line passing through (or incident with) two given distinct points.

The following theorem is important and is an immediate deduction from these axioms.

Theorem 3.1 *Two distinct lines meet in at most one point.*

Proof. We use a *reductio ad absurdum* argument. Assume that the two distinct lines meet in two or more distinct points. Take any two of these points; by Axiom E4, these two points define a unique line. This gives us a contradiction, and hence there can be at most one point common to two distinct lines. ∎

He gives a definition of parallel, and a single related axiom. The definition is that two lines are *parallel* if they have no common point. The axiom is:

Axiom E5 (Parallel Axiom) Given a line and a point not on the line, there is one and only one line containing the given point and parallel to the given line.

The following theorem is an immediate consequence of the fifth axiom.

Theorem 3.2 *Lines parallel to the same line are parallel to each other.*

Proof. Again we argue by contradiction. Let two distinct lines l and m be given parallel to a third line n, and assume that l is not parallel to m. Then l and m meet in a point P, say. Now, by Axiom E5, there is a unique line through P that is parallel to n. Therefore, l and m must be the same line, and we have our contradiction. Since, therefore, l and m are distinct and do not meet in any point, they are parallel to each other.   ■

It is the concept of parallel lines that, more than anything else, distinguishes the Euclidean from the projective plane. If we are traveling along a multilane highway that happens to be straight for a long way, then what our eyes see is all the lanes, apparently, converging. There is, therefore, a temptation to say that parallel lines "meet at infinity." Such a statement makes no sense in the Euclidean plane, but can take on a justifiable meaning in the projective plane. Let us construct the projective plane by the addition to the Euclidean plane of certain points and a special line. For each set of parallel lines in the Euclidean plane, let us add a single point to the plane. This point will represent the common intersection point of the set of parallel lines. Call this point an *ideal point*. There is exactly one such point for each set of parallel lines, and hence one and only one for each direction in the plane. Hence, if O is a point of the Euclidean plane and l is a line through O, we have added one ideal point to l. Let this point be denoted by P_l (Figure 3.2). We now introduce an additional line z into the Euclidean plane, which we call the *ideal line*. This line will pass through the point P_l and, in addition, will also pass through all the other ideal points, each one corresponding to a unique direction. This line is defined by the following axiom.

Incidence Axiom for Ideal Points The ideal point P_l of the line l is on a line m if and only if one of the following is true:

(a) the line m is parallel to the line l,
(b) the line m is identical to the line l, or
(c) the line m is identical to the line z.

We have now constructed the real projective plane. Two important results of this construction, which greatly simplify our later work, are that in the projective plane two distinct points uniquely determine a line and two distinct lines uniquely determine a point. Let A and B be two distinct points. If neither A

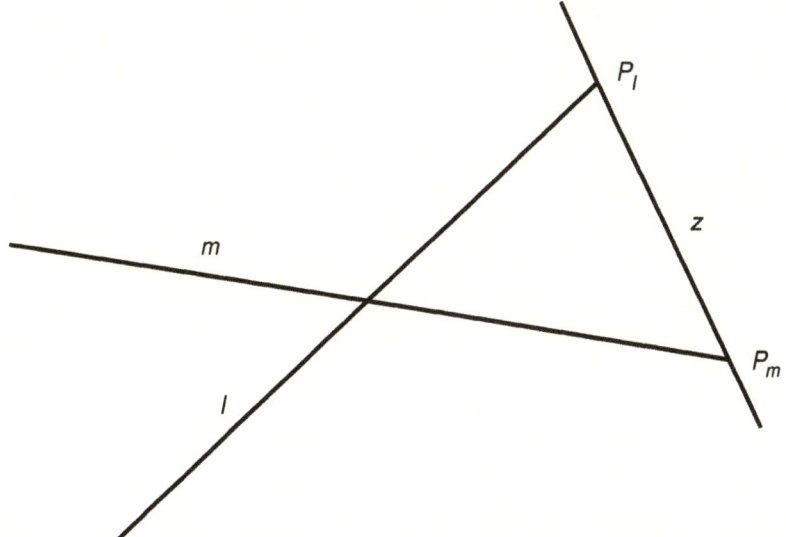

Figure 3.2. Each ordinary line contains a single point.

nor B is an ideal point then they uniquely determine a line in the usual Euclidean sense. If one of the points, say A, is an ideal point and B is not, then consider the pencil of lines through B. By construction of the projective plane there is a unique ideal point associated with each direction, and hence there is a unique direction such that a member of the pencil of lines through B will also pass through A. If both A and B are ideal points, then the line they define is the ideal line and no other, since a nonideal line contains exactly one ideal point. Clearly, the parallel axiom of the Euclidean plane is not valid in the projective plane.

Another concept that is important in the Euclidean plane but has no counterpart in the projective plane is the ordering of points on a line, that is, the concept that one point is "between" two others on a line. Hilbert addressed this idea in five axioms known as the axioms of order. We use the notation ABC to denote the concept that point B is between points A and C.

Axiom E6 If ABC, then A, B, and C are distinct and collinear.

Axiom E7 If ABC, then CBA.

Axiom E8 If A and C are distinct, there is a point B such that ABC and there is a point D such that ACD.

Axiom E9 Given three distinct collinear points, one and only one of the points is between the other two.

Axiom E10 Given four distinct collinear points, it is possible to name them A, B, C, and D in such a way that ABC, ABD, ACD, and BCD.

The first of these axioms simply tells us that the concept "between" only has meaning for distinct points and points that are on a line. The next one tells us that if B is between A and C, then it is also between C and A. Axiom E8 tells us that if we have two distinct points, then we can always find a third that is between them and one that is not between them. The ninth axiom gives us uniqueness of the relationship for three points, and the last one tells us that we can always order four points in a particular way.

3.4.1 Loss of Order in the Projective Plane

Let us now consider this concept of order or "betweenness" in a projective plane. It is reasonable and, indeed, the usual practice to call the ideal point associated with a set of parallel lines the *point at infinity*. We reiterate that it is a single point that has been added, not two points. The ideal line is usually called the *line at infinity*. If we think of the ideal point as not lying between any two other, ordinary points, we get into the following difficulties: Let A and B be ordinary points, and I the ideal point associated with the line defined by A and B. Then we must have ABI (or IAB). This would seem to be a way of saying that the ideal point does not lie between two ordinary points. However, Axiom E8 tells us that there is a point D (which must be an ordinary point, because there is only a single ideal point associated with each line) such that AID. Hence the intuitive concept of having the ideal point infinitely far to the right (or the left) leads to contradictions if we assume the axioms of order. The projective plane does not have an ordering system like that given by Hilbert's axioms. It is impossible to say whether a given point on a line lies to one or the other side of the point at infinity, and hence the ordering of points on the line, and all properties deduced therefrom, are lost.

Consider two lines meeting at a point P_0, and let A be another point on one of the lines. Rotating this line about A causes the intersection point P, say, with the other line to move along the line away from the original intersection point P_0. Continuing to rotate in the same direction, P continues to move in the same direction, moving further away from P_0. At some stage the two lines will become parallel, at which instant the intersection point P vanishes. Continuing to rotate in the same direction, P immediately appears somewhere on the opposite end of the line from that which it left at the instant of parallelism. The addition of the single ideal point in a sense closes the line. Any projective line can be chosen as the ideal line, the line at infinity, or whatever one wants to call it. There is no need to think of the projective plane as consisting of two kinds of points and two kinds of lines. There are simply points and lines. Only when related back to a Euclidean plane must we designate one of the lines to be the line at infinity, which then does not belong to the Euclidean plane. This is a very important point, and one we will make use of many times in our ensuing algebraic discussion: in the projective plane there is only one kind of point and only one kind of line.

The axioms of the real projective plane are very simple. They are:

Axiom P1 There exists at least one line.

Axiom P2 On each line there exist at least three points.

Axiom P3 Not all points lie on the same line.

Axiom P4 Two distinct points lie on one and only one line.

Axiom P5 Two distinct lines meet in one and only one point.

Axiom P6 There is a one-to-one correspondence between the real numbers and all but one point of a line.

3.5 Duality

When we examine the axioms of the real projective plane, and simple consequences thereof, we notice a remarkable *duality* between points and lines. There is at least one point, and there is at least one line. Two distinct lines determine a unique point, and two distinct points determine a unique line. Not all points lie on the same line, and not all lines pass through the same point, and so on. Thus, if we prove a theorem about points in relation to lines, we should be able to prove a dual theorem about lines in relation to points. The dual theorem is obtained by interchanging the words "point" and "line," "collinear" and "concurrent," "lying on" and "passing through," ("incident with" is self-dual), and so on.

Example 3.2 *Write the dual statements of each of the following:*

(a) The set of all points lying on a given line.
(b) Two lines and a point on neither line.
(c) The line determined by two given points, and the point of intersection of two given lines.

(a) The set of all lines passing through a given point.
(b) Two points and a line passing through neither point.
(c) The point determined by two given lines, and the line passing through two given points.

The ability to deduce a theorem from its dual is called the *principle of duality*. It is a very powerful principle in projective geometry. To be able to use it freely, we should prove the dual of each of the axioms of the real projective plane. In Chapter 2 we actually obtained dual results independently: there is a unique conic through five points, and there is a unique conic tangent to five lines; there are two conics through four points and tangent to one line, and there are two conics tangent to four lines and passing through one point; there are four conics through three points and tangent to two lines, and there are four conics tangent to three lines, and passing through two points.

3.6 The Cross Ratio

We have mentioned the concept of geometry as being the study of the properties that are invariant under certain transformations. We can ask ourselves what things remain unchanged as we move our object around using the allowable transformations. For example, we might ask how invariant the position of an object is. Rather immediately, we see that position is not invariant under even a Euclidean transformation, and so we would say that position is not a Euclidean property. However, distance, area, and angles are all preserved under Euclidean transformations, and so these are referred to as Euclidean properties. If we allow affine transformations (for example, the casting of shadows onto a plane by a parallel beam of light, while the object under study can be moved in three-space), we find that none of these three properties is invariant. The shadow of a rod of fixed length can be made to appear smaller by tilting the rod towards the axis of the beam of light. Doing the same thing with a closed planar figure will make its area become smaller, and angles can be made to appear larger or smaller, depending on how we tilt the object. Clearly, then, such properties are not affine properties. What properties are affine properties? We will not discuss this question, but simply state, and leave to the reader to prove, that one such property is the ratio of lengths measured along a line.

The projective transformation is even more general than the affine one, and hence we would expect there to be fewer projectively invariant properties than affine ones. Under an affine transformation, a circle will always look like a circle, an ellipse, or (in the degenerate case) a line. A circle can never look like a parabola under such a transformation. This is not so with projective tranformations. Consider the central projection, which is a special case of a projective transformation. Imagine, for example, that your eye is at the vertex of an infinite right circular cone and you are looking down the axis of the cone. From that position of your eye, all that you can see of the surface of the cone is, in fact, a circle. No matter what was drawn on the surface, from the viewpoint of the vertex, it would appear like a circle. Therefore, any section of the cone would result in a curve that, to you, would look like a circle. Depending on the particular section, this curve, *in the plane of the curve*, might be any of the conics (parabola, circle, ellipse, hyperbola, or line pair), but *projected from your eye* it will be a circle. Hence (somewhat intuitively) we see that all conics are equivalent in the projective plane. Hence, the property of being a particular kind of conic, which is, indeed, an affine property, is not a projective one. Once more, we will not dwell on the projective properties in general, but we will discuss one important one, the cross ratio.

Let A, B, C, and D be four collinear points (Figure 3.3). The *cross ratio* of the ordered pair of points (C, D) with respect to the ordered pair (A, B) is written $\{A, B; C, D\}$ and is defined by

$$\{A, B; C, D\} = \frac{AC/CB}{AD/DB} = \frac{AC}{CB} \times \frac{DB}{AD},$$

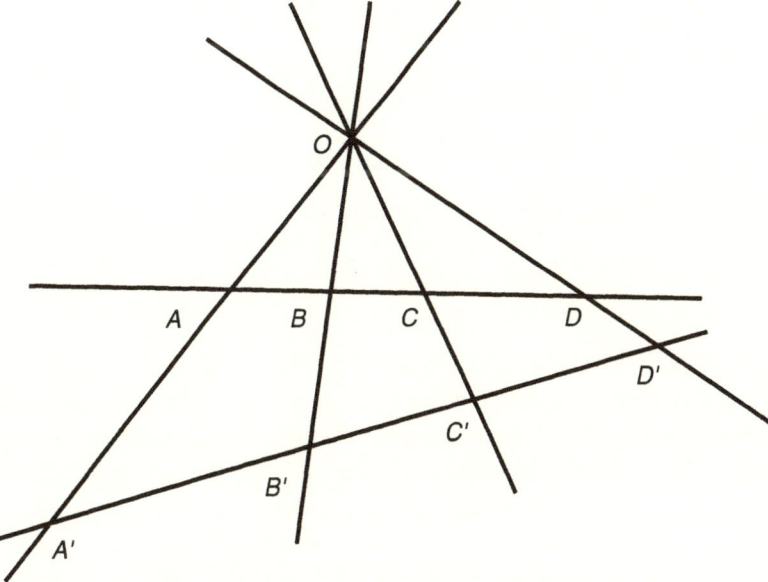

Figure 3.3. The cross ratio of four collinear points.

where the segments are considered as being directed segments, with DB, say, being the negative of BD. We have the following theorem.

Theorem 3.3 *If four collinear points A, B, C, and D are projected from a point O onto the four collinear points A', B', C', and D', respectively, then*

$$\{A, B; C, D\} = \{A', B'; C', D'\}.$$

Proof. Using the sine rule in the appropriate triangles, we obtain (Figure 3.3)

$$\frac{AC}{\sin AOC} = \frac{AO}{\sin ACO}, \qquad \frac{CB}{\sin COB} = \frac{BO}{\sin BCO}.$$

Therefore,

$$\frac{AC}{CB} = \frac{AO}{BO} \frac{\sin AOC}{\sin ACO} \frac{\sin BCO}{\sin COB}.$$

However, $\sin BCO = \sin ACO$: therefore,

$$\frac{AC}{CB} = \frac{AO}{BO} \frac{\sin AOC}{\sin COB}.$$

Similarly,

$$\frac{AD}{DB} = \frac{AO}{BO} \frac{\sin AOD}{\sin DOB}.$$

Hence,

$$\{A, B; C, D\} = \frac{\sin AOC \sin DOB}{\sin COB \sin AOD},$$

which is independent of the particular line. Hence

$$\{A, B; C, D\} = \{A', B'; C', D'\}. \qquad \blacksquare$$

The point O is referred to as the *center* of the projection, and the projection itself is what we have called a *central projection*. The cross ratio of four points is, therefore, invariant under a central projection. (A general projective transformation can be reduced to a sequence of central projections, and hence the cross ratio is invariant under a general projective transformation.) This result has been known for a very long time; certainly it was known to Pappus, and from Pappus's work one gets the impression that it was actually known to Euclid.

3.7 Desargues's Theorem

We now give an example of a synthetic argument. We choose the proof of a special case of Desargues's theorem. In the next chapter we will prove this theorem again, using algebraic methods. Here, we make no use of coordinates and manage a more general proof than that which we will give in the next chapter. Synthetic proofs tend to give more insight than algebraic ones, but do not lend themselves as easily to computation.

Two triangles ABC and $A'B'C'$ are said to be in *perspective* if the lines AA', BB', and CC' joining corresponding vertices are concurrent. The point of concurrency is called the *center of perspective*.

Theorem 3.4 (*Desargues*) *If two noncoplanar triangles without a common vertex are in perspective, then the points of intersection of corresponding sides are collinear.*

Proof. Let triangles ABC and $A'B'C'$ be in perspective from P (see Figure 3.4). Then AC and $A'C'$ lie in the plane defined by PA and PC and hence meet in a point B''. Now AC also lies in the plane ABC, and $A'C'$ lies in the plane $A'B'C'$, and hence the point B'' must lie on the line of intersection of these two planes. This line is unique, since the planes ABC and $A'B'C'$ are distinct. Similarly, AB and $A'B'$ meet at C'', and BC and $B'C'$ meet at A'', and these two

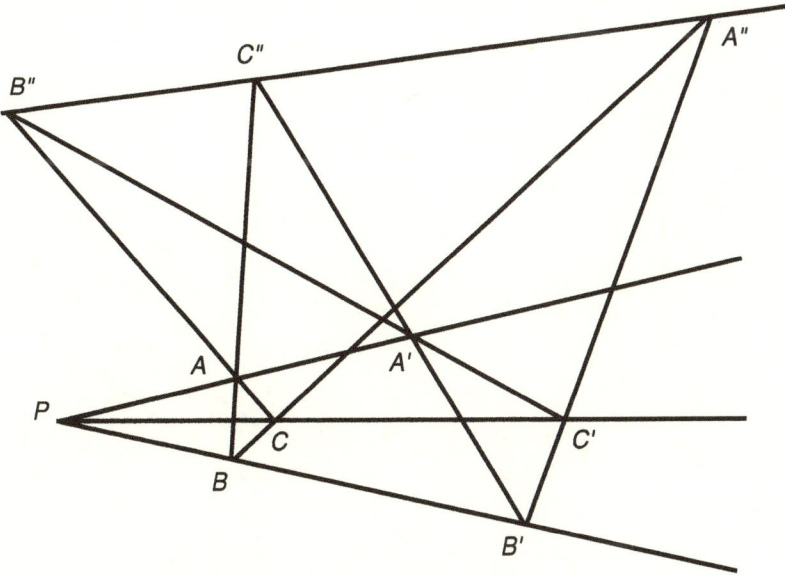

Figure 3.4. The theorem of Desargues.

points must also lie in the line of intersection of the planes ABC and $A'B'C'$. The points of intersection are therefore collinear. ∎

3.8 The Affine Plane: A Synthetic Introduction

Having discussed the projective plane and the ideal line and, in particular, having shown that the ideal line is just like any other line (that is, any line can be taken as the ideal line), we are now in a position to describe the affine plane from the synthetic viewpoint. The affine plane is defined as the projective plane with one line removed.

Consider three distinct projective lines a, b, and c. They must form a triangle. In Figure 3.5(a) these lines are not concurrent, and in Figure 3.5(b) the lines are concurrent. Let us now remove the line c from the projective plane to form an affine plane. The remaining two lines still intersect in their common vertex [Figures 3.5(a) and (b)]. However, if the three lines are concurrent and we remove c, the intersection point is also removed, and, by the definition of parallel lines as lines with no common point, the lines a and b become parallel in the

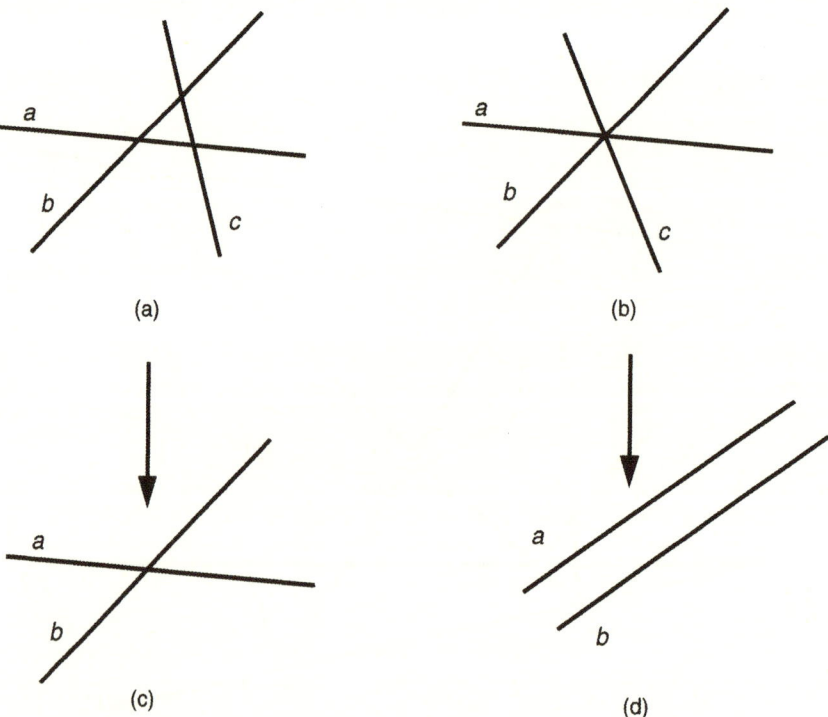

Figure 3.5. The removal of a projective line to form an affine plane.

corresponding affine plane. We constructed the projective plane by adding the ideal line to the Euclidean plane. Then we removed any special significance attached to this line, since all lines in the projective plane are the same. By removing any line from the projective plane we create an affine or Euclidean plane. There is no essential difference between the affine and the Euclidean planes. There is a difference between affine and Euclidean geometries. That difference is the difference between the transformations we use to study the various properties of interest.

3.9 Bibliographical Notes

Some texts that comment upon the development of the subject are *Euclidean and Non-Euclidean Geometries: Development and History* by Marvin Jay Greenberg, published by W. H. Freeman (1974), and *Foundations of Euclidean and Non-Euclidean Geometry* by Richard L. Faber, Marcel Dekker (1983). *An Outline of Projective Geometry* by Lynn E. Garner, Elsevier North Holland (1981), and *Projective Geometry and Algebraic Structures* by R. J. Mihalek, Academic Press (1972) treat the subject from a fairly abstract viewpoint.

Exercises

3.1.   Given four concurrent lines, construct a definition of the cross ratio of these lines that is consistent with the principle of duality.

3.2.   Show that
   (a) $\{A, B; C, D\} = \{C, D; A, B\}$,
   (b) $\{A, B; C, D\}\{B, A; C, D\} = 1$,
   (c) $\{A, B; C, D\}\{B, C; A, D\}\{C, A; B, D\} = -1$.

3.3.   If A, B, C, D, and E are collinear, show that
   (a) $\{A, B; C, D\}\{A, B; D, E\} = \{A, B; C, E\}$,
   (b) if $\{A, B; C, D\} = \{A, B; C, E\}$ then $D = E$.

3.4.   By considering Figure 3.1 and the fact that $B = C - A$, write out the algorithm for synthetic subtraction in the Euclidean plane.

3.5.   Using Hilbert's axioms of existence and incidence for the Euclidean plane, prove that
   (a) there exists at least one point;
   (b) there are at least two lines passing through each point;
   (c) not all lines pass through the same point.

3.6.   Using Axioms P1 through P5 of the projective plane, prove that
   (a) if P is a point of the plane, then there are at least three lines through P;

(b) there exists a set of four points, no three of which are collinear.

3.7. Let l_i, $i = 1, 2, \ldots, 7$, denote lines and P_i, $i = 1, 2, \ldots, 7$, denote points. Let the notation $\{P_i, P_j, P_k\} \wedge l_m$ mean that the points P_i, P_j, and P_k lie on the line l_m. Prove that, if

$$\{P_1, P_2, P_4\} \wedge l_1, \qquad \{P_2, P_3, P_5\} \wedge l_2, \qquad \{P_3, P_4, P_6\} \wedge l_3,$$
$$\{P_4, P_5, P_7\} \wedge l_4, \qquad \{P_5, P_6, P_1\} \wedge l_5, \qquad \{P_6, P_7, P_2\} \wedge l_6,$$
$$\{P_7, P_1, P_3\} \wedge l_7,$$

then the seven points and seven lines constitute a projective plane, that is, Axioms P1 through P5 are satisfied. We omit Axiom P6, since this omission allows us to consider projective planes other than the real one. The plane of this example is a finite one, that is, it has a finite number of points and lines. It can be "visualized" by taking P_1, P_4, and P_7 as the vertices of a triangle, P_6 the center of the inscribed circle that meets $P_1 P_4$ in P_2, $P_4 P_7$ in P_5 and $P_7 P_1$ in P_3. Then three of the lines are the sides of the triangle, three more lines connect a vertex through the center of the circle to the point on the opposite side, and the seventh line is the circle itself. It can be shown that a finite projective plane contains $N^2 + N + 1$ points and lines, where N is an integer greater than one. Our example is the smallest projective plane.

3.8. What are the dual statements to
    (a) four points, no three of which are collinear,
    (b) there exists at least one line,
    (c) on each line there exists at least three points,
    (d) distinct concurrent lines meet in a unique point.

3.9. Prove the converse of Theorem 3.4.

3.10. Construct two triangles that are in perspective *in the plane*, and verify the statement that Desargues's theorem is also true for coplanar triangles.

3.11. Construct two triangles in different planes, but such that the triangles are in perspective in three-space, and verify Desargues's theorem.

3.12. Let A, B, C and D, E, F be two sets of distinct collinear points. Let BF meet EC in P, CD meet FA in Q, and AE meet DB in R. Prove that P, Q, R are collinear. This result is known as the theorem of Pappus.

4

Algebraic Projective Geometry

4.1 Introduction

This is the chapter in which we will develop the tools needed for the subsequent discussion of algebraic curves. Although we are often interested in curves and surfaces in real space, we will gain our insight via a study in complex space, for in that context we can avail ourselves of more powerful tools. Furthermore, Euclidean and affine transformations are special cases of projective transformations. Projective geometry, where coordinates can have complex values, is therefore the most suitable framework for our study, and we develop this framework in this chapter.

We start with a motivation for homogeneous coordinates and then discuss such coordinate systems. The use of homogeneous coordinates is central to everything else we shall discuss, and it is important that the reader become familiar with them and also become facile at moving between them and affine coordinates. It is the change from homogeneous coordinates to affine coordinates that will enable us to deduce properties in affine space from those we have deduced in projective space.

One of the consequences of the axioms of projective geometry is the principal of duality. The duality is between points and lines, and by its use we can deduce a result concerning points and lines from one concerning lines and points; for example, knowing that there is, in general, a unique conic through five points, we can deduce that there is also, in general, a unique conic tangent to five lines. This duality is a fundamental property of projective geometry and not an artifact of any particular representation. Hence this duality will still be there in our coordinate representation of the geometry. The implication is that, if we deduce some equation that tells us something about points, there will be a corresponding equation that tells us about lines. Coordinates, therefore, are not only a representation of points; they can be a representation of

129

lines. There are both point and line coordinates, and we discuss them in this chapter.

Having developed coordinates, we then move on to the central theme of transformations. In this chapter we put this idea on a firm footing by introduction the algebraic structure of a *group*. From this point on we shall have a clear understanding of how the different transformations (rigid-motion, affine, similarity, and projective) relate to each other. The transformations will be described in terms of matrices, and the group properties are related to corresponding matrix properties.

One way to think about different geometrical planes is in terms of the number of points that can be mapped onto an equal number of points. In projective space this number is four, and this result is referred to as the *fundamental theorem* of projective (plane) geometry. We discuss this and a few other important properties, for example, the projective invariance of the conics.

The remainder of the chapter is devoted to a more detailed discussion of some of the special, but important cases of the projective transformation. We will consider central projections, affine transformations, similarity, and Euclidean transformations. Chapter 3 discussed the synthetic approach. This is important, as it is often the first step in a more detailed analysis. However, for our purposes the synthetic approach alone will not yield the equations, in terms of coordinates, that we need for computational purposes. It is the analytical approach that will enable us to study curves and surfaces in terms of equations, and it is the materials in the current chapter, the point and line coordinates and the transformations, that are the building blocks we need for our study of curves, which begins in Chapter 5.

4.2 More General Algebraic Methods

In the second chapter we discussed some important properties of conics and how they related to our topic. The mathematical tools we used were, in the main, those of analytical geometry in the real Euclidean plane. It will be necessary, however, to examine more general curves, and for this examination we shall need to use the tools of projective geometry. In the previous chapter we gave a brief introduction to the synthetic approach to projective geometry. Here we will develop an algebraic approach, the approach relying on coordinates and equations. Consider the following example.

Example 4.1 *Determine the points where the two conics $x^2 + y^2 - 1 = 0$ and $x^2 + y^2 - 2 = 0$ meet.*

The two conics would meet where

$$x^2 + y^2 = 1 \quad \text{and} \quad x^2 + y^2 = 2,$$

that is, where $1 = 2$. Hence, we conclude that the two conics do not meet at all.

This example is somewhat trivial, but, bearing in mind that so many of our results of the second chapter hinged on the fact that two conics meet in four points, we must be worried about the validity of that discussion. Let us look at another simple example and then pose some very important questions.

Example 4.2 *Sketch the parametric curves*

(a) $x = t^3, y = t^2,$
(b) $x = t^3 + t, y = t^2.$

(a): We notice that $x^2 = t^6$ and $y^3 = t^6$. Therefore, it is easy to eliminate the parameter to get

$$x^2 = y^3,$$

and so we see that the curve is symmetric about the y-axis. Clearly, the curve passes through the origin. Following the usual curve-sketching procedures, we differentiate to get

$$3y^2 y' = 2x,$$

$$y' = \frac{2x}{3y^2} = \frac{2}{3}x^{-1/3},$$

and hence the curve has a vertical slope at the origin. Also

$$y' > 0, \quad \text{for all } x > 0, \quad \text{and} \quad y'' < 0 \quad \text{for all } x > 0.$$

The curve must then look like that in Figure 4.1, which has a cusp at the origin.

(b): Eliminating the parameter, we obtain

$$x^2 = y(y+1)^2.$$

This curve also passes through the origin and is symmetric about the y-axis. Furthermore,

$$y' = \frac{2x}{(y+1)(3y+1)}.$$

It would be daunting to attempt to eliminate y from the right-hand side of this expression, from which we can see, however, that $y' = 0$ at the origin.

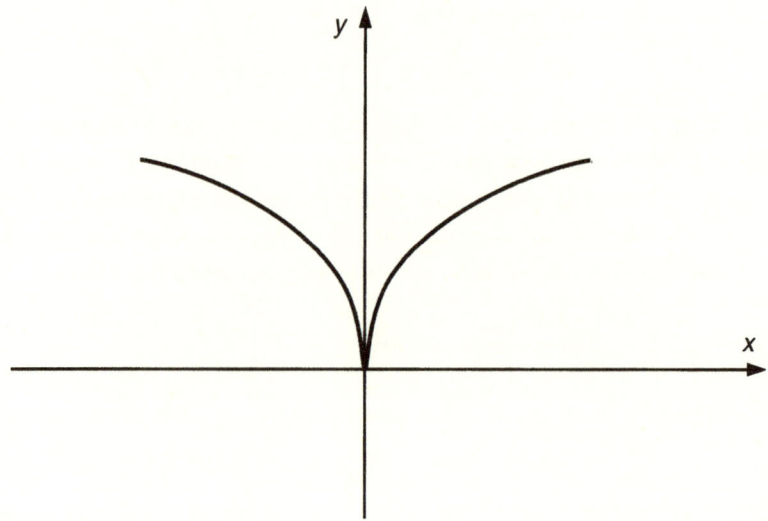

Figure 4.1. The polynomial cubic with a cusp at the origin.

Also,

$$y'' = \frac{2(1 - 3y)}{(3y + 1)^3}.$$

Therefore, $y'' = 0$ when $y = \frac{1}{3}$, and we have inflections at the points

$$x = \pm\frac{4}{3\sqrt{3}}, \qquad y = \frac{1}{3}.$$

We notice that $y > 0$, for all $x > 0$, and the curve must, therefore, look like that in Figure 4.2.

Both of the above parametrizations were polynomial. The y was t^2 in both cases, and the x-parametrizations differed only by a single term. Nonetheless, the curves actually look very different from each other. Both these curves are of the general form

$$x = a_3t^3 + a_2t^2 + a_1t + a_0,$$
$$y = b_3t^3 + b_2t^2 + b_1t + b_0,$$

a type of curve that occurs very frequently indeed. Do all curves of this latter form look like one of the two curves of Example 4.2? The answer to this question is clearly no, for the curve $x = t^3$, $y = t$ is unlike either. The following question is more difficult to answer. How many "different-looking" curves are contained in the general polynomial cubic form? In Example 4.2 we were lucky

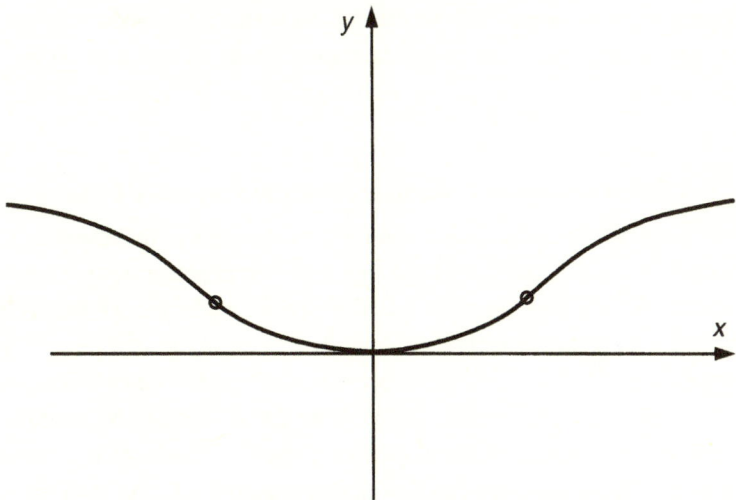

Figure 4.2. The polynomial cubic with two inflection points.

to be able to eliminate the parameter. How can we analyze the behavior of the general polynomial cubic curve? If we proceed as in Example 4.2, we rapidly encounter a mess of algebraic difficulties.

We saw in the second chapter that the conic can be parametrized in terms of rational functions. Can we eliminate the parameter in the polynomial cubic to obtain some curve of the form $f_n(x, y) = 0$, and if so, what will be the order n of the curve? If the mathematical methods we used to analyze conics were, in certain situations, suspect, is there any better method we can use? The answer is yes. All parametric curves with parametrizations that are rational functions are algebraic curves and can be written in the form $f_n(x, y) = 0$, although the converse is not true. In order to fully appreciate this we must develop a clear idea of what an algebraic curve is. We must, therefore, become familiar with homogeneous coordinates, the notion of a complex point of a curve, and the ramifications of working in projective, rather than Euclidean, space. The techniques we will develop will provide the tools necessary to analyze some very important "real" problems.

The adjectives "Euclidean" and "projective" are used to describe both spaces and geometries, and it is helpful to draw a clear distinction between a geometry and a space. We will think of a *geometry* as the study of a certain class of questions, and a *space* as the environment in which this study is carried out. We shall be studying a part of algebraic geometry, the part dealing with algebraic curves, and this is most appropriately done in projective space. An inherent part of algebraic geometry is the projective transformation of curves, and projective transformations are the central theme of projective geometry. This is an inherently physical study, the ramifications of which are witnessed, perhaps unconsciously, by everyone who is graced with the power of sight.

4.3 Homogeneous Coordinates for the Projective Plane

In view of our previous discussion about the plane itself in the absence of any particular coordinate system, we hope that its inherent properties are now well understood. We are now ready to introduce coordinates that we will need in order to use the algebraic methods.

Although a point in the Euclidean plane can be represented by two coordinates, it can also be represented by more. If we use more than two coordinates, then there must be some relationship among them. For example, if we have three nonconcurrent lines, we can describe the position of a point in the plane by the three directed distances of the point from the lines. These distances will be connected by some linear relationship of the form

$$ax + by + cz = d,$$

where x, y, and z represent the three distances.

The very striking difference between the Euclidean and projective planes is the absence, in the latter, of the parallel axiom and the fact that any two distinct lines determine a point. We are prompted to attempt a solution to the simultaneous equations

$$x + y = 0,$$

$$x + y = cz, \qquad c \neq 0.$$

We obtain the solution $z = 0$, $y = -x$. Hence, were we to use the three coordinates (x, y, z) to describe the unique point where the two distinct lines meet, this point would have coordinates $(x, -x, 0)$. However, the coordinate triple $(x, -x, 0)$ is certainly not unique, for x has not been specified. Let us assume that $x \neq 0$. Not only $(x, -x, 0)$ but any point with coordinates $(\alpha x, -\alpha x, 0)$, $\alpha \neq 0$, satisfies the equations of the two lines. We therefore associate all such sets of coordinates with the same point. However, the point $(0, 0, 0)$ also satisfies both equations, and we are faced with a dilemma, for now we have two points lying on two distinct lines and have therefore violated the axioms. Somehow we must patch up our coordinate system.

Consider the two lines

$$a_1 x + b_1 y + c_1 = 0,$$

$$a_2 x + b_2 y + c_2 = 0.$$

We replace x by $\bar{x}/\bar{z}$ and y by $\bar{y}/\bar{z}$ to obtain

$$a_1 \frac{\bar{x}}{\bar{z}} + b_1 \frac{\bar{y}}{\bar{z}} + c_1 = 0,$$

$$a_2 \frac{\bar{x}}{\bar{z}} + b_2 \frac{\bar{y}}{\bar{z}} + c_2 = 0.$$

Now multiply by $\bar{z}$ to get

$$a_1 \bar{x} + b_1 \bar{y} + c_1 \bar{z} = 0,$$

$$a_2 \bar{x} + b_2 \bar{y} + c_2 \bar{z} = 0.$$

If $\bar{z} \neq 0$, the Euclidean coordinates are given by the ratios

$$x = \frac{\bar{x}}{\bar{z}} \quad \text{and} \quad y = \frac{\bar{y}}{\bar{z}}.$$

If $\bar{z} = 0$, then there is no corresponding Euclidean point. Therefore, $\bar{z} = 0$ must be the equation of the line at infinity.

Solving these two equations, we get

$$\frac{\bar{x}}{b_1 c_2 - b_2 c_1} = \frac{\bar{y}}{c_1 a_2 - c_2 a_1} = \frac{\bar{z}}{a_1 b_2 - a_2 b_1}.$$

Hence, the coordinates of the point of intersection are

$$(b_1 c_2 - b_2 c_1, c_1 a_2 - c_2 a_1, a_1 b_2 - a_2 b_1).$$

These three coordinates are all zero if and only if

$$\frac{a_1}{a_2} = \frac{b_1}{b_2} = \frac{c_1}{c_2},$$

in which case the two lines are identical. Therefore, we exclude the coordinate triple $(0, 0, 0)$.

This approach provides an intuitive introduction to a coordinate system. All we need is a coordinate system for points that, together with an algebraic representation for lines, satisfies the axioms of the projective plane. Now "point" and "line" are undefined words. However, from the computational point of view we shall have to provide some definitions. Points will be defined in terms of a coordinate system, and lines will be defined in terms of equations involving the coordinates. Once having defined "point" and "line," we shall be in a position to determine if these definitions, together with the usual rules of algebra, will give us a system that satisfies the required axioms. If it does, then it is a valid algebraic model of the projective plane. If it does not, then we could consider alternative definitions and/or different rules of algebraic operation in order to achieve a model that fits the axioms. Such procedures are common in mathematics.

Definition 4.1 *A point will be represented by the ordered triple of real numbers* (x, y, z), *not all zero, and all triples of the form* $\alpha(x, y, z) = (\alpha x, \alpha y, \alpha z)$, $\alpha \neq 0$, *will represent the same point as* (x, y, z).

Definition 4.2 *Let* $A(x_1, y_1, z_1)$ *and* $B(x_2, y_2, z_2)$ *be two distinct points. We define the* **line** AB *to be the set of all points* $L(x, y, z)$ *with the property that*

$$x = \lambda x_1 + \mu x_2, \qquad y = \lambda y_1 + \mu y_2, \qquad z = \lambda z_1 + \mu z_2,$$

for some real numbers λ *and* μ *not both zero.*

Let us now examine the ramifications of these definitions. The reader may want to refer to Section 1.3 for the definition of a field.

Theorem 4.1 *The definitions of point and line, together with the algebraic operations on elements of a field, are consistent with the axioms of the real projective plane (Axioms P1 to P6, Section 3.4).*

Proof. There are six axioms we must consider. We lay out this proof in an informal style.

1. Axiom P6 Let us first try to show the one-to-one correspondence between the real numbers and all but one point of the line.

If $\mu \neq 0$, then L can be represented by the coordinates

$$\left(\frac{\lambda}{\mu} x_1 + x_2, \frac{\lambda}{\mu} y_1 + y_2, \frac{\lambda}{\mu} z_1 + z_2 \right).$$

Setting $\lambda/\mu = t$, we get the coordinates of L as

$$(t x_1 + x_2, t y_1 + y_2, t z_1 + z_2),$$

and the correspondence between a real number t and a point of the line is clear. However, there is precisely one point on the line, namely the point corresponding to $\mu = 0$, to which there does not correspond a real number. We have, therefore, shown consistency with Axiom P6.

2. Axiom P1 The three equations that define a line through the two distinct points $A(x_1, y_1, z_1)$ and $B(x_2, y_2, z_2)$ can be written in the form

$$\begin{pmatrix} x & x_1 & x_2 \\ y & y_1 & y_2 \\ z & z_1 & z_2 \end{pmatrix} \begin{pmatrix} -1 \\ \lambda \\ \mu \end{pmatrix} = \begin{pmatrix} 0 \\ 0 \\ 0 \end{pmatrix}.$$

For a nontrivial solution of this system we require the matrix on the left to be singular. Hence, its determinant or, equivalently, the determinant of its transpose must be zero, that is,

$$\det \begin{pmatrix} x & y & z \\ x_1 & y_1 & z_1 \\ x_2 & y_2 & z_2 \end{pmatrix} = 0,$$

which, upon expansion, becomes

$$(y_1 z_2 - y_2 z_1)x + (z_1 x_2 - z_2 x_1)y + (x_1 y_2 - x_2 y_1)z = 0.$$

This is an equation of the form

$$lx + my + nz = 0.$$

We note that

$$l = m = n = 0 \iff \frac{x_1}{x_2} = \frac{y_1}{y_2} = \frac{z_1}{z_2},$$

that is, the two points are the same. Since we assumed that the points are distinct, the numbers l, m, and n cannot all be zero.

Definition 4.3 *A polynomial, in any number of variables, is **homogeneous of degree** n if the sum of the degrees of the variables in each term of the polynomial is n.*

For example, the polynomial

$$2x^2 + 3xy + y^2 - xz - 5yz + z^2$$

is a homogeneous polynomial of degree 2, whereas the polynomial

$$2x^2 + 3xy + y^2 - x - 5y + z$$

is a polynomial of degree 2, but it is not homogeneous, because each of the terms $-x$, $-5y$, and z is of degree 1 whereas each of the other three terms has a total degree of 2.

We see, therefore, that our definition of a line implies that the coordinates are the zeros of a homogeneous polynomial of degree one, not all of whose coefficients can be zero.

The converse is also true, for if

$$\det \begin{pmatrix} x & y & z \\ x_1 & y_1 & z_1 \\ x_2 & y_2 & z_2 \end{pmatrix} = 0,$$

then the rows of the determinant must be dependent, which implies that there exist numbers α, β, and γ such that

$$\alpha x + \beta x_1 + \gamma x_2 = 0, \qquad \alpha y + \beta y_1 + \gamma y_2 = 0,$$

$$\alpha z + \beta z_1 + \gamma z_2 = 0.$$

Now, $\alpha \neq 0$; for, if it were zero, the points A and B would not be distinct. Setting $\lambda = -\beta/\alpha$, $\mu = -\gamma/\alpha$, we obtain

$$x = \lambda x_1 + \mu x_2, \qquad y = \lambda y_1 + \mu y_2, \qquad z = \lambda z_1 + \mu z_2,$$

which is the equation of a line.

This equivalence enables us to use the form of the line most suited to our needs.

Now, there certainly exist at least two points, for example, $(0, 0, 1)$ and $(1, 1, 1)$, and, hence, from our way of defining a line, there exists at least one line. Axiom P1 is, therefore, satisfied.

3. Axiom P2 Given a line defined by points A and B, any choice of λ and μ, both nonzero, will provide a third point distinct from A and B satisfying Axiom P2.

4. Axiom P3 The points $(0, 0, 1)$ and $(0, 1, 1)$ lie on the line $x = 0$, but the point $(1, 1, 1)$ does not lie on this line. This satisfies Axiom P3.

5. Axiom P4 This axiom is equivalent to the statement: A line is uniquely determined by any two of its points. We shall show that our definitions of point and line are consistent with this latter formulation of the axiom.

Let $A(x_1, y_1, z_1)$ and $B(x_2, y_2, z_2)$ be two given distinct points. Let C and D be two other points of the line AB. Let their coordinates be

$$(\lambda_1 x_1 + \mu_1 x_2, \lambda_1 y_1 + \mu_1 y_2, \lambda_1 z_1 + \mu_1 z_2)$$

and

$$(\lambda_2 x_1 + \mu_2 x_2, \lambda_2 y_1 + \mu_2 y_2, \lambda_2 z_1 + \mu_2 z_2),$$

respectively. Now, the line CD is the set of points $L(x, y, z)$ whose coordinates satisfy

$$x = \rho(\lambda_1 x_1 + \mu_1 x_2) + \eta(\lambda_2 x_1 + \mu_2 x_2),$$

and similarly for y and z. These equations can be rearranged to give

$$x = (\rho\lambda_1 + \eta\lambda_2)x_1 + (\rho\mu_1 + \eta\mu_2)x_2,$$

and similarly for y and z. However, this is the line AB. Therefore, the line is uniquely determined by any two of its points.

6. Axiom P5 We have one final axiom to check, namely Axiom P5, which states that two distinct lines meet in one, and only one, point.

Let the lines be

$$l_1 x + m_1 y + n_1 z = 0 \quad \text{and} \quad l_2 x + m_2 y + n_2 z = 0.$$

When we solve these equations we obtain the ratios

$$\frac{x}{m_1 n_2 - m_2 n_1} = \frac{y}{n_1 l_2 - n_2 l_1} = \frac{z}{l_1 m_2 - l_2 m_1},$$

and these equations are satisfied by the coordinates of the point

$$(m_1 n_2 - m_2 n_1, n_1 l_2 - n_2 l_1, l_1 m_2 - l_2 m_1),$$

which is unique up to a constant multiple, and hence the point is unique. ∎

We have shown, therefore, that our definitions of "point" and "line" are consistent with the axioms of the real projective plane. We will henceforth always assume that, once coordinates for points in the projective plane are given, a line is taken to be one of the two equivalent forms given above. The definition of the points gives us what is referred to as a *system of homogeneous coordinates* for the real projective plane. A set of lines that share a common point are said to be *concurrent*. The figure formed by three nonconcurrent lines is a *triangle*, the lines are called the *sides* of the triangle, and the three points where each pair of sides meet are called the *vertices* of the triangle. Recalling our discussion of the lack of ordering in the projective plane, we must emphasize that each entire line is a side of the triangle, not just a segment of the line, and that we cannot distinguish between the "inside" and the "outside" of a triangle. For a given homogeneous coordinate system, the points $X(1, 0, 0)$, $Y(0, 1, 0)$, and $Z(0, 0, 1)$ are the vertices of what is called the *triangle of reference*.

Axiom P6, which states the correspondence between all but one point of a projective line and the *real* numbers, is in fact the defining axiom for a *real* projective plane, and is often not included in the set of axioms. If we use complex numbers instead of real numbers, the projective plane obtained in this way is called the *complex projective plane*.

We are now finally in a position to define a real planar algebraic curve.

Definition 4.4 *A **real planar algebraic curve** is the locus of zeros, in the complex projective plane, of a homogeneous polynomial of degree at least one in three variables and with real coefficients. The three variables provide a coordinate system for the projective plane.*

Note that the qualifier "real" in the Definition 4.4 of the curve refers to the type of the coefficients, and not to the type of points on the curve, which may be complex.

4.3.1 Projective Coordinates: Examples

In this section we will work through several examples in the use of projective coordinates. We will assume throughout that we are working over the field of complex numbers. If the equation of a curve is given to us in nonhomogeneous coordinates x and y, we can associate with it an algebraic curve in the projective plane by replacing x by x/z and y by y/z, and multiplying the equation by the highest power of z that appears in the denominator. For example, $x^2 + y^2 - 1 = 0$ is associated with $x^2 + y^2 - z^2 = 0$ via the intermediate stages of writing $(x/z)^2 + (y/z)^2 - 1 = 0$ and multiplying by z^2.

Let us now repeat the first example of this chapter.

Example 4.1

Alternative solution. Determine the points where the two circles $x^2 + y^2 - 1 = 0$ and $x^2 + y^2 - 2 = 0$ meet.

The associated projective conics are, respectively, $x^2 + y^2 - z^2 = 0$ and $x^2 + y^2 - 2z^2 = 0$. This implies a repeated point at $z = 0$. When $z = 0$, $x = \pm i y$ and we obtain four points of intersection, namely

$$(i, 1, 0), \qquad (i, 1, 0), \qquad (-i, 1, 0), \quad \text{and} \quad (-i, 1, 0).$$

We have to count intersections with their appropriate multiplicities. This is consistent with the way zeros are counted when one applies the fundamental theorem of algebra, and this is one of the reasons why we have to work in the complex projective plane rather than the real one. It is also consistent with our concept (discussed in Section 2.9) of a tangent to a curve at a point as being a line that meets the curve at least twice at that point.

Example 4.3 *Find the points where the line through the origin is tangent to the circle* $x^2 + y^2 - 1 = 0$.

We use the circle in the form $x^2 + y^2 - z^2 = 0$. A line through the origin has equation $y - mx = 0$. These curves meet where $(1 + m^2)x^2 - z^2 = 0$, and the line is a tangent to the circle if and only if this equation has a repeated root. Therefore, $m^2 = -1$, and the two tangents have equations

$$y + ix = 0 \quad \text{and} \quad y - ix = 0.$$

These tangents meet the circle when $z = 0$, with $y + ix = 0$ meeting the circle at $(1, -i, 0)$ and $y - ix = 0$ meeting the circle at $(1, i, 0)$.

The points $(1, \pm i, 0)$ are rather special points, as we shall see from the next example.

Example 4.4 *If $z = 0$ is the line at infinity, show that all circles meet this line in the points* $(1, \pm i, 0)$.

If the Euclidean system is rectangular, then any circle can be written in the associated form

$$x^2 + y^2 + 2gxz + 2fyz + cz^2 = 0,$$

so that $z = 0$ implies $x = \pm iy$, and hence the line at infinity meets any circle in the points $(1, \pm i, 0)$. These points are called the *circular points at infinity*. We notice that if, say, $x = 0$ is the equation of the line at infinity, then the circle will be

$$y^2 + z^2 + 2hxy + 2gzx + ax^2 = 0,$$

and the circular points at infinity will be $(0, 1, \pm i)$. We also notice that if the Euclidean coordinate system is not rectangular, then the equation of the circle includes an xy term, and the coordinates of the points where the line at infinity meets the circle are changed.

The point $(1, 1, 1)$ is referred to as the *unit point*. This can be chosen to be any point not on the triangle of reference, for if a point P not on the triangle of reference has coordinates (α, β, γ) with $\alpha\beta\gamma \neq 0$, the transformation

$$x' = \frac{x}{\alpha}, \qquad y' = \frac{y}{\beta}, \qquad z' = \frac{z}{\gamma}$$

gives P the coordinates $(1, 1, 1)$ but leaves the triangle of reference unchanged.

Similarly, the line

$$x + y + z = 0,$$

is referred to as the *unit line*, and the line $lx + my + nz = 0$ can be transformed into the unit line by the transformation

$$x' = lx, \qquad y' = my, \qquad z' = nz.$$

4.3.2 Algebraic Duality and Projective Line Coordinates

We have seen that points and lines are duals in the projective plane. We have also seen that our algebraic definition of the projective plane in terms of homogeneous point coordinates, and either of the equivalent definitions of lines, are indeed consistent with the axioms of the projective plane. There must, therefore, be an algebraic duality in the way in which we represent points and lines in our homogeneous coordinate system. We have seen that any line can be expressed in the form

$$lx + my + nz = 0. \tag{4.1}$$

Hence, a line is uniquely determined by the triple (l, m, n), and clearly the triple $(\alpha l, \alpha m, \alpha n)$, $\alpha \neq 0$, determines the same line. We can therefore use (l, m, n) as homogeneous coordinates for the line. These coordinates are called *line coordinates*. Now, if (l, m, n) are line coordinates and (x, y, z) are point coordinates, it is usual to interpret the equation (4.1) as the set of points lying on the line determined by the triple (l, m, n). We could, however, consider it as the set of lines passing through the point determined by the triple (x, y, z). These two statements are duals of each other. The particular interpretation will depend upon which triple is fixed and which is variable. Thus,

$$2x + 3y - z = 0$$

must clearly represent the set of all points that lie on the particular line defined by the triple $(2, 3, -1)$. Likewise,

$$2l + 3m - n = 0$$

represents the set of all lines that pass through the particular point defined by the triple $(2, 3, -1)$. Therefore, $2l + 3m - n = 0$ is the equation of the point $(2, 3, -1)$. We will use the same ordered-triple notation for both point and line coordinates, but to avoid confusion we will use x, y, and z only for point coordinates.

From duality and the definition of a line (Definition 4.2) we see that if

$$x_1 l + y_1 m + z_1 n = 0 \quad \text{and} \quad x_2 l + y_2 m + z_2 n = 0$$

are the equations of the points (x_1, y_1, z_1) and (x_2, y_2, z_2), respectively, then the equation of any point on the line determined by the two given points must be

$$\lambda(x_1 l + y_1 m + z_1 n) + \mu(x_2 l + y_2 m + z_2 n) = 0.$$

The duality allows us to construct, in line coordinates, a pencil of conics tangential to four given lines in exactly the same fashion as we constructed, in point coordinates, a pencil of conics through four points. In Section 2.10 we did not actually do this. We used a rather cumbersome method when constructing conics tangential to given lines (Example 2.18). Let us now repeat the construction of such a conic using the power of duality.

Example 4.5 *Find the equation of the conic that touches the five lines*

$$x = 0, \qquad y = 0, \qquad x - 2y + 1 = 0,$$

$$2x - y - 1 = 0 \quad \text{and} \quad x + y - 3 = 0.$$

The five given lines have coordinates $(1, 0, 0)$, $(0, 1, 0)$, $(1, -2, 1)$, $(2, -1, -1)$, and $(1, 1, -3)$, respectively. Working now in line coordinates, these triples take the place of points. If we take the first two lines, then

$$\det \begin{pmatrix} l & m & n \\ 1 & 0 & 0 \\ 0 & 1 & 0 \end{pmatrix} \equiv n = 0$$

represents a point on the two lines $(1, 0, 0)$ and $(0, 1, 0)$. The first two lines then give us the point $n = 0$. Similarly, the third and fourth lines give us the point

$$\det \begin{pmatrix} l & m & n \\ 1 & -2 & 1 \\ 2 & -1 & -1 \end{pmatrix} \equiv l + m + n = 0,$$

the first and third lines determine the point $m + 2n = 0$, and the second and fourth lines determine the point $l + 2n = 0$. Therefore, the pencil of conics with tangential equation (line equation)

$$n(l + m + n) + \alpha(m + 2n)(l + 2n) = 0,$$

is, by construction, tangent to the four lines whose coordinates are $(1, 0, 0)$, $(0, 1, 0)$, $(1, -2, 1)$, and $(2, -1, -1)$.

We now wish to force a member of the pencil to be tangent to the line $(1, 1, -3)$. Therefore,

$$(-3)(-1) + \alpha(-5)(-5) = 0,$$

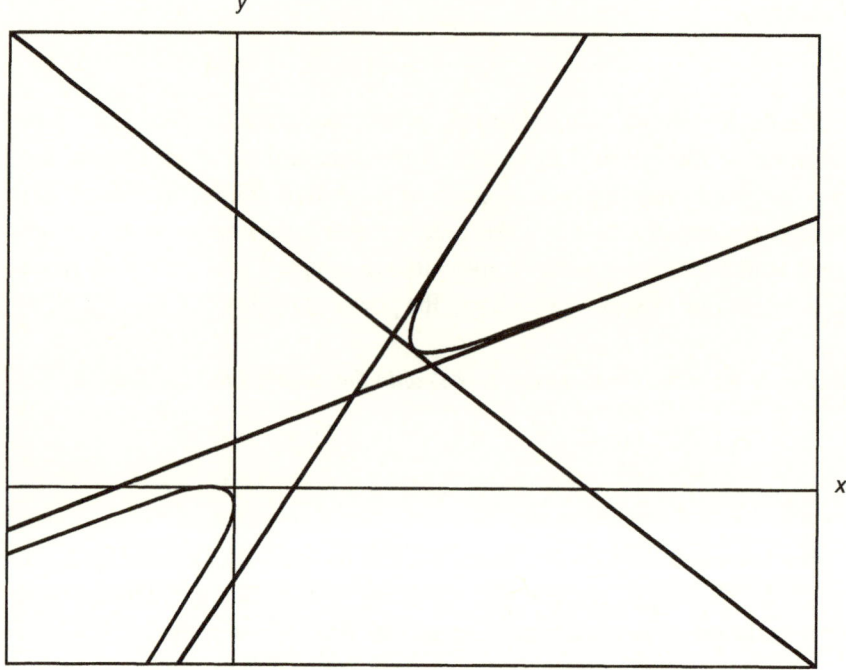

Figure 4.3. Conic tangent to five lines.

whence

$$\alpha = -\tfrac{3}{25}.$$

Hence, the unique conic in line coordinates, that is, the tangential equation of the unique conic, is

$$13n^2 + 19nl + 19nm - 3ml = 0,$$

which is what we obtained before by a less systematic method. The point equation associated with this tangential equation is (Figure 4.3)

$$361x^2 - 878xy + 361y^2 + 114xz + 114yz + 9z^2 = 0,$$

where, this time, we have written the equation in projective coordinates.

Example 4.6 *Find the coordinates of the line determined by the points* $(2, -1, 3)$ *and* $(1, 0, 2)$, *and the coordinates of the point determined by the lines* $(4, 2, -1)$ *and* $(1, -3, 1)$.

The line determined by the two given points is

$$\det \begin{pmatrix} x & y & z \\ 2 & -1 & 3 \\ 1 & 0 & 2 \end{pmatrix} = -2x - y + z = 0,$$

and hence the coordinates of the line are $(-2, -1, 1)$.

Similarly, the equation of the point determined by the two given lines is

$$\det \begin{pmatrix} l & m & n \\ 4 & 2 & -1 \\ 1 & -3 & 1 \end{pmatrix} = -l - 5m - 14n = 0,$$

and hence the coordinates of the point are $(-1, -5, -14)$.

4.3.3 Other Projective Coordinate Systems

In the examples so far we have used the most common coordinate system for the projective plane, that being the one associated with the rectangular Cartesian coordinates in the Euclidean plane, the association being made in the following manner: If (x, y, z) represents a point in the real projective plane then, for $z \neq 0$, there is a corresponding point $(x/z, y/z)$ in the Euclidean plane. This particular coordinate system is, however, only one of infinitely many which could be used. It is the most common, but in special cases there may be some particular physical interpretation, such as direction from a point or distance from a line, that another coordinate system will highlight to advantage. Two examples of such coordinate systems are given in Exercises 4.18 and 4.19. In the next example we show how to construct a one-dimensional projective coordinate system.

Consistent with the coordinate system we have used for the real projective plane, we require that a one-dimensional real projective space S_1 be the set of points that can be represented by ordered pairs of real numbers (a, b) with the following three properties:

1. For each $s \in S_1$ there exists at least one associated ordered pair (a, b) with $a^2 + b^2 \neq 0$.
2. For each ordered pair (a, b) with $a^2 + b^2 \neq 0$ there exists a unique $s \in S_1$.
3. If two ordered pairs (a_1, b_1) and (a_2, b_2) represent the same point $s \in S_1$, then there exists a real number $\rho \neq 0$ such that $(a_1, b_1) = \rho(a_2, b_2) = (\rho a_2, \rho b_2)$.

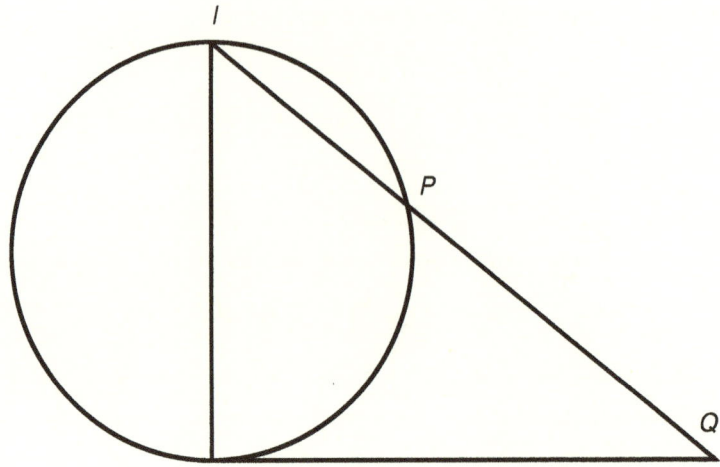

Figure 4.4. Projective line coordinates.

Example 4.7 *Let S be the set of points on the Euclidean circle*

$$x^2 + (y - 1)^2 = 1.$$

Let I be the point with coordinates $(0, 2)$ *(Figure 4.4). Associate with the point I the ordered pairs* $(\rho, 0)$, $\rho \neq 0$. *Let P be a point other than I on the circle, and construct a line through I and P to meet the x-axis in the point Q, which has coordinates* $(x, 0)$, *say. Associate with P the ordered pairs* $(\rho x, \rho)$, $\rho \neq 0$. *We have then associated ordered pairs with each point of S. Show that this association specifies a one-dimensional projective coordinate system in S over the real numbers.*

By construction, we have produced ordered pairs for each point of S. The first requirement for a coordinate system is, therefore, satisfied.

Consider the ordered pair of real numbers (a, b), not both zero. If $b = 0$, then (a, b) is associated with I. If $b \neq 0$, then (a, b) is associated with the point on S other than I where the line, in the Euclidean plane, through I and the point with Euclidean coordinates $(a/b, 0)$ meets the circle. Specifically, this line has the equation

$$ay + 2bx - 2a = 0,$$

and meets the circle at the point I $(0, 2)$ and at the point P of S with coordinates

$$\left(\frac{4ab}{a^2 + 4b^2}, \frac{2a^2}{a^2 + 4b^2} \right). \tag{4.2}$$

Hence, if $b \neq 0$, we associate the unique point of S with the coordinates (4.2) with the ordered pair (a, b). The second property of a one-dimensional projective coordinate system is, therefore, verified.

Let (a, b) and (c, d) correspond to the same point of S. If this point is I then $b = d = 0$, and a and c are nonzero. Therefore,

$$(a, b) = \frac{a}{c}(c, d).$$

If the point of correspondence is not I, then

$$\frac{a}{b} = \frac{c}{d} \quad \text{and} \quad (a, b) = \frac{b}{d}(c, d).$$

Hence, if (a, b) and (c, d) correspond to the same point, then there exists a $\rho \neq 0$ such that

$$(a, b) = \rho(c, d).$$

Finally, let $(a, b) = \rho(c, d)$ for $\rho \neq 0$. If $d = 0$, then $b = 0$ and both ordered pairs correspond to I. If $d \neq 0$, then (c, d) corresponds to

$$\left(\frac{4cd}{c^2 + 4d^2}, \frac{2c^2}{c^2 + 4d^2} \right).$$

However, since $a = \rho c$ and $b = \rho d$, this point is the same as (4.2) This verifies the third required property, and hence we do indeed have a one-dimensional projective coordinate system in S over the reals.

4.4 Projective Transformations

Let us now illustrate a very important fact, which will later be proved as a theorem (Theorem 4.4), namely that any four points, no three of which are collinear, can be mapped by a projective transformation onto any four points, no three of which are collinear.

Example 4.8 *Find a transformation of coordinates which takes the points* $(2, 1, 0)$, $(0, 1, 2)$, $(2, 1, 1)$ *and* $(3, 4, -1)$ *into the points* $(1, 0, 0)$, $(0, 1, 0)$, $(0, 0, 1)$ *and* $(1, 1, 1)$, *respectively.*

We first recall that the triangle of reference is given by the three points $(1, 0, 0)$, $(0, 1, 0)$, $(0, 0, 1)$ and that the unit point has coordinates $(1, 1, 1)$.

We will do the mapping with a linear transformation. We seek a 3×3 matrix A such that

$$A \begin{pmatrix} 2 \\ 1 \\ 0 \end{pmatrix} = \pi \begin{pmatrix} 1 \\ 0 \\ 0 \end{pmatrix}, \qquad A \begin{pmatrix} 0 \\ 1 \\ 2 \end{pmatrix} = \rho \begin{pmatrix} 0 \\ 1 \\ 0 \end{pmatrix}, \qquad A \begin{pmatrix} 2 \\ 1 \\ 1 \end{pmatrix} = \sigma \begin{pmatrix} 0 \\ 0 \\ 1 \end{pmatrix},$$

and

$$A \begin{pmatrix} 3 \\ 4 \\ -1 \end{pmatrix} = \tau \begin{pmatrix} 1 \\ 1 \\ 1 \end{pmatrix}.$$

The first three of these equations can be written in matrix form as

$$A \begin{pmatrix} 2 & 0 & 2 \\ 1 & 1 & 1 \\ 0 & 2 & 1 \end{pmatrix} = \begin{pmatrix} \pi & 0 & 0 \\ 0 & \rho & 0 \\ 0 & 0 & \sigma \end{pmatrix}.$$

The noncollinearity of any subset of three points guarantees the nonsingularity of any 3×3 matrix formed by their coordinates. Therefore,

$$A = \frac{1}{2} \begin{pmatrix} \pi & 0 & 0 \\ 0 & \rho & 0 \\ 0 & 0 & \sigma \end{pmatrix} \begin{pmatrix} -1 & 4 & -2 \\ -1 & 2 & 0 \\ 2 & -4 & 2 \end{pmatrix}.$$

The final equation

$$A \begin{pmatrix} 3 \\ 4 \\ -1 \end{pmatrix} = \tau \begin{pmatrix} 1 \\ 1 \\ 1 \end{pmatrix},$$

then gives

$$\tfrac{15}{2}\pi = \tfrac{5}{2}\rho = -6\sigma = \tau.$$

Therefore,

$$A = \tau \begin{pmatrix} -\frac{1}{15} & \frac{4}{15} & -\frac{2}{15} \\ -\frac{1}{5} & \frac{2}{5} & 0 \\ -\frac{1}{6} & \frac{1}{3} & -\frac{1}{6} \end{pmatrix}.$$

We can choose τ to be any nonzero real number. For $\tau = 1$ we note that

$$\begin{pmatrix} -\frac{1}{15} & \frac{4}{15} & -\frac{2}{15} \\ -\frac{1}{5} & \frac{2}{5} & 0 \\ -\frac{1}{6} & \frac{1}{3} & -\frac{1}{6} \end{pmatrix} \begin{pmatrix} 2 & 0 & 2 & 3 \\ 1 & 1 & 1 & 4 \\ 0 & 2 & 1 & -1 \end{pmatrix} = \begin{pmatrix} \frac{2}{15} & 0 & 0 & 1 \\ 0 & \frac{2}{5} & 0 & 1 \\ 0 & 0 & -\frac{1}{6} & 1 \end{pmatrix},$$

and, hence, the transformation satisfies the required conditions.

As our next example we use algebra to prove a special case of Desargues's Theorem. In Section 3.7 we used synthetic methods to prove this theorem for noncoplanar triangles.

Theorem 4.2 *(Desargues) If two coplanar triangles without a common vertex are in perspective, then the points of intersection of corresponding sides are collinear.*

Proof. Let the two triangles be ABC and $A'B'C'$, and let ABC be the triangle of reference (Figure 4.5), with vertex coordinates $A(1, 0, 0)$, $B(0, 1, 0)$, and $C(0, 0, 1)$. Let $P(\alpha, \beta, \gamma)$ be the *center of perspective*, where the corresponding sides AA', BB', and CC' meet. The coordinates of A' can be written in the form $(\rho + \sigma\alpha, \sigma\beta, \sigma\gamma)$, but since this point is not the same as A, we know that $\sigma \neq 0$. Hence, we can assume that $\sigma = 1$ [since $(\sigma\alpha, \sigma\beta, \sigma\gamma) = (\alpha, \beta, \gamma)$ if $\sigma \neq 0$]. Therefore, the points A', B', and C' can be expressed in the forms

$$A'(\alpha + \lambda, \beta, \gamma), \qquad B'(\alpha, \beta + \mu, \gamma), \qquad C'(\alpha, \beta, \gamma + \nu).$$

The line $B'C'$ has equation

$$\det \begin{pmatrix} x & y & z \\ \alpha & \beta + \mu & \gamma \\ \alpha & \beta & \gamma + \nu \end{pmatrix} = 0,$$

which meets BC, i.e., the line $x = 0$, when

$$\alpha (\nu y + \mu z) = 0.$$

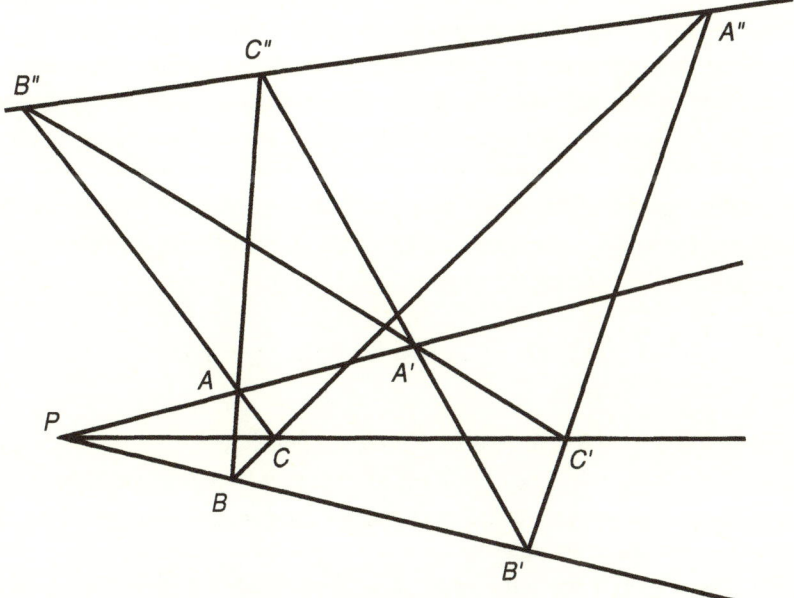

Figure 4.5. The theorem of Desargues.

If $\alpha = 0$, the lines BC and $B'C'$ are not distinct. The required intersection therefore corresponds to $vy + \mu z = 0$, namely the point $(0, \mu, -v)$. Similarly, the lines CA and $C'A'$ meet when $vx + \lambda z = 0$ at the point $(-\lambda, 0, v)$, and the lines AB and $A'B'$ meet when $\mu x + \lambda y = 0$ at the point $(\lambda, -\mu, 0)$. These three points lie on the line

$$\mu v x + \lambda v y + \lambda \mu z = 0. \qquad \blacksquare$$

4.4.1 Groups

In Example 4.8 above we made use of a linear transformation to map four noncollinear points onto four noncollinear points. This transformation was represented by a nonsingular 3×3 matrix and is an example of what we will call a *projective transformation*. It is often a key step in finite-element computations to map an arbitrary element to some standard element and then perform the computations on this standard element, mapping the results back to the original element. The transformations that perform this task are called *affine transformations*. In computer-aided design, transformations are used to change the size of objects but leave the shape unaltered (*similarity transformations*), to obtain the mirror image of a figure (*reflections*), and to reposition (rotate or simply translate) an object (*rigid motions*). All these various transformations are special cases of the projective transformations, and hence each of them can be represented by some nonsingular 3×3 matrix. Recall, too, our classification of geometries as the study of the properties which are invariant under certain transformations. We now make this definition more precise by classifying a geometry as the study of the properties that are invariant under a *group* of transformations. We shall see that when the transformations are represented by matrices it is particularly easy to determine whether or not they form a group. This fact then enables us to study all these important transformations of computer-aided design and the finite-element method in a particularly incisive fashion.

Let us begin by defining a group.

Definition 4.5 *A **group** is a set S and a binary law of operation $\otimes$ that satisfy the following properties:*

1. *For $a, b \in S$, $a \otimes b \in S$, that is, S is closed under $\otimes$.*
2. *For $a, b, c \in S$, $a \otimes (b \otimes c) = (a \otimes b) \otimes c$, that is, $\otimes$ is associative.*
3. *There exists an element $u \in S$ such that $u \otimes a = a \otimes u = a$ for all $a \in S$, that is, there exists an identity.*
4. *For $a \in S$ there exists an element $b \in S$ such that $a \otimes b = b \otimes a = u$, that is, each element of S has an inverse in S.*

The notation is usually simplified by calling the binary law "multiplication" and omitting the symbol $\otimes$, and by writing 1 for the identity element and a^{-1} for the inverse of a. The conditions above can then be written in the following simpler form:

1. For $a, b \in S$, $ab \in S$.
2. For $a, b, c \in S$, $a(bc) = (ab)c$.
3. $1a = a1 = a$ for all $a \in S$.
4. For $a \in S$ there exists $a^{-1} \in S$ such that $aa^{-1} = a^{-1}a = 1$.

Example 4.9 *Let S be the set of ordered pairs of real numbers (a, b) where $a \neq 0$, and define multiplication by $(a, b)(c, d) = (ac, bc + d)$. Show that S, together with this multiplication, forms a group.*

1: For $a, b, c,$ and d real, ac and $bc + d$ are real, and hence the set is closed under multiplication.

2: For $a, b, c, d, e,$ and f real,

$$
\begin{aligned}
(a, b)\,[(c, d)(e, f)] &= (a, b)(ce, de + f) \\
&= (ace, bce + de + f) \\
&= (ace, (bc + d)e + f) \\
&= (ac, bc + d)(e, f) \\
&= [(a, b)(c, d)]\,(e, f),
\end{aligned}
$$

and hence the multiplication is associative.

3: The element $(1, 0)$ belongs to S, and

$$
(1, 0)(a, b) = (a, b)(1, 0) = (a, b).
$$

Therefore, $(1, 0)$ is the identity for multiplication.

4: For $(a, b) \in S$, we have $a \neq 0$ and

$$
(a, b) \left(\frac{1}{a}, -\frac{b}{a} \right) = \left(\frac{1}{a}, -\frac{b}{a} \right) (a, b) = (1, 0);
$$

hence,

$$
(a, b)^{-1} = \left(\frac{1}{a}, -\frac{b}{a} \right).
$$

Therefore, S, together with the given multiplication, forms a group.

We usually define a set of transformations to be a group if the following conditions are satisfied.

1. The application of two consecutive transformations (i.e., the *product* of two transformations) is also a transformation of the set.
2. The product of three transformations is associative.
3. Every transformation has an inverse.

The third property implies the existence of an identity, and the first property guarantees that this identity lies in the set.

Geometrical transformations acting upon an object can be thought of as changing the object in some fashion. If T is a transformation and f is some geometrical object, the notation Tf will denote the resulting object when T transforms f. The identity transformation will be denoted by I, so $If = f$. A transformation may be defined directly in terms of what it does to a particular object, in which case the product of two transformations may not have a meaning without considering the particular object. In other words, if T and S are transformations, we must define TS by the statement

$$(TS)f \equiv T(Sf).$$

Example 4.10 *Show that I and the reflection of a circle in a fixed diameter form a group.*

Let T denote the reflection of the circle in a fixed diameter. Then the set contains only two elements, namely T and I. Closure is easily established, since we have such a small number of possibilities for products. These are I^2, IT, TI, and T^2. Now,

$$(I^2)f = I(If) = If,$$
$$(IT)f = I(Tf) = Tf,$$
$$(TI)f = T(If) = Tf,$$

and from the definition of T it is clear that $T^2 = I$. Therefore, any product of three transformations also reduces to either T or I. Hence, it is easy to show that we have associativity. We also notice that T is its own inverse and hence, somewhat trivially, we have a transformation group.

Rather than define transformations in this fashion, we will define all our transformations in terms of matrices, in which case we can define the products of transformations by the multiplication of matrices. We will rarely draw any distinction between the transformation and the matrix which represents it. That is, if A and B are matrix representations of transformations T and S, say, we will often say that "T equals A," "S equals B," and "TS equals AB." Our analysis of transformations is now the analysis of matrices.

Example 4.11 *Repeat Example 4.10 using matrices.*

The identity transformation I is represented by any nonzero multiple of the 3×3 identity matrix

$$A = \begin{pmatrix} \alpha & 0 & 0 \\ 0 & \alpha & 0 \\ 0 & 0 & \alpha \end{pmatrix} = \alpha I, \qquad \alpha \neq 0.$$

We choose the circle to be positioned with its center at $(0, 0, 1)$, that is, at one of the vertices of the triangle of reference, and let the fixed diameter be the line $y = 0$. The reflection T in this line maps the point (a, b, c) to the point $(\rho a, -\rho b, \rho c)$, $\rho \neq 0$, and is represented by any matrix of the form

$$B = \begin{pmatrix} \beta & 0 & 0 \\ 0 & -\beta & 0 \\ 0 & 0 & \beta \end{pmatrix} = \beta \begin{pmatrix} 1 & 0 & 0 \\ 0 & -1 & 0 \\ 0 & 0 & 1 \end{pmatrix} = \beta R, \qquad \beta \neq 0.$$

It is immediately clear that the set of matrices of these types forms a group under matrix multiplication: the product of any two matrices is again one of the given types, matrix multiplication is associative, and the inverse matrices $\alpha^{-1} I$ and $\beta^{-1} R$ also belong to the set.

Matrix multiplication is associative. Furthermore, the identity matrix clearly represents the identity transformation. Hence, a set of matrix transformations forms a group if it is closed under multiplication and if each matrix in the set has an inverse in the set. This, in turn, implies that the matrices of transformation groups are nonsingular.

Theorem 4.3 *The set of nonsingular $n \times n$ matrices forms a group under matrix multiplication.*

Proof. Let A and B be nonsingular $n \times n$ matrices. Then AB is an $n \times n$ matrix. Also, since $\det AB = \det A \det B$, the matrix AB is nonsingular. Since A is nonsingular, A^{-1} exists, is nonsingular, and is an $n \times n$ matrix. The $n \times n$ identity matrix I is nonsingular and hence belongs to the set. Matrix multiplication is associative, that is, $A(BC) = (AB)C$. Therefore, the set of nonsingular $n \times n$ matrices forms a group. ∎

Such a group of transformations is called the *projective group* for $(n - 1)$-space. In our particular applications the matrices will be 3×3 when we are working in the plane, and 4×4 when we are in three-space.

Definition 4.6 *A **projective transformation** in* $(n-1)$*-space is an* $n \times n$ *non-singular matrix.*

4.4.2 The Fundamental Theorem

We can now ask ourselves which properties are, and which are not, invariant under such a transformation. In this regard, Example 4.8 is very instructive. Let us generalize it into a theorem.

Theorem 4.4 *(**Fundamental Theorem**) There exists a projective transformation, unique up to a scalar multiple, which transforms four given points in a plane, no three of which are collinear, into four given points in a plane, no three of which are collinear.*

We will prove this theorem in three simple steps, the first two of which we will regard as lemmas. All the work is done in the projective plane.

Lemma 4.1 *A projective transformation takes noncollinear points into non-collinear points.*

Proof. Let A be a projective transformation, and let $P_j(x_{1j}, x_{2j}, x_{3j})$, $j = 1, 2, 3$, be three noncollinear points that are transformed to $Q_j(y_{1j}, y_{2j}, y_{3j})$, $j = 1, 2, 3$. Let X and Y be the 3×3 matrices whose ijth entries are x_{ij} and y_{ij}, respectively. Then $Y = AX$. Since the points P_j are noncollinear, the matrix X is nonsingular. Since A is a projective transformation, A is nonsingular. Therefore, Y is nonsingular, and the points Q_j are therefore noncollinear. ∎

Lemma 4.2 *There exists a projective transformation, unique up to a scalar multiple, that transforms four given points, no three of which are collinear, into the vertices of the triangle of reference and the unit point.*

Proof. Let the four given points be $P_j(x_{1j}, x_{2j}, x_{3j})$, $j = 1, 2, 3, 4$. Then X, as defined in Lemma 4.1, is nonsingular. Therefore, $C = X^{-1}$ is a projective transformation, and C maps P_1, P_2, and P_3 onto the vertices of the triangle of reference.

Let C map P_4 to the point $Q(\alpha, \beta, \gamma)$. Then, from Lemma 4.1, Q cannot be on the triangle of reference, and therefore $\alpha\beta\gamma \neq 0$. The transformation

$$B = \begin{pmatrix} 1/\alpha & 0 & 0 \\ 0 & 1/\beta & 0 \\ 0 & 0 & 1/\gamma \end{pmatrix},$$

is therefore a projective transformation. The transformation represented by B leaves the triangle of reference unchanged but maps Q to the unit point $(1, 1, 1)$. Since C and B are projective transformations, so is $A = BC$, and this is the transformation we seek.   ∎

We can now prove Theorem 4.4.

Proof. Let $\{P_i\}$ and $\{Q_i\}$, $i = 1, 2, 3, 4$, be the two sets of points, and suppose that we wish to map P_i to Q_i, $i = 1, 2, 3, 4$. From Lemma 4.2, let A_1 be the projective transformation that maps the set $\{P_i\}$ onto the vertices of the triangle of reference and the unit point, and let A_2 be the projective transformation that maps the set $\{Q_i\}$ onto the vertices of the triangle of reference and the unit point. Then $A_2^{-1}A_1$ is a projective transformation that maps P_i to Q_i, $i = 1, 2, 3, 4$.   ∎

Example 4.12 *Find a projective transformation that maps the four points* $(1, 2, -1), (-1, 1, 0), (1, 1, 0),$ *and* $(0, 4, -1)$ *to* $(2, 0, -1), (1, 4, 2), (1, 1, 1),$ *and* $(4, 5, 2)$, *respectively.*

Let

$$X_1 = \begin{pmatrix} 1 & -1 & 1 \\ 2 & 1 & 1 \\ -1 & 0 & 0 \end{pmatrix};$$

then

$$C_1 = X_1^{-1} = \frac{1}{2} \begin{pmatrix} 0 & 0 & -2 \\ -1 & 1 & 1 \\ 1 & 1 & 3 \end{pmatrix}$$

and

$$C_1 \begin{pmatrix} 0 \\ 4 \\ -1 \end{pmatrix} = \begin{pmatrix} 1 \\ \frac{3}{2} \\ \frac{1}{2} \end{pmatrix}.$$

Therefore, the matrix

$$A_1 = \frac{1}{2} \begin{pmatrix} 1 & 0 & 0 \\ 0 & \frac{2}{3} & 0 \\ 0 & 0 & 2 \end{pmatrix} \begin{pmatrix} 0 & 0 & -2 \\ -1 & 1 & 1 \\ 1 & 1 & 3 \end{pmatrix} = \begin{pmatrix} 0 & 0 & -1 \\ -\frac{1}{3} & \frac{1}{3} & \frac{1}{3} \\ 1 & 1 & 3 \end{pmatrix}$$

maps the first set of four points to $(1, 0, 0)$, $(0, 1, 0)$, $(0, 0, 1)$ and $(1, 1, 1)$. Let

$$X_2 = \begin{pmatrix} 2 & 1 & 1 \\ 0 & 4 & 1 \\ -1 & 2 & 1 \end{pmatrix};$$

then

$$C_2 = X_2^{-1} = \frac{1}{7}\begin{pmatrix} 2 & 1 & -3 \\ -1 & 3 & -2 \\ 4 & -5 & 8 \end{pmatrix}$$

and

$$C_2\begin{pmatrix} 4 \\ 5 \\ 2 \end{pmatrix} = \begin{pmatrix} 1 \\ 1 \\ 1 \end{pmatrix}.$$

Therefore, the matrix

$$A_2 = I \times \frac{1}{7}\begin{pmatrix} 2 & 1 & -3 \\ -1 & 3 & -2 \\ 4 & -5 & 8 \end{pmatrix} = \frac{1}{7}\begin{pmatrix} 2 & 1 & -3 \\ -1 & 3 & -2 \\ 4 & -5 & 8 \end{pmatrix} = X_2^{-1}$$

maps the second set of four points to $(1, 0, 0)$, $(0, 1, 0)$, $(0, 0, 1)$ and $(1, 1, 1)$. The required transformation is, therefore,

$$A_2^{-1}A_1 = X_2A_1 = \begin{pmatrix} 2 & 1 & 1 \\ 0 & 4 & 1 \\ -1 & 2 & 1 \end{pmatrix}\begin{pmatrix} 0 & 0 & -1 \\ -\frac{1}{3} & \frac{1}{3} & \frac{1}{3} \\ 1 & 1 & 3 \end{pmatrix} \equiv \begin{pmatrix} 2 & 4 & 4 \\ -1 & 7 & 13 \\ 1 & 5 & 14 \end{pmatrix}.$$

We introduced this discussion by saying that a transformation moves objects. Suppose that we were performing a dynamic simulation of an agricultural tractor driving along rough ground. We could model the tractor's motion by a system of differential equations that is being solved to determine each new position of the tractor, and then apply a transformation to each major component, such as the wheels, to move them into the new position. We could display the whole situation on a screen, and the viewer would see the tractor "drive" along the road, bumping up and down. In this situation, the transformation indeed appears to be moving the object. Imagine now that we are examining, not the dynamic stability of the tractor, but the visibility from the driver's seat. We could look at the tractor from any point by leaving the tractor stationary and moving the "eye" to some desired position. That is, we select a point and make a central projection (projective transformation) from that point. In this way the "eye" can be placed inside the tractor's cab, and we shall then be able to determine if there is any structural piece, for example, a thick vertical exhaust pipe, that is obstructing a clear view. In this situation the transformation has not moved the object, but has, in effect, mapped a figure from one plane to another. A transformation can, therefore, be interpreted in several ways. If $\underline{x} = (x \ \ y \ \ z)^T$, we can think of the equation $\underline{x}' = A\underline{x}$ as a change of coordinates in the plane; as a transformation that does not change the coordinates but instead moves objects [that is, the point with coordinates (x, y, z) is moved to a new position where its new coordinates are (x', y', z')], or as a mapping from one plane to a different plane altogether. The appropriate interpretation of the transformation depends on the context.

4.4.3 The Cross Ratio

In Chapter 3 we discussed the cross ratio from a synthetic viewpoint. Here we look at it in an algebraic framework. There are some very obvious consequences of Theorem 4.4. Clearly, distance is not preserved under a projective transformation. Neither is angle, area, or order, nor the concept of being inside a triangle. From Lemma 4.1, we see that collinear points will remain collinear, and, from duality, concurrent lines will remain concurrent, but there seems to be very little that remains constant in comparison with what changes. A projective transformation can hence be expected to greatly alter the appearance of an object. In Section 3.2 we mentioned central projections as being related to the way we see things. We have defined a plane projective transformation as a nonsingular 3×3 matrix (Definition 4.6). Now we must establish an equivalence between these two in order to verify the consistency of our argument. To do this we will use the cross ratio.

If $\underline{a}$ and $\underline{b}$ are coordinate triples of two distinct points in the projective plane, then the line determined by these two points was defined to be the set of points with coordinates of the form

$$x = \lambda a + \mu b, \quad \text{with } \lambda, \mu \text{ not both zero.}$$

The point is determined by the ratio μ/λ (or λ/μ). If we set $\theta = \mu/\lambda$ and use the symbol ∞, which we read as "infinity," when $\lambda = 0$, then we can write

$$\underline{x} = \underline{a} + \theta\underline{b},$$

and the point $\underline{x}$ is uniquely determined by the value of θ. We refer to θ as the *projective parameter* of the line.

The value of θ associated with each point $\underline{x}$ will be unique once specific triples representing $\underline{a}$ and $\underline{b}$, called the *base points* of the line, are given. Furthermore, if a projective transformation maps $\underline{a}$ to $\underline{a}'$ and $\underline{b}$ to $\underline{b}'$, then the point $\underline{x}$ will be mapped to $\underline{x}' = \underline{a}' + \theta \underline{b}'$ and hence the projective parameter of a point on a line will remain invariant under a projective transformation if we use the images of the base points $\underline{a}$ and $\underline{b}$ as the base points $\underline{a}'$ and $\underline{b}'$ for determining the projective parametrization of the transformed line.

Now, let P_i be the point $\underline{a} + \theta_i \underline{b}$, $i = 1, 2, 3, 4$. Then we define the cross ratio by

$$\{P_1, P_2; P_3, P_4\} \equiv \frac{\left(\dfrac{\theta_3 - \theta_1}{\theta_2 - \theta_3}\right)}{\left(\dfrac{\theta_4 - \theta_1}{\theta_2 - \theta_4}\right)} = \left(\frac{\theta_3 - \theta_1}{\theta_2 - \theta_3}\right)\left(\frac{\theta_2 - \theta_4}{\theta_4 - \theta_1}\right).$$

This definition results in precisely the same number as our previous definition, but applies as well in the general projective plane as it does in any of its special cases. To see this, all we need to do is to map the points $\underline{a}$ and $\underline{b}$ to $(0, 0, 1)$ and $(1, 0, 1)$. The point P_i now has coordinates

$$(x_i, 0, 1) = (1 - x_i)\left((0, 0, 1) + \frac{x_i}{1 - x_i}(1, 0, 1)\right),$$

and its projective parameter is $\theta_i = x_i/(1 - x_i)$. This parameter is unchanged by the transformation. If we choose $z = 1$ for all the points, the numbers x_i give the Cartesian coordinates of the points, so that P_i has Cartesian coordinates $(x_i, 0)$. Then

$$\{P_1, P_2; P_3, P_4\} = \left(\frac{x_3 - x_1}{x_2 - x_3}\right)\left(\frac{x_2 - x_4}{x_4 - x_1}\right),$$

which, upon substitution for x_i in terms of θ_i, becomes

$$\{P_1, P_2; P_3, P_4\} = \left(\frac{\theta_3 - \theta_1}{\theta_2 - \theta_3}\right)\left(\frac{\theta_2 - \theta_4}{\theta_4 - \theta_1}\right).$$

If we change parametrization, using $\underline{a}'$ and $\underline{b}'$ as the new base points of the parametrization, then

$$\underline{a} = \underline{a}' + \mu \underline{b}', \qquad \underline{b} = \underline{a}' + \nu \underline{b}',$$

and

$$P_i = (1 + \theta_i)\underline{a}' + (\mu + \theta_i \nu)\underline{b}',$$

and hence the new projective parameter θ_i' is given by

$$\theta_i' = \frac{\mu + \theta_i \nu}{1 + \theta_i}.$$

This change, however, does not affect the cross ratio (see Exercise 4.29). The cross ratio of four collinear points is then invariant under a projective transformation. In dual fashion, we also know that the cross ratio of concurrent lines remains invariant.

This redefinition of the cross ratio enables us to extend our concept of a central projection. Look again at Figure 3.3. We see that the point O is now any point of the projective plane and hence can be a point on the ideal line where a set of parallel lines in the Euclidean plane meet. Central projections, therefore, include parallel projections. Figure 3.3 indicates a central projection of a line onto another line in the plane defined by the initial line and the center of perspective. By taking a sequence of central projections, we can map a line onto any desired plane, and the sequence of central projections will preserve the cross ratio of collinear points. For example, the application of two central projections can map four collinear points P_i, $i = 1, \ldots, 4$, in one plane to four collinear points R_i, $i = 1, \ldots, 4$, in a different plane, with

$$\{P_1, P_2; P_3, P_4\} = \{Q_1, Q_2; Q_3, Q_4\} = \{R_1, R_2; R_3, R_4\}.$$

We depict this in Figure 4.6.

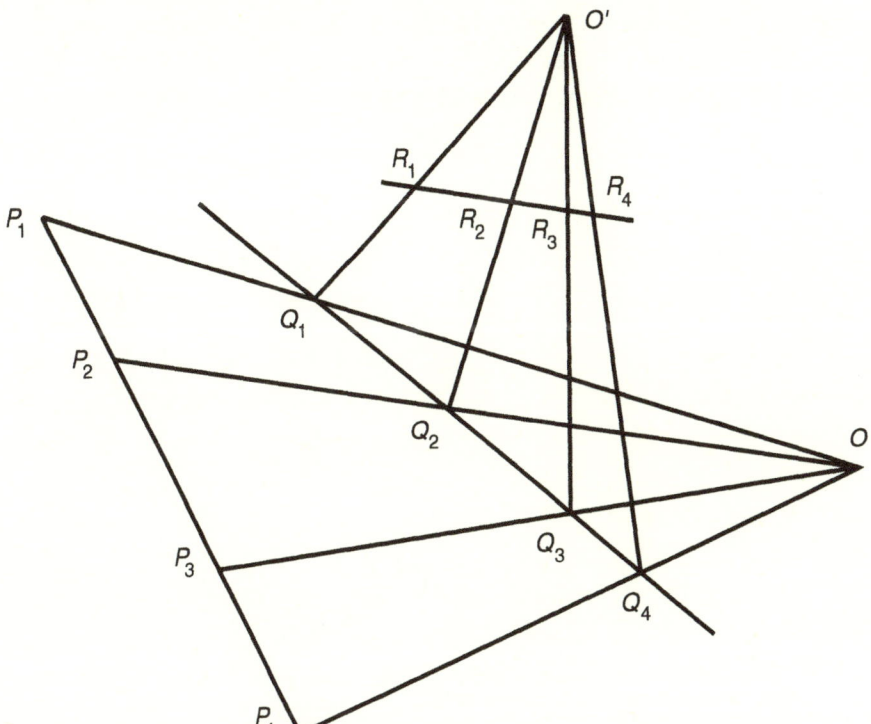

Figure 4.6. Three ranges of four points related by central projections.

4.4.4 Central Projections

We are now in a position to develop the equivalence between central projections and the nonsingular matrices of projective transformations. The equivalence is actually between a sequence of central projections and the projective transformations. We begin with a theorem similar to Theorem 4.4.

Theorem 4.5 *There is a sequence of central projections that transforms four given points in a plane, no three of which are collinear, into four given points in the same plane, no three of which are collinear.*

Proof. Let the two sets of points be $\{P_1, P_2, P_3, P_4\}$ and $\{Q_1, Q_2, Q_3, Q_4\}$. We seek to relate these sets by a sequence of central projections. Let the two sets lie in a plane π_1 and project the points P_1, P_2, P_3, and P_4 from some point O_1 not in π_1 to points R_1, R_2, R_3, and R_4, respectively, which lie in a plane π_2 distinct from π_1 (Figure 4.7). Since the sets $\{P_i\}$ and $\{R_i\}$ are related by a central projection, all we need to show is that the sets $\{R_i\}$ and $\{Q_i\}$ are related by a sequence of central projections.

Choose a point O_2 on Q_1R_1, and project π_2 onto a plane π_3 that passes through Q_1 but is distinct from π_1 and does not contain any of the points Q_i, $i = 2, 3, 4$ (Figure 4.8). Then R_1, R_2, R_3 and R_4 are projected from O_2 to Q_1, S_2, S_3, and S_4, respectively. Let the lines Q_1Q_2 and Q_3Q_4 meet at A_1 on π_1,

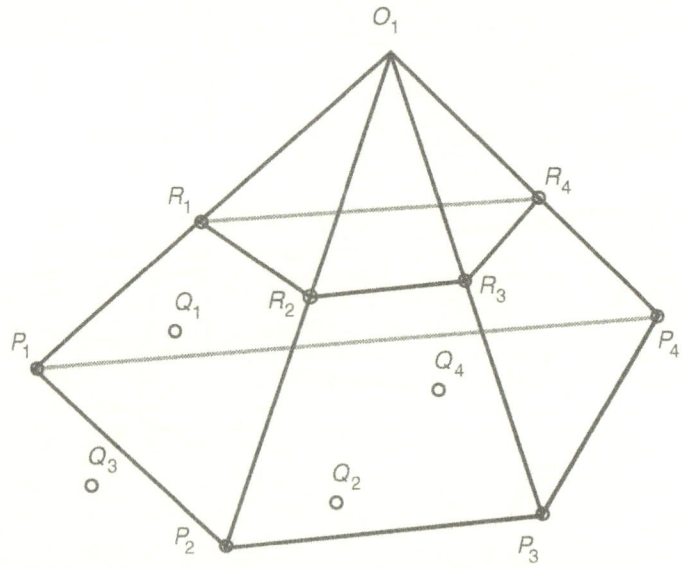

Figure 4.7. Projecting P_1, P_2, P_3, P_4 to R_1, R_2, R_3, R_4.

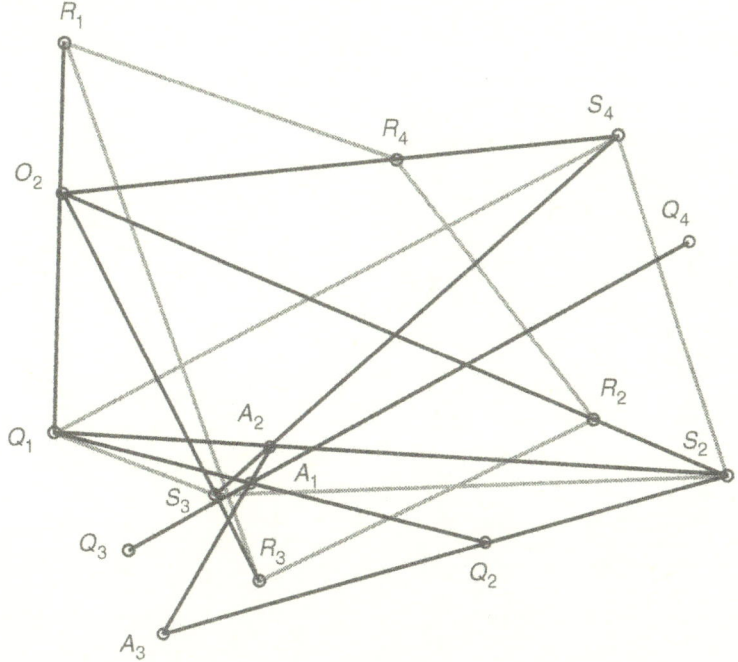

Figure 4.8. Projecting R_1, R_2, R_3, R_4 to Q_1, S_2, S_3, S_4.

and let the lines $Q_1 S_2$ and $S_3 S_4$ meet at A_2 on π_3. Now, the lines $Q_1 Q_2$ and $Q_1 S_2$ form a plane, and hence the lines $A_1 A_2$ and $Q_2 S_2$ are coplanar and hence meet at a point, say A_3.

Now, from A_3, project π_3, the plane of Q_1, S_2, S_3, S_4, and A_2, onto a plane π_4 that passes through Q_1 and Q_2 but is distinct from π_1 and does not contain A_3 (Figure 4.9). The projections of Q_1, S_2, S_3, S_4, and A_2 are Q_1, Q_2, T_3, T_4, and A_1. Since S_3, S_4, and A_2 are collinear, their images T_3, T_4, and A_1 must also be collinear. This line and the line $Q_3 Q_4 A_1$ form a plane, and hence $Q_3 T_3$ and $Q_4 T_4$ must intersect. Let the point of intersection be A_4.

Finally, we project from A_4 onto π_1 and the points Q_1, Q_2, T_3, and T_4 to Q_1, Q_2, Q_3, and Q_4, respectively. The sequence of central projections has then given the following sequence of transformations:

$$[P_1, P_2, P_3, P_4] \quad \text{to} \quad [R_1, R_2, R_3, R_4] \quad \text{to} \quad [Q_1, S_2, S_3, S_4]$$

$$\text{to} \quad [Q_1, Q_2, T_3, T_4] \quad \text{to} \quad [Q_1, Q_2, Q_3, Q_4]. \quad \blacksquare$$

Theorem 4.6 *In terms of a fixed coordinate system, any sequence of central projections that maps the plane onto itself is a projective transformation.*

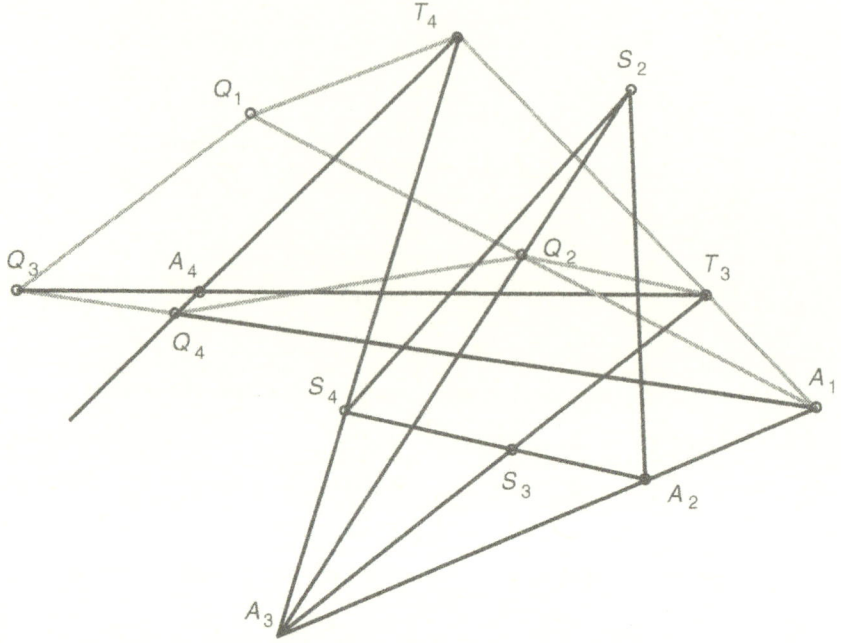

Figure 4.9. Projecting Q_1, S_2, S_3, S_4, A_2 to Q_1, Q_2, T_3, T_4, A_1.

Proof. Let X, Y, Z, and U be the vertices of the triangle of reference and the unit point. Then the sequence of central projections will map these points to X', Y', Z', and U', no three points being collinear. Let these new points define a coordinate system, and let P' be the image of P under the sequence of central projections. If (x, y, z) are the coordinates of P in the system based upon X, Y, Z, and U, and (x', y', z') are the coordinates of P' in the system based upon X', Y', Z', and U', then

$$\frac{y}{z} = \{U, P; Y, Z\} = \{U', P'; Y', Z'\} = \frac{y'}{z'},$$

since the cross ratio is invariant under a central projection. Repeating this step for the other ratios, we get

$$(x, y, z) = \rho(x', y', z'),$$

and hence P and P' represent the same point in different coordinate systems. Now, from Theorem 4.4, two coordinate systems are related by a projective transformation and, hence, P is related to P' by a projective transformation. ∎

Theorems 4.5, 4.6, and 4.4, applied in that order, show that a projective transformation is a sequence of central projections. This is because Theorem 4.5 tells us that there is a sequence of central projections that maps any four points to any four points. Then Theorem 4.6 assures us that such a sequence is a projective transformation, and Theorem 4.4 tells us that this transformation is unique. Hence, we have established the equivalence between central projections and projective transformations.

4.4.5 Fixed Points and Lines

If a projective transformation A has an eigenvector $\underline{v}$, then, since $A\underline{v} = \lambda\underline{v}$ for some nonzero λ, both the point and the line represented by $\underline{v}$ will remain fixed under the transformation A. The fixed points and lines of a particular transformation are then determined by the eigenstructure of the matrix A.

If the vector $\underline{l} = (l \ m \ n)^T$ represents line coordinates and the vector $\underline{x} = (x \ y \ z)^T$ represents point coordinates, then a line is given by an equation of the form

$$\underline{l}^T \underline{x} = 0.$$

If $\underline{x}$ is transformed to $\underline{x}'$ by a projective transformation A, we have $\underline{x}' = A\underline{x}$ and hence

$$\underline{l}^T \underline{x} = \underline{l}^T A^{-1} \underline{x}' = \underline{l}'^T \underline{x}',$$

which implies that the new line coordinates are related to the old ones by the transformation

$$\underline{l}'^T = \underline{l}^T A^{-1}.$$

This fact enables us to find the fixed lines in a similar fashion to that of finding the fixed points. Note that the concept of a "fixed line" does not imply that all the points on the line are fixed, but merely that points on the line are mapped to points on the same line.

Example 4.13 *Find the fixed points and lines of the projective transformation*

$$A = \begin{pmatrix} -1 & 0 & 0 \\ 2 & 1 & 0 \\ 1 & -1 & 1 \end{pmatrix}.$$

The eigenvalues are -1 and 1 with corresponding eigenvectors $(-1 \ 1 \ 1)^T$ and $(0 \ 0 \ 1)^T$, respectively. There is only a single eigenvector associated with the repeated eigenvalue 1. The points $(1, -1, -1)$ and $(0, 0, 1)$ are therefore fixed points. From duality there must be two fixed lines, and clearly the line

$$\det \begin{pmatrix} x & y & z \\ 1 & -1 & -1 \\ 0 & 0 & 1 \end{pmatrix} = -x - y = 0$$

determined by the two fixed points must be a fixed line.

The eigenvalues of the inverse of a matrix are the reciprocals of the eigenvalues of the original matrix. Since

$$A^{-1} = \begin{pmatrix} -1 & 0 & 0 \\ 2 & 1 & 0 \\ 3 & 1 & 1 \end{pmatrix},$$

the fixed lines are given by the eigenvectors of its transposed matrix, namely $(1 \ 0 \ 0)^T$ and $(1 \ 1 \ 0)^T$. The first corresponds to the line $x = 0$, whereas the second corresponds to $x + y = 0$, which is the line determined above by the fixed points. The given transformation, therefore, has two fixed points $(1, -1, -1)$ and $(0, 0, 1)$ and two fixed lines $x + y = 0$ and $x = 0$.

4.4.6 The Projective Invariance of Conics

Since a projective transformation is linear, the image of an algebraic curve is another algebraic curve of the same order. In particular, a conic will be mapped to a conic under a projective transformation. If we think of the different conics as sections of a right circular cone, we are led to expect that all conics can be transformed into circles; for if our vantage point were the vertex of the cone, then any curve on the lateral surface of the cone would appear as a circle, irrespective of whether the curve is planar or not. The proof of the projective equivalence of conics is contained in the following theorem, which is proved by the use of pencils of conics together with Theorem 4.4.

The equation of the general conic, in projective coordinates, can be written in the form

$$\underline{x}^T A \underline{x} = 0,$$

where A is the symmetric matrix

$$\begin{pmatrix} a & h & g \\ h & b & f \\ g & f & c \end{pmatrix}.$$

We showed in Section 2.5 that the conic is a line pair if and only if A is singular, and hence a nondegenerate conic has an equation $\underline{x}^T A \underline{x} = 0$, where

A is nonsingular. Furthermore, if $\underline{x}_0$ is a point of the nondegenerate conic $\underline{x}^T A \underline{x} = 0$, then the equation of the tangent to the conic at $\underline{x}_0$ is $\underline{x}_0^T A \underline{x} = 0$ (see Exercise 4.36).

Theorem 4.7 *Two nondegenerate conics are projectively equivalent.*

Proof. Let A and B be two points of a conic S_1, and let the tangents to S_1 at A and B meet at C. Let D be a point of the conic different from A and B (Figure 4.10). Then no three of A, B, C, and D are collinear. Take a coordinate system such that A, B, and C are the vertices $(1, 0, 0)$, $(0, 1, 0)$, and $(0, 0, 1)$, respectively, of the triangle of reference, and D is the unit point. Then the pencil of conics tangential to AC at A and tangential to BC at B has the equation

$$xy + \alpha z^2 = 0.$$

This pencil must contain S_1. Now, S_1 passes through $D(1, 1, 1)$; hence the unique member of the pencil which passes through D is given by $\alpha = -1$. Therefore, S_1 has the equation $xy - z^2 = 0$. If S_2 is any other conic, we can similarly write its equation in the form

$$x'y' - z'^2 = 0$$

in a suitably chosen coordinate system. From the fundamental theorem, these two coordinate systems are related by a projective transformation. Therefore, the two conics $xy - z^2 = 0$ and $x'y' - z'^2 = 0$ are projectively related and thus the conics S_1 and S_2 are projectively equivalent. ∎

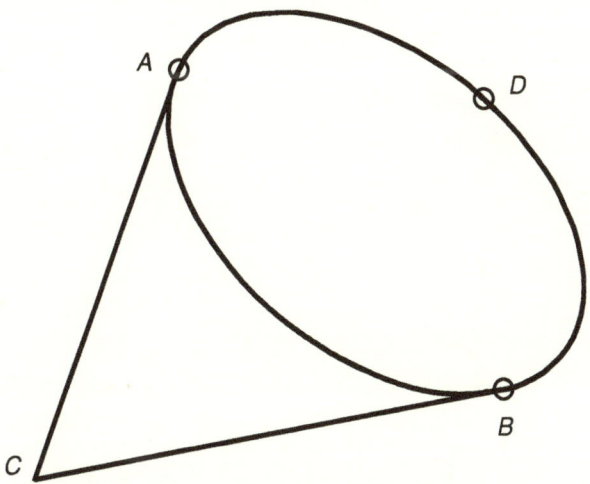

Figure 4.10. Tangents and chord as coordinate system.

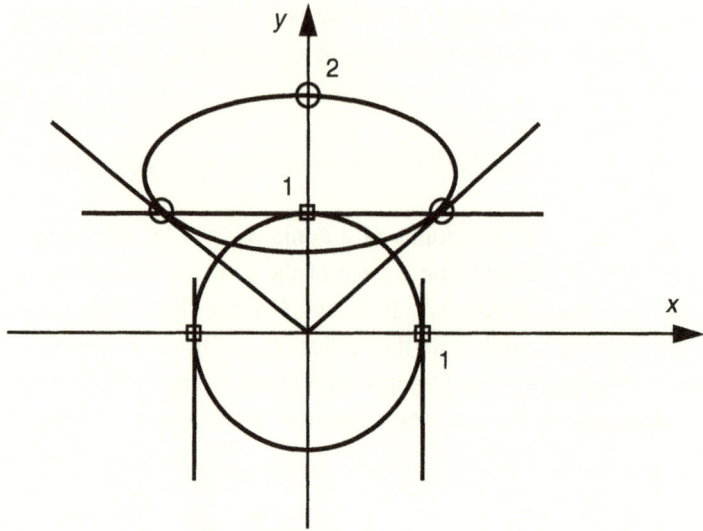

Figure 4.11. Transformation of an ellipse to a circle.

Example 4.14 *Determine a projective transformation which maps the ellipse* $x^2 + 3y^2 - 8yz + 4z^2 = 0$ *to the circle* $x^2 + y^2 - z^2 = 0$.

We take the points $(-1, 1, 1)$, $(1, 1, 1)$, and $(0, 2, 1)$ on the given ellipse (Figure 4.11). The tangent at $(-1, 1, 1)$ has equation

$$(-1\,1\,1) \begin{pmatrix} 1 & 0 & 0 \\ 0 & 3 & -4 \\ 0 & -4 & 4 \end{pmatrix} \begin{pmatrix} x \\ y \\ z \end{pmatrix} = 0,$$

which is $x + y = 0$. Similarly, the tangent at $(1, 1, 1)$ has equation $x - y = 0$. The line determined by the points $(-1, 1, 1)$ and $(1, 1, 1)$ is

$$\det \begin{pmatrix} x & y & z \\ -1 & 1 & 1 \\ 1 & 1 & 1 \end{pmatrix} = 0,$$

which is $y - z = 0$. Let $x' = \rho(x + y)$, $y' = \sigma(x - y)$, $z' = \tau(y - z)$; then the point $(0, 2, 1)$ maps to $(1, 1, 1)$ if $\rho : \sigma : \tau = 1 : -1 : 2$, which implies

$$\begin{pmatrix} x' \\ y' \\ z' \end{pmatrix} = \begin{pmatrix} 1 & 1 & 0 \\ -1 & 1 & 0 \\ 0 & 2 & -2 \end{pmatrix} \begin{pmatrix} x \\ y \\ z \end{pmatrix} = A\underline{x},$$

say. Therefore, replacing $\underline{x}$ by $A^{-1}\underline{x}$ will map the ellipse to $xy - z^2 = 0$. Now,

$$A^{-1} = \frac{1}{2}\begin{pmatrix} 1 & -1 & 0 \\ 1 & 1 & 0 \\ 1 & 1 & -1 \end{pmatrix}.$$

Taking the tangents to the circle at the points $(-1, 0, 1)$ and $(1, 0, 1)$, using $(0, 1, 1)$ as the third point, and proceeding as above, we find that replacing $\underline{x}$ by $B\underline{x}$, where

$$B = \begin{pmatrix} -1 & 0 & 1 \\ 1 & 0 & 1 \\ 0 & 1 & 0 \end{pmatrix},$$

maps $xy - z^2 = 0$ to $x^2 + y^2 - z^2 = 0$. Therefore, replacing $\underline{x}$ by $A^{-1}B\underline{x}$ will map the given ellipse to the required circle, where

$$A^{-1}B = \begin{pmatrix} -1 & 0 & 0 \\ 0 & 0 & 1 \\ 0 & -\frac{1}{2} & 1 \end{pmatrix}.$$

Hence, replacing x by $-x$, y by z, and z by $-\frac{1}{2}y + z$ will map $x^2 + 3y^2 - 8yz + 4z^2 = 0$ to $x^2 + y^2 - z^2 = 0$.

The fact that all conics can be written in the form $xy - z^2 = 0$ (or $xz - y^2 = 0$, or any other desired quadratic form) enables us to produce parametrizations of the curve in a very simple fashion. We can write $(x, y, z) = (s^2, t^2, st)$ in terms of the projective coordinates s and t of the line, which is clearly a parametrization of $xy - z^2 = 0$.

Example 4.15 *Construct a parametrization of*

$$3x^2 + 2xy + 3y^2 - 18xz - 18yz + 36z^2 = 0$$

based upon the point $(3, 3, 1)$ and tangents at $(3, 1, 1)$ and $(1, 3, 1)$.

The conic is

$$(x \quad y \quad z)\begin{pmatrix} 3 & 1 & -9 \\ 1 & 3 & -9 \\ -9 & -9 & 36 \end{pmatrix}\begin{pmatrix} x \\ y \\ z \end{pmatrix} = 0.$$

The tangent at $(3, 1, 1)$ is

$$(3 \quad 1 \quad 1)\begin{pmatrix} 3 & 1 & -9 \\ 1 & 3 & -9 \\ -9 & -9 & 36 \end{pmatrix}\begin{pmatrix} x \\ y \\ z \end{pmatrix} = x - 3y = 0.$$

Similarly, the tangent at $(1, 3, 1)$ has equation $3x - y = 0$. The line determined by the two points $(3, 1, 1)$ and $(1, 3, 1)$ has the equation $x + y - 4z = 0$. Hence

$$x' = \rho(x - 3y), \qquad y' = \sigma(3x - y), \qquad z' = \tau(x + y - 4z).$$

The point $(3, 3, 1)$ will be mapped to $(1, 1, 1)$ if $\rho : \sigma : \tau = -1 : 1 : 3$, and hence the transformation is

$$\underline{x}' = \begin{pmatrix} -1 & 3 & 0 \\ 3 & -1 & 0 \\ 3 & 3 & -12 \end{pmatrix} \underline{x} = \begin{pmatrix} s^2 \\ t^2 \\ st \end{pmatrix},$$

which leads to

$$\underline{x} = \begin{pmatrix} 3(s^2 + 3t^2) \\ 3(3s^2 + t^2) \\ 3s^2 + 3t^2 - 2st \end{pmatrix}$$

as the required parametrization.

The parametric points $(0, 1)$, $(1, 0)$, and $(1, 1)$ are mapped to $(3, 1, 1)$, $(1, 3, 1)$, and $(3, 3, 1)$, respectively.

4.5 Affine Transformations

A projective transformation is equivalent to a sequence of central projections and thus has important connections to how we actually see the world around us. Any four planar points, no three of which are collinear, can be mapped to any four planar points, no three of which are collinear, and they can be used to map any nondegenerate conic to any nondegenerate conic. The cross ratio of four collinear points or four concurrent lines is invariant under such transformations, but little else is invariant. There is no concept of parallel lines. As mentioned in the previous section, the other transformations useful in computer-aided design and the finite-element method (rigid motions, affine transformations, etc.) are simply special cases of the projective transformations. Just as central projections have important applications, so also have parallel projections, but since there are no parallel lines in the projective plane, we shall not be able to effect a parallel projection in the projective plane. How can this difficulty be surmounted? To answer this question, recall the way we constructed the projective plane by the addition of an ideal line to the Euclidean plane, and the synthetic discussion of the affine plane in Section 3.8.

Let P_2 denote the projective plane, and let l denote a line of P_2. We define an *affine plane* A_2 by $A_2 = P_2 \backslash l$, that is, A_2 is the complement of l with respect to P_2. An affine plane is, therefore, the projective plane from which one line has been removed. From all projective transformations, let G denote the set of those transformations under which l is invariant, that is, for which l is a fixed line. Then G is a group and also maps A_2 onto A_2 (see Exercise 4.39). We call this group the *affine group* corresponding to A_2. If we think of this selected line l as the "ideal" line, we can define two lines in the plane as being parallel if they meet at a point of l. This definition is completely consistent with our previous definition of parallel lines as lines that do not meet, for if the two parallel lines meet at a point of l, then, since $A_2 = P_2 \backslash l$, these lines do not meet in A_2. An immediate consequence of this is that a line in the affine plane divides the plane into two mutually exclusive sets of points, one on each "side" of the line, thus reintroducing the concept of order, which does not exist in the projective plane. Note that the affine plane is not Euclidean, since we have not reintroduced Euclidean properties such as invariance of distance and angles.

The most common transformation in the finite-element method is the mapping of some arbitrary triangle onto some standard triangle. This transformation is justified by the following theorem.

Theorem 4.8 *The vertices of a triangle can be mapped onto the vertices of any other triangle by means of an affine transformation, that is, any two triangles in the affine plane are affinely equivalent.*

Proof. Let m_1, m_2, m_3 and n_1, n_2, n_3, respectively, be the sides of the two triangles, and let l be the line that was removed from P_2 to produce the affine plane under consideration. Then no three of the lines m_1, m_2, m_3, and l are concurrent. Similarly, no three of the lines n_1, n_2, n_3, and l are concurrent. Therefore, from the dual of Theorem 4.4, they are projectively equivalent. In particular, there is a projective transformation that maps m_i to n_i, $i = 1, 2, 3$, and leaves l invariant. This transformation must, therefore, belong to the subgroup of affine transformations, and hence the two triangles are affinely equivalent. ∎

For any specified line l in the projective plane, we can always choose a coordinate system such that $l = 0$ is a side of the triangle of reference. Denote the other two sides by $x = 0$ and $y = 0$. Then we can use (x, y, l) as coordinates in the projective plane—which, of course, includes the affine plane, so that (x, y, l) serve as coordinates in the affine plane. There is, however, a redundancy in this notation. If the affine plane under consideration is the one from which the line l has been removed, then there are no points in this affine plane for which $l = 0$ or, equivalently, $l \neq 0$ for all points in the affine plane. This means that any point in the affine plane has projective coordinates of the form

$(a, b, l) \equiv (l^{-1}a, l^{-1}b, 1) = (\alpha, \beta, 1)$. With this normalization, we can use the first two coordinates as the affine coordinates of points in the affine plane. A point in the affine plane is now described by (x, y). We note that (x, y) and $(\rho x, \rho y)$ represent the same point in the affine plane if and only if $\rho = 1$, since their projective coordinates are $(x, y, 1)$ and $(\rho x, \rho y, 1)$, respectively.

With this choice of coordinate system, a projective transformation which leaves l invariant must be of the form

$$A = \begin{pmatrix} a_{11} & a_{12} & a_{13} \\ a_{21} & a_{22} & a_{23} \\ 0 & 0 & 1 \end{pmatrix}.$$

We define a nonsingular matrix of this form to be an *affine transformation*.

Example 4.16 *Find the affine transformation that maps* $A(1, 1)$, $B(0, 2)$, *and* $C(1, 0)$ *to the points* $P(0, 0)$, $Q(1, 1)$, *and* $R(-1, 2)$, *respectively.*

The lines AB and BC must be mapped to PQ and QR, respectively. Hence, if (x, y) are the original and (x', y') the transformed coordinates, we must have

$$-x' + y' = \alpha(-x - y + 2),$$
$$-x' - 2y' + 3 = \beta(2x + y - 2).$$

This guarantees that B is mapped to Q. Imposing the other conditions, we get $\alpha = \beta = 3$. This yields the transformation

$$\begin{pmatrix} x' \\ y' \\ 1 \end{pmatrix} = \begin{pmatrix} 0 & 1 & -1 \\ -3 & -2 & 5 \\ 0 & 0 & 1 \end{pmatrix} \begin{pmatrix} x \\ y \\ 1 \end{pmatrix},$$

which is the required affine transformation.

In this section, we have defined the general affine transformations, with the property that they keep one specified line, namely the "ideal" line, invariant. It is now left to the reader (see Exercise 4.43) to show that the type of a conic remains invariant under any affine transformation, and therefore the property of a conic of being an ellipse, a parabola, or a hyperbola is an affine property.

Since any triangle can be mapped to any other triangle, it is clear that area is not preserved. If $\tilde{x} = f(x, y)$, $\tilde{y} = g(x, y)$ is any differentiable transformation, then

$$\iint d\tilde{x}\, d\tilde{y} = \iint J\left(\frac{f, g}{x, y}\right) dx\, dy,$$

where

$$J\left(\frac{f,g}{x,y}\right) = \det \begin{pmatrix} \dfrac{\partial f}{\partial x} & \dfrac{\partial f}{\partial y} \\[2ex] \dfrac{\partial g}{\partial x} & \dfrac{\partial g}{\partial y} \end{pmatrix}$$

denotes the Jacobian of the transformation. Therefore, area elements are multiplied by the absolute value of the Jacobian of the transformation. Affine transformations whose Jacobian determinants have value ± 1 are called *equiaffine transformations* or *equiareal transformations*, and they form a subgroup of the affine transformations (see Exercise 4.44).

4.5.1 Similarity and Euclidean Transformations

Since we have removed a line from the projective plane to produce an affine plane, we have removed a point from each projective line to form the corresponding affine line.

Let P and Q have affine coordinates (x_1, y_1) and (x_2, y_2), respectively. A point R on the line through P and Q is defined to be a member of the set of points with projective coordinates

$$(x, y, l) = v(x_1, y_1, 1) + \rho(x_2, y_2, 1),$$

and hence the point on the ideal line $l = 0$ corresponds to $v + \rho = 0$. Therefore, the affine points on the line through P and Q correspond to $v + \rho \neq 0$, and the affine line consists of the set of points

$$(x, y, l) = (vx_1 + \rho x_2, vy_1 + \rho y_2, v + \rho), \qquad v + \rho \neq 0$$

$$\equiv \left(\frac{v}{v + \rho}x_1 + \frac{\rho}{v + \rho}x_2, \frac{v}{v + \rho}y_1 + \frac{\rho}{v + \rho}y_2, 1\right).$$

Therefore, in affine coordinates, the line through P and Q is given by the set of points such that

$$(x, y) = \left(\frac{v}{v + \rho}x_1 + \frac{\rho}{v + \rho}x_2, \frac{v}{v + \rho}y_1 + \frac{\rho}{v + \rho}y_2\right)$$

$$= \frac{v}{v + \rho}(x_1, y_1) + \frac{\rho}{v + \rho}(x_2, y_2)$$

$$= (1 - \mu)(x_1, y_1) + \mu(x_2, y_2),$$

that is,

$$R = (1 - \mu)P + \mu Q = P + \mu(Q - P).$$

We can think of μ as a parameter defining a *distance* from the point P relative to the point Q. The value μ is then the *affine line parameter* based upon P and Q. If collinear points P_i, $i = 1, 2, 3$, correspond to affine parameters μ_i, $i = 1, 2, 3$, we define the ratio $\{P_1, P_2, P_3\}$ by

$$\{P_1, P_2, P_3\} = \frac{P_1 P_2}{P_2 P_3} = \frac{\mu_2 - \mu_1}{\mu_3 - \mu_2}.$$

This is an affine invariant. That is, the ratio of distances on a line or on parallel lines is invariant under an affine transformation (see Exercise 4.45). Elimination of the line parameter shows that the equation of the affine line through distinct points (x_1, y_1) and (x_2, y_2) can be expressed in the form

$$\det \begin{pmatrix} x & y & 1 \\ x_1 & y_1 & 1 \\ x_2 & y_2 & 1 \end{pmatrix} = 0.$$

This is a linear equation of the form $ax + by + c = 0$, with $a^2 + b^2 \neq 0$, in terms of the coordinates of a general point (x, y) on the line.

We now define the measure of *distance d* between two affine points (x_1, y_1) and (x_2, y_2) to be

$$d(P_1, P_2) \equiv \sqrt{(x_2 - x_1)^2 + (y_2 - y_1)^2},$$

and we define the measure of the *angle θ* between the two affine lines $l_1 \equiv a_1 x + b_1 y + c_1 = 0$ and $l_2 \equiv a_2 x + b_2 y + c_2 = 0$ to be

$$\theta(l_1, l_2) \equiv \arccos \left(\frac{|a_1 a_2 + b_1 b_2|}{\sqrt{a_1^2 + b_1^2} \sqrt{a_2^2 + b_2^2}} \right),$$

where the function $\arccos(u)$ is defined by its series expansion. These definitions are consistent with those used in the Euclidean plane with a rectangular coordinate system. The angle θ is a measure of the "difference in direction" between lines. Parallel lines correspond to $\theta(l_1, l_2) = \arccos(1) = 0$. When $a_1 a_2 + b_1 b_2 = 0$, we have $\theta(l_1, l_2) = \arccos(0) = \pi/2$, and we say that such lines are *perpendicular* or *orthogonal*.

For an arbitrary affine transformation A, let $\underline{x}' = A\underline{x}$. Then

$$(x_2' - x_1')^2 + (y_2' - y_1')^2 = \left(a_{11}^2 + a_{21}^2\right)(x_2 - x_1)^2 + \left(a_{12}^2 + a_{22}^2\right)(y_2 - y_1)^2$$

$$+ 2(a_{11}a_{12} + a_{21}a_{22})(x_2 - x_1)(y_2 - y_1).$$

Therefore, distances are scaled equally in all directions if

$$a_{11}^2 + a_{21}^2 = a_{12}^2 + a_{22}^2,$$

$$a_{11}a_{12} + a_{21}a_{22} = 0.$$

The general solution of these equations is $a_{22} = \pm a_{11}$, $a_{21} = \mp a_{12}$, yielding a transformation of the form

$$\begin{pmatrix} a & b & c \\ \mp b & \pm a & d \\ 0 & 0 & 1 \end{pmatrix}, \qquad a^2 + b^2 = k^2 > 0.$$

The value $|k|$ is the *scale factor*. Such transformations are called *similarity transformations*. Two planar figures are said to be similar if one can be transformed into the other by a similarity transformation. Similarity transformations form a group (a proper subgroup of the affine transformations), and we can therefore talk of similarity geometry. Distances are not invariant under similarity transformations, but the ratios of distances in any direction will be preserved. For example, the eccentricity of a conic will be invariant under a similarity transformation, and angles between lines will be invariant.

If $|k| = 1$, the transformations are called *motions* or *Euclidean transformations*. The group of motions (see Exercise 4.48) contains the translations, the rotations, the reflections, and the glide reflections. The glide reflections and the reflections are *indirect* motions in that they reverse the sense of angles. The translations and rotations are *direct* motions. Let us now investigate some of these transformations.

The translations of points are represented by transformations of the form

$$x' = x + c, \qquad y' = y + d,$$

and hence are represented in 3×3 matrix form as

$$T = \begin{pmatrix} 1 & 0 & c \\ 0 & 1 & d \\ 0 & 0 & 1 \end{pmatrix}.$$

A rotation of a point about the origin through an angle θ is given by (Figure 4.12)

$$x' = r\cos(\theta + \alpha) = x\cos\theta - y\sin\theta,$$

$$y' = r\sin(\theta + \alpha) = x\sin\theta + y\cos\theta,$$

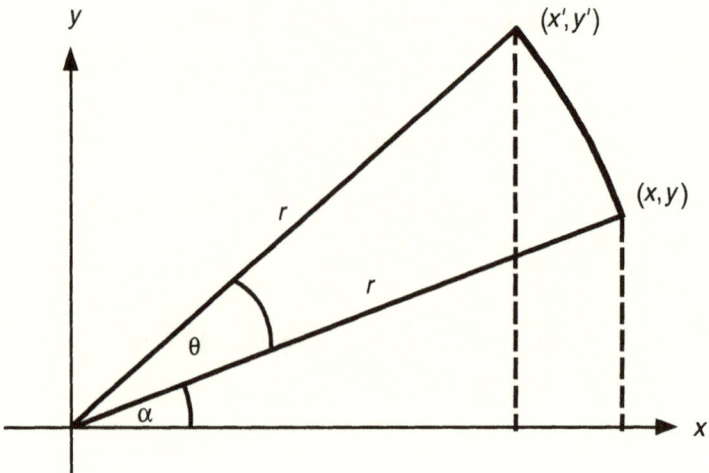

Figure 4.12. Rotation around the origin.

which is represented by the matrix

$$R = \begin{pmatrix} \cos\theta & -\sin\theta & 0 \\ \sin\theta & \cos\theta & 0 \\ 0 & 0 & 1 \end{pmatrix}.$$

A rotation about the point (p, q) through an angle θ is given by

$$x' = (x - p)\cos\theta - (y - q)\sin\theta + p,$$

$$y' = (x - p)\sin\theta + (y - q)\cos\theta + q.$$

This can be interpreted as a translation of the point to the origin followed by a rotation about the origin through an angle θ and then followed by the inverse translation, that is, by the matrix

$$\begin{pmatrix} 1 & 0 & p \\ 0 & 1 & q \\ 0 & 0 & 1 \end{pmatrix} \begin{pmatrix} \cos\theta & -\sin\theta & 0 \\ \sin\theta & \cos\theta & 0 \\ 0 & 0 & 1 \end{pmatrix} \begin{pmatrix} 1 & 0 & -p \\ 0 & 1 & -q \\ 0 & 0 & 1 \end{pmatrix}.$$

However, we can also write the transformation in the form

$$x' = x\cos\theta - y\sin\theta + c,$$

$$y' = x\sin\theta + y\cos\theta + d,$$

where

$$c = -p \cos \theta + q \sin \theta + p,$$
$$d = -p \sin \theta - q \cos \theta + q.$$

Hence, we can think of the transformation as a rotation about the origin through an angle θ followed by a translation, the matrix being written in the form

$$\begin{pmatrix} 1 & 0 & c \\ 0 & 1 & d \\ 0 & 0 & 1 \end{pmatrix} \begin{pmatrix} \cos \theta & -\sin \theta & 0 \\ \sin \theta & \cos \theta & 0 \\ 0 & 0 & 1 \end{pmatrix}.$$

The general rotation can also be interpreted as a translation followed by a rotation, as we can see from the following argument. Let R be a rotation and T a translation, as defined above. Then

$$R^{-1} = \begin{pmatrix} \cos \theta & \sin \theta & 0 \\ -\sin \theta & \cos \theta & 0 \\ 0 & 0 & 1 \end{pmatrix}$$

and

$$R^{-1}TR = \begin{pmatrix} 1 & 0 & c \cos \theta + d \sin \theta \\ 0 & 1 & -c \sin \theta + d \cos \theta \\ 0 & 0 & 1 \end{pmatrix},$$

which is a translation T', and hence $TR = RT'$.

The collection of translations and rotations are referred to as *displacements* or *rigid motions*, and have equations of the general form

$$x' = x \cos \theta - y \sin \theta + c,$$
$$y' = x \sin \theta + y \cos \theta + d.$$

The matrix representation is therefore

$$\begin{pmatrix} a & -b & c \\ b & a & d \\ 0 & 0 & 1 \end{pmatrix}, \qquad \text{where} \quad a^2 + b^2 = 1.$$

The rigid motions of a plane form a group called the *displacement group* (see Exercise 4.55). Displacement transformations are extremely important in geometric modeling, where simple primitives are used to build a model of a complex structure by moving the primitives to different positions during an assembly process, the final structure being a collection of displaced primitives. Most of the common properties, such as distance, angle, and order, are preserved under rigid motions. The properties invariant under this group are called *Euclidean properties*.

Many planar objects are symmetrical about some given line, the points on one side of the line being mirror images of points on the other side of the line. The transformation that maps points in the plane into their mirror images with respect to a fixed line is called a *reflection*. From Figure 4.13 we see that, if P is reflected in a line OQ to the point P', then

$$x' = x - PP'\sin\theta, \qquad y' = y + PP'\cos\theta,$$

where θ is the angle of inclination of the line OQ. Now, $PP' = 2PQ$, and $PQ = OP\sin(\theta - \alpha)$, where $x = OP\cos\alpha$ and $y = OP\sin\alpha$. Substitution into the equations for x' and y' yields

$$x' = x\cos 2\theta + y\sin 2\theta,$$

$$y' = x\sin 2\theta - y\cos 2\theta.$$

The line OQ is referred to as the *axis of reflection*.

We can reflect in an arbitrary line by first reflecting in a line through the origin and then performing a translation through a distance $2n$ in a direction perpendicular to the given line, where n is the length of the directed perpendicular *from* the origin *to* the given line, which defines the direction of the translation (Figure 4.14). If the angle of inclination of the given line is θ, and the angle of

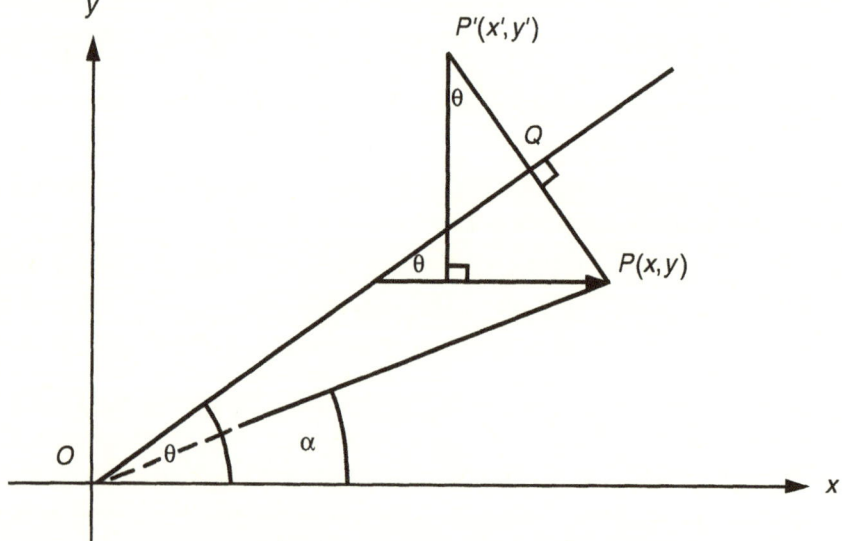

Figure 4.13. Reflection in a line through the origin.

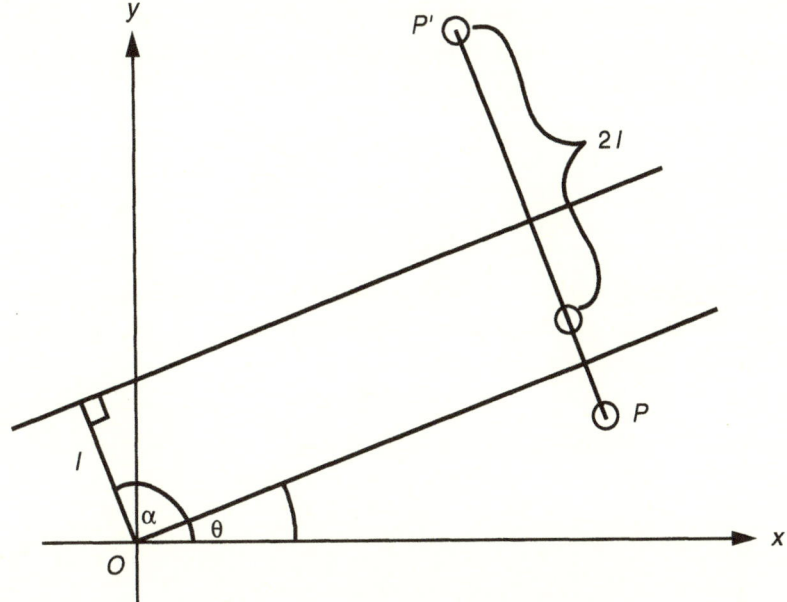

Figure 4.14. Reflection in a general line.

inclination of the directed perpendicular is α, then the reflection is given by

$$x' = x \cos 2\theta + y \sin 2\theta + 2n \cos \alpha,$$
$$y' = x \sin 2\theta - y \cos 2\theta + 2n \sin \alpha.$$

Clearly, $\alpha = \theta + \pi/2$. The corresponding matrix is of the general form

$$\begin{pmatrix} a & b & c \\ b & -a & d \\ 0 & 0 & 1 \end{pmatrix}, \tag{4.3}$$

where, for example, $a = \cos 2\theta$ and $c = 2n \cos \alpha$, with $a^2 + b^2 = 1$ and $b(c^2 - d^2) = 2acd$. This latter expression can be more simply expressed as $ac + bd + c = 0$ (see Exercise 4.60). Distances are preserved under such transformations, but the sense of angles is reversed. Of course, any point on the axis of reflection is invariant under this transformation (see Exercise 4.61).

Disregarding any restriction on c and d, we can still write the matrix (4.3) in the form

$$\begin{pmatrix} 1 & 0 & c \\ 0 & 1 & d \\ 0 & 0 & 1 \end{pmatrix} \begin{pmatrix} a & b & 0 \\ b & -a & 0 \\ 0 & 0 & 1 \end{pmatrix},$$

and hence we can see that any matrix of this form with $a^2 + b^2 = 1$ can be thought of as a reflection in a line through the origin followed by a translation. Of course, the matrix can also be written in the form

$$\begin{pmatrix} a & b & 0 \\ b & -a & 0 \\ 0 & 0 & 1 \end{pmatrix} \begin{pmatrix} 1 & 0 & \tilde{c} \\ 0 & 1 & \tilde{d} \\ 0 & 0 & 1 \end{pmatrix},$$

where $\tilde{c} = ac + bd$ and $\tilde{d} = bc - ad$, with $a^2 + b^2 = 1$. Hence, such matrices can also be regarded as a translation followed by a reflection in a line through the origin.

Example 4.17 *Show that the transformation*

$$5x' = -3x + 4y + 15,$$

$$5y' = 4x + 3y + 5$$

does not represent a general reflection but can be written either as the product of a reflection and a displacement or as the product of another displacement and the same reflection.

The associated matrix is

$$\begin{pmatrix} -\frac{3}{5} & \frac{4}{5} & 3 \\ \frac{4}{5} & \frac{3}{5} & 1 \\ 0 & 0 & 1 \end{pmatrix}.$$

The condition $a^2 + b^2 = 1$ is satisfied, but $ac + bd + c = 2$, and hence the transformation does not represent a reflection. We can, however, write it in the form

$$\begin{pmatrix} 1 & 0 & 3 \\ 0 & 1 & 1 \\ 0 & 0 & 1 \end{pmatrix} \begin{pmatrix} -\frac{3}{5} & \frac{4}{5} & 0 \\ \frac{4}{5} & \frac{3}{5} & 0 \\ 0 & 0 & 1 \end{pmatrix},$$

in which case it represents a reflection in the line $y = x \tan \theta$, where $\tan 2\theta = -\frac{4}{3}$, followed by a translation given by

$$x' = x + 3, \qquad y' = y + 1.$$

Alternatively, the transformation can be written in the form

$$\begin{pmatrix} -\frac{3}{5} & \frac{4}{5} & 0 \\ \frac{4}{5} & \frac{3}{5} & 0 \\ 0 & 0 & 1 \end{pmatrix} \begin{pmatrix} 1 & 0 & -1 \\ 0 & 1 & 3 \\ 0 & 0 & 1 \end{pmatrix},$$

in which case it represents a translation given by

$$x' = x - 1, \qquad y' = y + 3,$$

followed by a reflection in the line $y' = x' \tan \theta$, where $\tan 2\theta = -\frac{4}{3}$.

Example 4.18 *Show that the transformation*

$$x' = 2x - 3y + 1,$$

$$y' = 3x + 2y - 4$$

is a similarity transformation, and describe its geometrical significance.

The associated matrix is

$$\begin{pmatrix} 2 & -3 & 1 \\ 3 & 2 & -4 \\ 0 & 0 & 1 \end{pmatrix},$$

which is of the form of a similarity transformation. This matrix can be factored as

$$\begin{pmatrix} 1 & 0 & 1 \\ 0 & 1 & -4 \\ 0 & 0 & 1 \end{pmatrix} \begin{pmatrix} 2 & -3 & 0 \\ 3 & 2 & 0 \\ 0 & 0 & 1 \end{pmatrix},$$

and thus represents a rotation around the origin through an angle θ where $\tan \theta = \frac{3}{2}$, and a change of scale by a factor of $\sqrt{13}$, followed by a translation of the origin to the point $(1, -4)$.

4.6 Bibliographical Notes

We particularly recommend *Projective and Euclidean Geometry* (2nd edition) by W. T. Fishback, Wiley (1962). All but the last two chapters of this thirteen-chapter text are pertinent to our own discussion and form a most suitable background for the study of algebraic curves.

Some texts approach geometry directly from the coordinate point of view and place much more emphasis on group transformations than we have done. *Transformations and Geometries* by David Gans, Appleton-Century-Crofts (1969), is one such example. *A Course in Modern Geometries* by Judith N. Cederberg, Springer-Verlag (1989, reprinted 1991), and C. R. Wylie, Jr.'s *Introduction to Projective Geometry*, McGraw-Hill (1970), may also be useful.

Exercises

4.1. Where, in the complex projective plane, does the line $x - y - 4 = 0$ meet the ellipse $x^2 + 4y^2 - 1 = 0$?

4.2. Where, in the complex projective plane, does the line $x + y - 1 = 0$ meet the conic $2x^2 + 4xy + 2y^2 - x + 3y + 1 = 0$?

4.3. Find the tangents to $x^2 + 4y^2 - z^2 = 0$ that pass through the point $(0, 0, 1)$. Find also the points of tangency.

4.4. Show that the points (x_1, y_1, z_1), (x_2, y_2, z_2), and (x_3, y_3, z_3) are collinear if and only if the determinant whose rows are these three triples is zero.

4.5. Prove that three points can be transformed, by a linear transformation, into the reference points if and only if they are noncollinear. Hence, produce a transformation that maps $(3, 2, 1)$, $(0, 1, 2)$, and $(1, -3, 0)$ onto $(1, 0, 0)$, $(0, 1, 0)$, and $(0, 0, 1)$, respectively.

4.6. Find a matrix that maps the points $(1, 2, 3)$, $(0, 2, 1)$, $(1, -1, 0)$, and $(1, -1, 1)$ onto the points $(1, 0, 0)$, $(0, 1, 0)$, $(0, 0, 1)$, and $(1, 1, 1)$, respectively.

4.7. Show that the point $(-1, 3, 4)$ lies on the line $(6, 2, 0)$.

4.8. What are the general line coordinates of a line through the intersection of two given lines?

4.9. Prove that the four lines $(1, 1, 2)$, $(3, -1, 4)$, $(5, 1, 8)$, and $(2, 0, 3)$ are concurrent.

4.10. Using line coordinates and proceeding in a manner similar to Example 4.5, find the tangential pencil of conics touching the lines $3x - y + z = 0$, $x - y + 2z = 0$, $y - z = 0$, and $4x - 2y + z = 0$.

4.11. Proceeding as in Example 4.5, find the conic that is tangent to the lines $x - 2y + 2 = 0$, $2x - y + 2 = 0$, $x - 2y - 2 = 0$, $2x - y - 2 = 0$, and $x + y - 2 = 0$.

4.12. Find the coordinates of the line determined by the points $(0, 1, -3)$ and $(2, 1, 7)$.

4.13. Find the coordinates of the points determined by the line pairs
   (a) $(2, -1, 1)$ and $(3, 1, 0)$,
   (b) $(-1, 4, 2)$ and $(2, -8, 7)$.

4.14. Find the vertices of the triangle formed by the three lines $x + y + z = 0$, $y = 0$, and $3x - 4y + 5z = 0$.

4.15. Find the equations of the lines that join the point of intersection of the lines $2x + y + 4z = 0$ and $x - y + z = 0$ to the vertices of the triangle of reference.

4.16. Find the coordinates of the sides of the triangle whose vertices have equations $2l + m + n = 0$, $l - m + 2n = 0$, and $4l + m - 3n = 0$.

4.17. The vertices of a triangle are $A(-1, 1, 1)$, $B(1, -1, 1)$, and $C(1, 1, -1)$, and D is the unit point. The lines AD and BC meet in P, BD and CA meet in Q, and CD and AB meet in R. The lines QR and BC meet in X, RP and CA meet in Y, and PQ and AB meet in Z. Show that X, Y, and Z lie on a line, and find its equation.

4.18. Let S comprise the two sets of "points" defined in the following way. S_1 is the set of Euclidean points inside the circle $x^2 + y^2 = 1$, and S_2 is the set of pairs $[(x, y), (-x, -y)]$ of Euclidean points that lie on the circle $x^2 + y^2 = 1$. Associate with S the triple $[\rho x, \rho y, \rho(1 - x^2 - y^2)]$. Show that this defines a two-dimensional projective coordinate system on S over the reals.

4.19. Let S be the set of all lines through a point in three-dimensional Euclidean space. Associate with a line in S the ordered triple comprising its direction cosines. Show that this defines a two-dimensional projective coordinate system on S over the reals.

4.20. Prove that there is exactly one parabola tangent to four lines in general position. Find, and sketch, the parabola that is tangent to the lines $x + y = 0$, $y = 0$, $x - y + 1 = 0$, and $x - 3 = 0$. (Be careful. This exercise is sneaky; cf. exercise 2.28.)

4.21. Find the loci of the centers of all the circles that are tangent to the two circles $x^2 + y^2 - 1 = 0$ and $(x - 5)^2 + y^2 - 4 = 0$.

4.22. Show that the following are groups:
(a) the set of real numbers under addition,
(b) the set of nonzero real numbers under multiplication,
(c) the set of complex numbers of the form $e^{i\theta} \equiv \exp(i\theta)$ under multiplication,
(d) the n complex nth roots of unity under multiplication.

4.23. Which of the following are groups?
(a) I and the reflection of a circle in its center,
(b) I and the two transformations equivalent to the reflections of a circle in two different fixed diameters,
(c) rotations of a circle through $\pi/3$, $2\pi/3$, π, $4\pi/3$, $5\pi/3$, and 2π.

4.24. Consider a square centered at the origin and with vertices at $(\pm 1, \pm 1)$. Consider rotations by $\pi/2$, π, $3\pi/2$, and 2π, and reflections in the lines $x = 0$, $y = 0$, $x = y$, and $x = -y$. Do these transformations form a group?

4.25. What is the dual of Theorem 4.4?

4.26. State and prove the generalization of Theorem 4.4 to n dimensions.

4.27. Find a projective transformation that maps $(1, 2, 1)$, $(0, 1, 1)$, $(2, 1, 1)$, and $(-1, 1, 4)$ onto $(1, 5, 7)$, $(1, 4, 2)$, $(-1, 4, 10)$, and $(1, 13, 1)$, respectively.

4.28. Prove that concurrent lines will remain concurrent under a projective transformation.

4.29. Show that a change of parametrization does not affect the cross ratio.

4.30. Prove that if a projective transformation A has three distinct nonzero eigenvalues, then there are three distinct fixed points and lines. Prove that the fixed points are noncollinear (and hence the fixed lines are nonconcurrent), and sketch the situation.

4.31. Find the fixed points and lines of the projective transformation

$$\begin{pmatrix} 0 & 6 & 2 \\ 1 & -1 & -1 \\ 2 & 6 & 0 \end{pmatrix}.$$

4.32. If a projective transformation A is such that $A^p = I$, and $A^q = I$ implies $q \geq p$, then we say that the transformation has *period p*. Prove that no projective transformation of period 2 can have three noncollinear fixed points without being the identity. Under what conditions could such a transformation have period 3?

4.33. Find the projective transformations that have $(2, 1, 0)$, $(-1, 1, 1)$, and $(1, 2, -2)$ as fixed points.

4.34. Find the fixed points of the transformations
(a)

$$\begin{pmatrix} 3 & 1 & -2 \\ -6 & -2 & 6 \\ -2 & -1 & 3 \end{pmatrix},$$

(b)

$$\begin{pmatrix} -4 & 6 & -6 \\ -2 & 4 & -5 \\ 1 & -1 & 0 \end{pmatrix}.$$

4.35. Discuss the fixed points and lines of a projective transformation that has a single eigenvalue.

4.36. Show that the tangent to the nondegenerate conic $\underline{x}^T A \underline{x} = 0$ at the point $\underline{x}_0$ has the equation $\underline{x}_0^T A \underline{x} = 0$.

4.37. Find a transformation that maps the conic $2xy - xz - yz + z^2 = 0$ to $x^2 + 2y^2 - z^2 = 0$.

4.38. Find a parametrization of the conic

$$(2a - 5)(x^2 + y^2) + 14xy - (2a - 3)(xz + yz) + 2z^2 = 0$$

based on tangents at the two points $(0, 1, 1)$ and $(1, 0, 1)$, and using the third point $(-1, -1, 1)$.

4.39. If P_2 denotes the projective plane and l is a line of P_2, show that the set G of all transformations under which l is invariant forms a group, and $A_2 = P_2 \setminus l$ is invariant under G.

4.40. Prove that the set of affine transformations forms a group.

4.41. Show that an affine transformation maps intersecting lines onto intersecting lines, and parallel lines onto parallel lines.

4.42. Find the affine transformation that maps $(2, -1)$, $(3, 1)$, and $(-1, -1)$ to the points $(0, 0)$, $(1, 0)$, and $(0, 1)$, respectively.

4.43. Show, both synthetically (using properties on the ideal line) and algebraically, that an affine transformation maps a conic onto one of the same type, that is, a hyperbola maps to a hyperbola, and so on.

4.44. Prove that the equiaffine transformations form a proper subgroup of the affine transformations.

4.45. Prove that the ratio of distances on a line or on parallel lines is an affine invariant.

4.46. Prove that the similarity transformations form a proper subgroup of the affine transformations but are not a subgroup of the equiaffine transformations.

4.47. Show that the eccentricity of a conic is invariant under a similarity transformation.

4.48. Prove that the set of motions forms a group.

4.49. Let T be the translation defined by $x' = x - 1$, $y' = y + 2$. Find the image under T of
(a) $x - 4y + 1 = 0$,
(b) $x^2 + y^2 - 2x + 4y + 4 = 0$,
(c) $x^2 - 2y = 0$.

4.50. Find the translation matrix that maps the point $(4, -1)$ to the point $(3, 2)$.

4.51. Prove that the property of a line being tangent to a conic is invariant under a translation.

4.52. Prove that rotations about the origin form a group.

4.53. Express a rotation about $(2, 1)$ by $30°$ as a matrix.

4.54. Find a rotation about the origin that, when followed by a translation, is the same as a rotation about the point $(2, 1)$ by $\pi/3$.

4.55. Prove that the set of rigid motions forms a group.

4.56.   Prove that distances are preserved under the group of displacements.

4.57.   Find the displacement matrix that maps the points $(2, -1)$ and $(3, 4)$ onto the points $(-6, 3)$ and $(-11, 2)$.

4.58.   Show that the property of being a circle is invariant under a displacement.

4.59.   Construct the matrices that represent a reflection in each of the lines $x - y\sqrt{3} = 2$ and $x - y\sqrt{3} = -2$.

4.60.   Show that the condition $b(c^2 - d^2) = 2acd$ is equivalent to the condition $ac + bd + c = 0$ when $a^2 + b^2 = 1$.

4.61.   Show, by substitution into the equations representing a general reflection, that points on the axis of reflection are invariant under the reflection transformation.

4.62.   Show that the transformation

$$5x' = -3x + 4y + 2,$$
$$5y' = 4x + 3y - 1$$

is a reflection, and find the axis of reflection. Write the transformation as the product of a translation and a reflection, and also as the product of a reflection and a translation.

4.63.   Show that the transformation

$$2x' = \sqrt{3}x + y + 2,$$
$$2y' = x - \sqrt{3}y + 2$$

does not represent a reflection. Write the transformation as a translation followed by a reflection.

4.64.   Show that

$$x' = x + 4y + 1,$$
$$y' = 4x - y + 2$$

is a similarity transformation, and give two equivalent geometrical interpretations of the transformation.

4.65.   Show that not all numbers of the form $a + ib\sqrt{5}$, with a and b integers, can be factored in a unique way into products of numbers of the same type.

4.66.   Express the curve $x^3 - x^2z - x^2y - yz^2 + 2xyz = 0$ in a coordinate system where $(1, 0, 0)$, $(2, 1, 1)$, and $(0, -1, 1)$ are the coordinates of the triangle of reference and $(3, 0, 2)$ is the unit point.

4.67.   Express the curve of the previous exercise in an affine plane that has been constructed from the projective plane by removal of the line $x + z = 0$.

5

Algebraic Curves

5.1 Introduction

Now that we have become thoroughly familiar with the projective plane, with the use of homogeneous coordinates over the complex field, and with linear transformations, we can proceed with a discussion of algebraic curves. We will consider general properties of algebraic curves and investigate the special properties of curves with rational parametrizations. We will learn how to count curve intersections and how to resolve the singularities of curves by using special transformations. We will prove that every rational curve is an algebraic curve, and investigate the necessary and sufficient conditions under which an algebraic curve has a rational parametrization.

We start by repeating the definition of an algebraic curve which we gave in Section 4.2. An algebraic curve is the locus of zeros, in the complex projective plane, of a homogeneous polynomial in three variables of degree at least one. The coefficients of the polynomial are, in general, complex numbers. If the coefficients happen to be real, we refer to the curve as a real algebraic curve. However, the points of a real algebraic curve are still the points in the complex projective plane whose coordinates are zeros of the homogeneous polynomial, that is, whose coordinates satisfy an equation of the form $f(x, y, z) = 0$, where $f(x, y, z)$ is a homogeneous polynomial with real coefficients.

We often omit the dependence upon the variables x, y, and z, writing f instead of $f(x, y, z)$, and refer to "the curve f" when we mean "the curve whose equation is $f(x, y, z) = 0$." If f is of degree n, we will say that the curve is of *order n*. If the polynomial f is irreducible, then the curve is said to be an *irreducible curve*. A full treatment of the meaning of irreducibility and factorization would divert us from our main theme. However, to fully appreciate both the power and limitations of the specific applications, we must discuss a minimal amount of geometry. Our goal is to explore the interplay

185

between algebraic definitions of curves and surfaces (i.e., definitions in terms of implicit equations) and parametric representations.

We still have not solved the problem about the number of intersections of two conics. We will do this in the present chapter, placing the conics in the framework of algebraic curves and commenting on some of the main qualitative properties of such curves, such as multiple points, loops, cusps, tangents, and inflection points.

The most usual procedure for analyzing a curve in the affine plane will be to introduce homogeneous coordinates in order to associate the affine curve with a curve in the projective plane, which we will call the *projective completion* of the affine curve, perform some analysis on this curve, and, finally, transfer back to the affine plane. Let $f(\underline{x}) = 0$ be a curve, and let $l(\underline{x}) = 0$ be a line. If l is not a component of the curve f, we can make a change of coordinates such that, in the new coordinate system, the equation of l is $z = 0$. In this coordinate system, z is not a factor of f. We can then associate with the curve $f(x, y, z) = 0$ in the projective plane the curve $f_A(x, y) \equiv f(x, y, 1) = 0$, which has been obtained from the projective plane by removal of the line l. The affine curve $f_A(x, y)$ will, in general, no longer be homogeneous. The coordinates (x, y) will be affine coordinates of the curve. We will call $f_A(x, y)$ the *affine specialization* of $f(x, y, z)$. We will often omit the subscript A, writing $f(x, y) = f(x, y, 1)$, with the understanding that we are referring to the affine specialization. It is thus very easy for us to move from an algebraic curve to an affine specialization.

Example 5.1 *Sketch the curve* $xy - z^2 = 0$ *in the real projective plane.*

We have, in effect, already done this in Section 2.11 when we discussed conics through two points with prescribed slopes at these points. The lines $x = 0$, $y = 0$, and $z = 0$ are shown in Figure 5.1. Since the curve meets the line $x = 0$ at a double root when $z = 0$, the curve must be tangent to the line $x = 0$ at the point $(0, 1, 0)$. Similarly, it is tangent to the line $y = 0$ at $(1, 0, 0)$. The curve is a conic through the points $(0, 1, 0)$ and $(1, 0, 0)$ and tangent to $x = 0$ and $y = 0$, respectively, at these points. If we remove the line $z = 0$, the curve is a hyperbola in the affine plane, with affine specialization $xy - 1 = 0$.

From the fundamental theorem of algebra we know that a polynomial in one variable can be factored into linear factors. For example, $x^2 + 1$ has the unique factorization $(x + i)(x - i)$. In homogeneous coordinates we could write this as $x^2 + y^2 = (x + iy)(x - iy)$. The unique factorization generalizes to more variables, but the unique factors are not always linear. If that were the case, all algebraic curves would be the product of lines, which would be somewhat uninteresting. In the case of conics we have already discussed the conditions

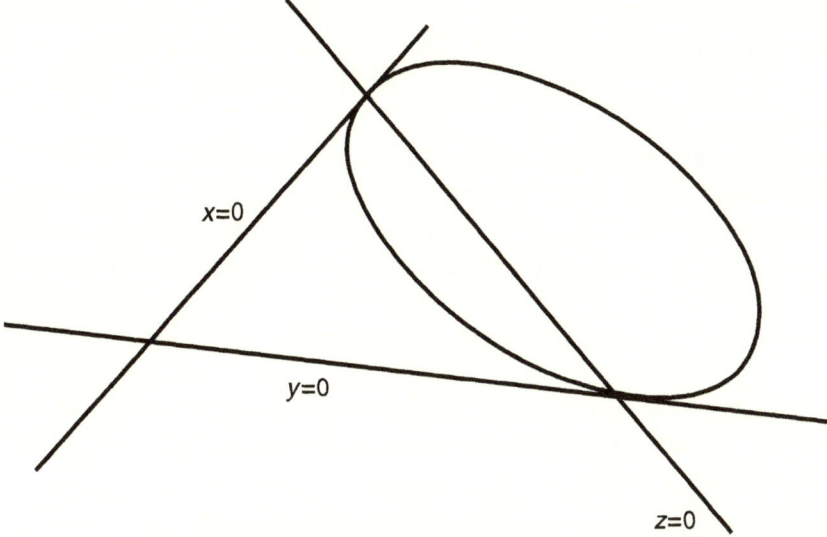

Figure 5.1. The projective curve $xy - z^2 = 0$.

that must be satisfied in order for the conic to reduce to a line pair. Let us simply state, without proof, that all polynomials can be factored in a unique way.

If f has a factorization $\prod_{i=1}^{m} f_i$, then each member of the set of curves $\{f_i = 0, \ i = 1, \ldots, m\}$ is called a *component* or *factor* of the curve $f = 0$.

Example 5.2 *Show that $x^2 + y^2 - 1$ is irreducible.*

It is easily shown that $x^2 + y^2 - 1$ is not the product of two real factors. Let us assume that

$$x^2 + y^2 - 1 = (ax + by + c)(lx + my + n),$$

where the coefficients a, b, c, l, m, and n are complex. That is, we assume that the complex locus of zeros of $x^2 + y^2 - 1$ is a complex line pair. Let

$$a = a_1 + ia_2, \qquad b = b_1 + ib_2, \qquad c = c_1 + ic_2.$$

If (x, y) is a real point of the complex line $ax + by + c = 0$, then x and y must satisfy both the equations

$$a_1 x + b_1 y + c_1 = 0,$$

$$a_2 x + b_2 y + c_2 = 0.$$

Not all of the a_1, b_1, a_2, and b_2 can be zero, and hence x and y lie on at least one real line. In other words, the real points of a complex line lie on a real line.

Therefore, the assumption that $x^2 + y^2 - 1 = 0$ is a complex line pair implies that the real points of this curve lie on a real line pair, which is false, since $x^2 + y^2 - 1$ is not the product of two real factors. Hence, $x^2 + y^2 - 1$, is irreducible.

Is it possible that any other nonhomogeneous quadratic equation in two variables will be reducible? We know that all conics are projectively equivalent to the circle $x^2 + y^2 - 1 = 0$. The previous example shows that this circle is not the product of two lines. The following theorem now answers our question.

Theorem 5.1 *Irreducibility is a projective invariant.*

Proof. Let $\underline{x}$ denote the triple (x, y, z) and assume that $f(\underline{x})$ is irreducible. Let $\underline{x} = A\underline{x}'$ be a projective transformation of coordinates, and let

$$g(\underline{x}') = f(A\underline{x}').$$

Then, applying the inverse transformation, we obtain

$$g(A^{-1}\underline{x}) = f(\underline{x}).$$

If g is reducible, then $g = g_1 g_2$ and

$$g(A^{-1}\underline{x}) = g_1(A^{-1}\underline{x})g_2(A^{-1}\underline{x}) = f(\underline{x}),$$

which means that f is reducible, which contradicts our assumption. Therefore, g is irreducible. ∎

Therefore, since the circle $x^2 + y^2 - 1 = 0$ can never be reduced to a line pair, neither can any other nondegenerate conic.

Since we will be discussing algebraic curves in some detail in this chapter, the following theorem reassures us about the existence of such curves.

Theorem 5.2 *There are infinitely many points on any algebraic curve of order n.*

Proof. Assume that x appears in the homogeneous equation of the curve $f(x, y, z) = 0$. Then the equation can be written in the form

$$\sum_{i=0}^{n} a_i(y, z)x^{n-i} = 0,$$

where the coefficients $a_i(y, z)$ are homogeneous polynomials of degree i. If the only term in the polynomial is $a_0 x^n$, then all the infinitely many points $(0, y, z)$ lie on the curve. Therefore, assume that $a_m(y, z)$ is not identically zero for at least one $m > 0$. From the fundamental theorem of algebra there are m ratios

y/z such that $a_m(y, z) = 0$. There are, therefore, infinitely many ratios y_j/z_j such that $a_m(y_j, z_j) \neq 0$. For each of these infinitely many ratios the equation

$$\sum_{i=0}^{n} a_i(y_j, z_j)x^{n-i} = 0, \qquad j = 1, 2, \ldots,$$

has n roots. Let x_j be one of these roots. Then the infinitely many points (x_j, y_j, z_j), $j = 1, 2, \ldots$, are on the curve. ∎

5.2 Multiple Points of a Curve

We now know that algebraic curves of order n exist. How much freedom do we have in constructing them? The number of terms in the general nth-degree homogeneous polynomial in three variables is $\frac{1}{2}(n+1)(n+2)$. Since one of the nonzero coefficients in the equation can be arbitrarily prescribed to have any nonzero value, the corresponding algebraic curve will have one fewer degrees of freedom. That is, the algebraic curve of order n has

$$\tfrac{1}{2}(n + 1)(n + 2) - 1 = \tfrac{1}{2}n(n + 3)$$

degrees of freedom and will, in general, be determined by this number of linear conditions or their equivalent. In general, there is a unique conic through five points, a unique cubic through nine points, and a unique quartic through fourteen points.

The points referred to here are "simple" points, and we now have to investigate what we mean by simple. Another of our stated goals is to be able to count the points in which two algebraic curves intersect, as this knowledge will provide us with a tool for constructing curves, as well as for deriving information about the qualitative behaviour of curves. Therefore, we also have to know how to find the multiplicity of each intersection.

Imagine a curve shaped like the figure eight. If we draw a line through the crossover point, the curve crosses this line twice at the intersection point, and we shall have to count this point at least twice (more if the line is tangent to the curve) when we count intersections of the line with the curve. The crossover point is an example of a multiple point of a curve. We shall have to define such points in a rigorous manner before we can formulate any theory related to counting the number of intersections of two curves. As a preliminary step, we investigate the number of intersections of a line with a general algebraic curve.

Theorem 5.3 *A line meets a curve of order n either in n points or in infinitely many points, in which case the line is a component of the curve.*

Proof. Let $f_n(\underline{x}) = 0$ be the given curve, and let the line be given by

$$\underline{x} = \lambda\underline{a} + \mu\underline{b}.$$

The line meets the given curve where $f_n(\lambda\underline{a} + \mu\underline{b}) = 0$. The expression $f_n(\lambda\underline{a} + \mu\underline{b})$ either is identically zero or is a homogeneous polynomial of degree n in λ and μ. Clearly, if the line is a component of the curve, then $f_n(\lambda\underline{a} + \mu\underline{b}) \equiv 0$. Conversely, if the line meets the curve in $n + 1$ or more points, then $f_n(\lambda\underline{a} + \mu\underline{b})$ has $n + 1$ zeros and must therefore be identically zero. Therefore, any point of the line must also be a point of the curve, that is, the line is a component of the curve.

Alternatively, $f_n(\lambda\underline{a} + \mu\underline{b})$ has n zeros, in which case the line meets the curve in n points, which are obtained by substituting the n solutions for the ratio λ/μ in the equation of the line. ∎

In the proof of Theorem 5.3, it was implicit that we counted the intersections with their appropriate multiplicities. In other words, the expression $f_n(\lambda\underline{a} + \mu\underline{b})$ will be of the form

$$\prod_{i=1}^{k}(\mu_i\lambda - \lambda_i\mu)^{r_i}, \qquad \text{where} \quad \sum_{i=1}^{k}r_i = n,$$

and this is interpreted by saying that the line meets the curve at the points P_i, $i = 1, 2, \ldots, k$, whose line coordinates are (λ_i, μ_i). It is clear, therefore, that the intersection multiplicity of the line and the curve at the point with coordinates (λ_i, μ_i) on the line is the maximum power of $\mu_i\lambda - \lambda_i\mu$ that can be factored from the polynomial $f_n(\lambda\underline{a} + \mu\underline{b})$. In Example 5.1 the conic $xy - z^2 = 0$ meets the line $x = 0$ twice at the point $(0, 1, 0)$. Their intersection occurs at one point, which must be counted twice because it is a point where the line is tangent to the curve.

If P is a point of a curve f, we can always make an affine specialization such that the curve has the equation $f(x, y) = 0$ and (a, b) are the affine coordinates of P. Then $f(a, b) = 0$, and any line through (a, b) has parametric equations

$$x = a + \lambda t, \qquad y = b + \mu t.$$

We use the Taylor expansion and find that this line meets the curve when

$$t\left[\lambda f_x(a, b) + \mu f_y(a, b)\right]$$

$$+ \frac{t^2}{2!}\left[\lambda^2 f_{xx}(a, b) + 2\lambda\mu f_{xy}(a, b) + \mu^2 f_{yy}(a, b)\right]$$

$$+ \cdots + \frac{t^n}{n!}\left[\lambda^n f_{xx\ldots x}(a, b) + \cdots + \mu^n f_{yy\ldots y}(a, b)\right] = 0,$$

where the subscripts x and y on f denote partial differentiation with respect to x and y, respectively. Obviously, any line through (a, b) meets the curve at (a, b), and this corresponds to the fact that $t = 0$ is a solution of the above equation. In general, $t = 0$ is a simple root of this equation, and there will be precisely one line, given by the ratio λ/μ that satisfies

$$\lambda f_x(a, b) + \mu f_y(a, b) = 0,$$

for which $t = 0$ is at least a double root. We call this line the *tangent* to the curve f at the point (a, b). We say, therefore, that every line through (a, b) meets the curve at least once there and that one line, the tangent, meets the curve at least twice there. It may happen that not only $f(a, b) = 0$, but also $f_x(a, b) = f_y(a, b) = 0$, in which case $\lambda f_x(a, b) + \mu f_y(a, b) = 0$ for any λ and μ. In this case $t = 0$ is at least a double root, and we say that now every line through (a, b) meets the curve at least twice there and that there are precisely two lines, called tangents, and given by the ratios λ/μ that satisfy the equation

$$\lambda^2 f_{xx}(a, b) + 2\lambda\mu f_{xy}(a, b) + \mu^2 f_{yy}(a, b) = 0,$$

that meet the curve at least three times at (a, b). This leads to the following definitions of the multiplicity of a point of a curve and tangents to a curve at a point.

Definition 5.1 *A point $P(a, b)$ of a curve $f(x, y) = 0$ is said to be a **point of multiplicity** r if $f(a, b) = 0$ and all the partial derivatives of f, up to and including those of order $r - 1$, are zero at (a, b), but there is at least one partial derivative of order r whose value at (a, b) is nonzero.*

*The **tangents** at a point $P(a, b)$ of multiplicity r are the lines through (a, b) whose slopes are given by the ratios λ/μ that satisfy the equation*

$$\sum_{k=0}^{r} \binom{r}{k} \lambda^{r-k} \mu^k \frac{\partial^r f(a, b)}{\partial r^{-k}x \, \partial^k y} = 0,$$

where all the partial derivatives are of order r and are evaluated at (a, b).

A point of multiplicity one is called *simple*. At a simple point there is one tangent line. A point of multiplicity two is called a *double point*, and so on. If the tangents at a multiple point are distinct, then the point is said to be *ordinary*. At the double point C in Figure 5.2 the tangents coincide and, therefore, this point is not an ordinary double point. Points of multiplicity two or more are said to be *singular*, and a curve with one or more singular points is called a *singular curve*.

The definition of the multiplicity of a point of a curve is easily extended to the projective plane by the use of the following lemma.

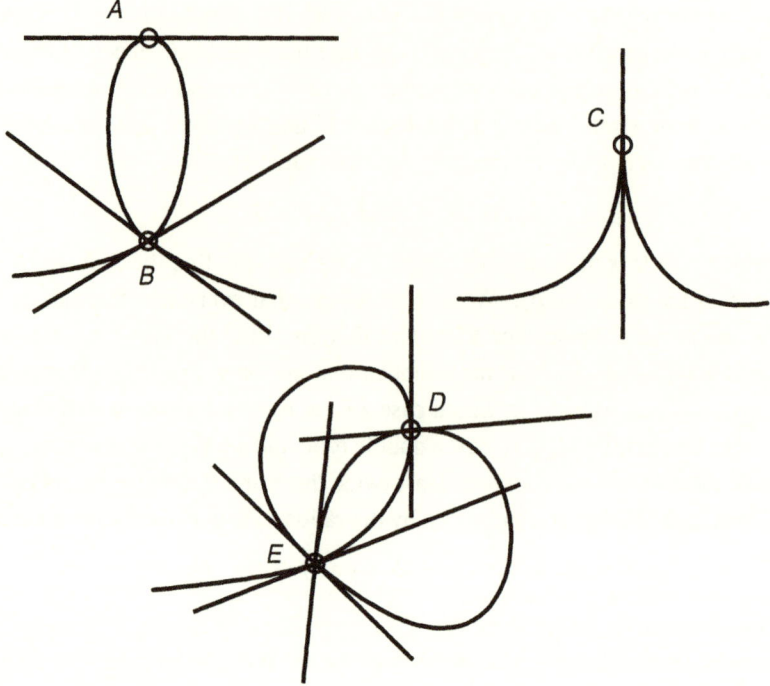

Figure 5.2. Examples of multiple points: simple point (A), double points (B, C, D), triple point (E), ordinary multiple points (B, D, E).

Lemma 5.1 *If $f(x_1, x_2, \ldots, x_m)$ is homogeneous of degree n, then*

$$\sum_{i=1}^{m} x_i \frac{\partial f}{\partial x_i} = nf.$$

Proof. Since f is homogeneous and of degree n,

$$f(tx_1, tx_2, \ldots, tx_m) = t^n f(x_1, x_2, \ldots, x_m).$$

Let $u_i = tx_i$. Then differentiation with respect to the parameter t yields

$$\sum_{i=1}^{m} x_i \frac{\partial f(u_1, u_2, \ldots, u_m)}{\partial u_i} = nt^{n-1} f(x_1, x_2, \ldots, x_m).$$

Substituting $t = 1$, we obtain the required result. ∎

From Lemma 5.1 we have, in the projective plane,

$$xf_x + yf_y + zf_z = nf,$$

where n is the order of the curve and f is homogeneous. Therefore,

$$f(a, b, 1) = f_x(a, b, 1) = f_y(a, b, 1) = 0 \quad \Rightarrow \quad f_z(a, b, 1) = 0.$$

Hence the conditions $f(a, b) = f_x(a, b) = f_y(a, b) = 0$ for the affine specialization are equivalent to the conditions

$$f(\rho a, \rho b, \rho) = f_x(\rho a, \rho b, \rho) = f_y(\rho a, \rho b, \rho) = f_z(\rho a, \rho b, \rho) = 0$$

for the projective completion. Continuing in this way, it is easy to see that if (a, b) are the affine coordinates of a point $\rho(a, b, 1)$ in the projective plane and if $f_A(x, y) \equiv f(x, y, 1)$ is the affine specialization of a curve $f(x, y, z) = 0$, then $f_A(a, b) = 0$, and all its partial derivatives of orders up to and including $r - 1$ are zero at (a, b) if and only if $f(\rho a, \rho b, \rho) = 0$ and all its partial derivatives of orders up to and including $r - 1$ are zero at $(\rho a, \rho b, \rho)$. Therefore, a point (a, b, c) of the curve $f(x, y, z) = 0$ is a point of multiplicity r if and only if f and all its partial derivatives of orders up to and including $r - 1$ are zero at (a, b, c).

There are some immediate and simple consequences of the definitions of multiple points, such as the following.

Theorem 5.4 *An irreducible curve of order $n > 1$ cannot have a point of multiplicity n.*

Proof. Let P be a point of multiplicity n. By Theorem 5.2 there exists a point Q of the curve that is distinct from P. By Theorem 5.3, the line determined by the points P and Q has at least $n + 1$ intersections with the curve; hence, it must be a component. Therefore, the curve is reducible, which is a contradiction. ■

5.3 An Introduction to Rational Curves

A curve is said to be *rational* if the homogeneous coordinates of all but a finite number of points of the curve can be expressed in the form

$$x = X(s, t), \qquad y = Y(s, t), \qquad z = Z(s, t),$$

where X, Y, and Z are homogeneous polynomials in the projective line coordinates s and t. The maximum of the degrees of X, Y, and Z is called the *degree of the parametrization*. In affine coordinates, with $\tilde{x} = x/z$, $\tilde{y} = y/z$, and $\tilde{s} = s/t$, the curve has the rational parametrization

$$\tilde{x} = \frac{X(\tilde{s}, 1)}{Z(\tilde{s}, 1)}, \qquad \tilde{y} = \frac{Y(\tilde{s}, 1)}{Z(\tilde{s}, 1)}.$$

If the denominator $Z(\tilde{s}, 1)$ is unity, the curve is said to be *polynomial*. This definition must not be confused with curves of the form $y = p_n(x)$, where $p_n(x)$ is a polynomial in x. The curve given by $x = f_n(t)$, $y = g_n(t)$ is, by definition, a polynomial curve, yet it cannot, in general, be written in the form $y = p_n(x)$.

Theorem 5.5 *An irreducible curve of order n with a point of multiplicity $n - 1$ is rational.*

Proof. Let P be the point of multiplicity $n - 1$, and choose a coordinate system such that P has homogeneous coordinates $(0, 0, 1)$. Let the affine specialization of the given homogeneous curve $f(x, y, z) = 0$ be given by $f(x, y) = 0$. Then the point P has affine coordinates $(0, 0)$, and the pencil of lines through P has the affine equation $y = mx$. This line meets $f(x, y) = 0$ when $f(x, mx) = 0$. By Theorem 5.3, this line meets the curve in n points, but, from the definition of multiple points, $n - 1$ of these points must be at $(0, 0)$, or equivalently, where $x = 0$. Therefore,

$$f(x, mx) = x^{n-1}[Z(m)x - X(m)],$$

where $Z(m)$ and $X(m)$ are polynomials of degrees at most n, in m. For any m, $n - 1$ of the possible intersections are at the origin. The remaining intersection point is given in parametric form by

$$x = \frac{X(m)}{Z(m)}, \qquad y = \frac{mX(m)}{Z(m)}.$$

The curve is, therefore, rational.   ∎

Example 5.3 *Show that the curve $x^3 + 3xy^2 + x^2 - 4y^2 = 0$ is rational, and produce a parametrization.*

Let $f(x, y) = x^3 + 3xy^2 + x^2 - 4y^2$. Then

$$\frac{\partial f}{\partial x} = 3x^2 + 3y^2 + 2x, \qquad \frac{\partial f}{\partial y} = 6xy - 8y,$$

and, since we have to look for a multiple point of the cubic curve, we have to find simultaneous solutions of $f = f_x = f_y = 0$. The simplest of the three equations shows that

$$\frac{\partial f}{\partial y} = 0 \quad \Longleftrightarrow \quad y = 0 \text{ or } x = \frac{4}{3}.$$

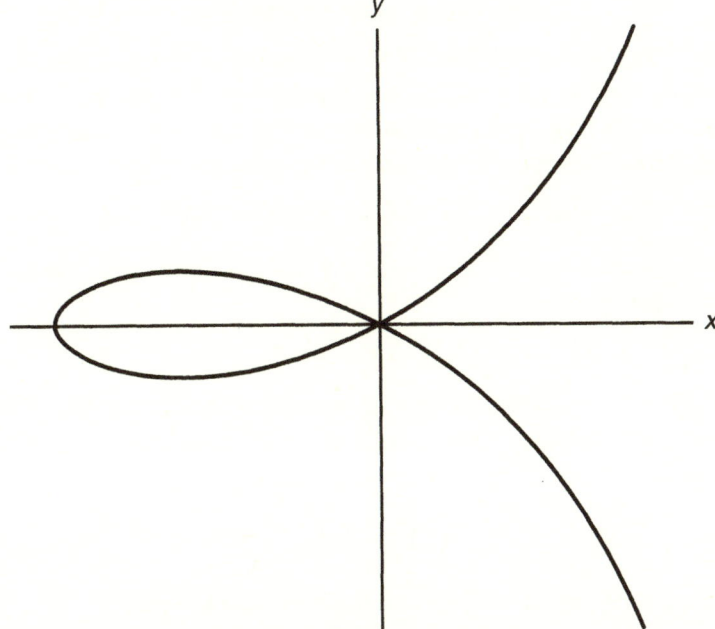

Figure 5.3. A rational cubic.

For $x = \frac{4}{3}$, we find that neither f nor f_x is zero. However, $y = 0$ implies that

$$\frac{\partial f}{\partial x} = x(3x + 2) \quad \text{and} \quad f = x^2(x + 1),$$

so that the only solution is $x = 0$, $y = 0$. Therefore, in affine coordinates, $(0, 0)$ is a double point (Figure 5.3). As expected, not all the partial derivatives of order two are zero at this point – the cubic cannot have a point of multiplicity greater than two. In projective coordinates, we could write

$$f(x, y, z) = x^3 + 3xy^2 + x^2z - 4y^2z,$$

and then f, f_x, f_y, and f_z would all be zero at the point $(0, 0, 1)$. Since $(0, 0)$ is a double point and the curve is a cubic, we know that it must be rational.

Substitute $y = mx$ in the equation of the curve to obtain

$$x^2[x(3m^2 + 1) + 1 - 4m^2] = 0.$$

Since $x = 0$ corresponds to the double point, the variable point of the curve (i.e., the third intersection with the curve of an arbitrary line through the double point) is given by

$$x = \frac{4m^2 - 1}{3m^2 + 1}, \qquad y = \frac{m(4m^2 - 1)}{3m^2 + 1},$$

which is a parametrization of the curve.

5.4 Tangents and Asymptotes

Another immediate consequence of our definitions of tangents and multiple points is the equation of a tangent at a simple point of a curve. We use the homogeneous form of the equation of the curve and assume that the point $\underline{a}$ is a simple point of the curve f and that $\underline{b}$ is any point whatsoever. Then

$$f(\underline{a} + \lambda \underline{b}) = f(\underline{a}) + \lambda \underline{b} \cdot \nabla f + O(\lambda^2),$$

and therefore, the line whose points are $\underline{a} + \lambda \underline{b}$ is a tangent to the curve if and only if the point $\underline{b}$ satisfies

$$\underline{b} \cdot \nabla f = 0,$$

for then the line has at least a double intersection with the curve at $\underline{a}$. We thus have a tangent if and only if the coordinates of the point $\underline{b} = (x, y, z)$ satisfy the condition

$$x \frac{\partial f}{\partial x}\Big|_{\underline{a}} + y \frac{\partial f}{\partial y}\Big|_{\underline{a}} + z \frac{\partial f}{\partial z}\Big|_{\underline{a}} = 0, \tag{5.1}$$

which must, therefore, be the equation of the tangent to f at $\underline{a}$.

If the affine specialization of a curve f is of the form

$$\sum_{i+j=r}^{n} a_{ij} x^i y^j = 0,$$

then the curve has a point of multiplicity r at the origin, since f and all its partial derivatives up to order $r - 1$ are zero at the origin. It is therefore a trivial matter to determine the multiplicity at the origin.

Can we investigate the multiplicity of any other point by first transferring it to the origin? Theorem 5.6 justifies this, but let us first demonstrate that checking the multiplicity at any of the vertices of the triangle of reference, not only at $(0, 0, 1)$, is equally simple.

Example 5.4 *Shows that the curve* $2x^5 + x^2y^2 - xy^2 - 3y^2 + 1 = 0$ *has a triple point at* $(0, 1, 0)$, *and that the curve* $y^3 - xy^2 + 2y^2 + x + 1 = 0$ *has a double point at* $(1, 0, 0)$.

In homogeneous form the first curve becomes

$$2x^5 + x^2y^2z - xy^2z^2 - 3y^2z^3 + z^5 = 0.$$

Its affine representation, where the line $y = 0$ has been removed from the projective plane, is

$$2x^5 + z^5 + x^2z - xz^2 - 3z^3 = 0.$$

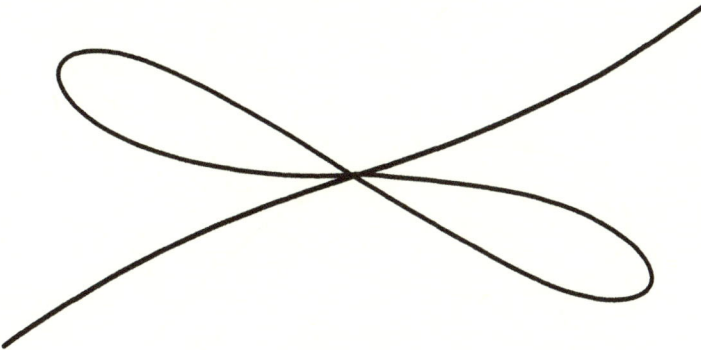

Figure 5.4. Quintic with a triple point.

Since this equation involves no terms of degree less than three, the point $(0, 1, 0)$ is a triple point (Figure 5.4)

For the second curve the projective completion is

$$y^3 - xy^2 + 2y^2z + xz^2 + z^3 = 0,$$

and when the line $x = 0$ is removed, we have the affine representation

$$y^3 + z^3 + 2y^2z + z^2 - y^2 = 0,$$

so that $y = 0$, $z = 0$ is a double point (Figure 5.5).

Theorem 5.6 *The multiplicity of a point of an algebraic curve is a projective invariant.*

Proof. This proof is left to the reader (Exercise 5.4). ∎

Example 5.5 *Show that the curve*

$$x^3 + 2x^2z - x^2y + xz^2 - 2xyz + y^3 + y^2z - yz^2 = 0$$

has a double point at $(-1, 0, 1)$.

Once we have checked that $(-1, 0, 1)$ does indeed lie on the curve, we map $(-1, 0, 1)$ to any vertex of the triangle of reference, for example, to $(0, 0, 1)$. Any mapping that does this will suffice for our purposes, so let us leave the other vertices unchanged. Then

$$A = \begin{pmatrix} 1 & 0 & 1 \\ 0 & 1 & 0 \\ 0 & 0 & 1 \end{pmatrix}$$

Algebraic Curves

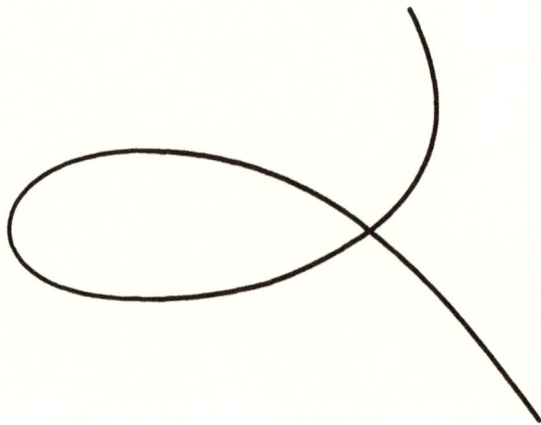

Figure 5.5. Cubic with a double point.

is a suitable projective transformation. We set $\underline{x} = A^{-1}\underline{\tilde{x}}$ and substitute into the equation to obtain

$$\tilde{x}^3 - \tilde{x}^2\tilde{z} - \tilde{x}^2\tilde{y} + \tilde{y}^3 + \tilde{y}^2\tilde{z} = 0.$$

The affine specialization that we get by removing the line $\tilde{z} = 0$ is

$$\tilde{x}^3 - \tilde{x}^2 - \tilde{x}^2\tilde{y} + \tilde{y}^3 + \tilde{y}^2 = 0,$$

so that $(0, 0, 1)$ is a double point. Hence, in the original coordinate system the point $(-1, 0, 1)$ is a double point. This curve is similar in appearance to the curve in Figure 5.5.

It is useful to be able to say something about the behavior of a curve in the vicinity of a singular point. Knowing the tangents at the singular point is a help in this regard. The following theorem, the proof of which we also leave as an exercise (Exercise 5.6), gives us the equations of the tangents.

Theorem 5.7 *If $f(x, y) = 0$ has a point of multiplicity r at the origin, then the components of the curve obtained from f by equating the terms of degree r to zero are the tangents to f at the origin.*

Example 5.6 *Find the tangents to $x^3 + 3xy^2 + x^2 - 4y^2 = 0$ at the origin.*

The origin is a double point, and the tangents are given by $x^2 - 4y^2 = 0$, that is, $x = \pm 2y$.

A double point for which the tangents are distinct is called a *node*. If the tangents at a double point are coincident and meet the curve in precisely three points there, the double point is called a *cusp*. Nodes at which the tangents are real are called *crunodes*. When the tangents at a node are complex the node is called an *acnode*.

Example 5.7 *Sketch the curves* $y^3 + x^2 - y^2 = 0$ *and* $y^3 + x^2 + y^2 = 0$.

Each of the curves has a double point at the origin (therefore, both curves are rational cubics). The tangents at the origin to the first curve are $x = \pm y$ and to the second $x = \pm iy$.

Setting $x = r \cos \theta$ and $y = r \sin \theta$ in the second curve, we find that

$$\sin^3 \theta = -\frac{1}{r},$$

and hence we see that there is no solution if $r < 1$. Hence, there are no points of the curve within a unit radius of the double point, that is, the double point is *isolated*. This is always the case when all the tangents at a multiple point are complex. The origin is a crunode of the first curve and an acnode of the second. The curves are depicted in Figure 5.6.

In the complex projective plane these two curves both have ordinary double points at the origin. However, the curves look very different from each other when viewed in the affine plane. This example highlights the care we must take

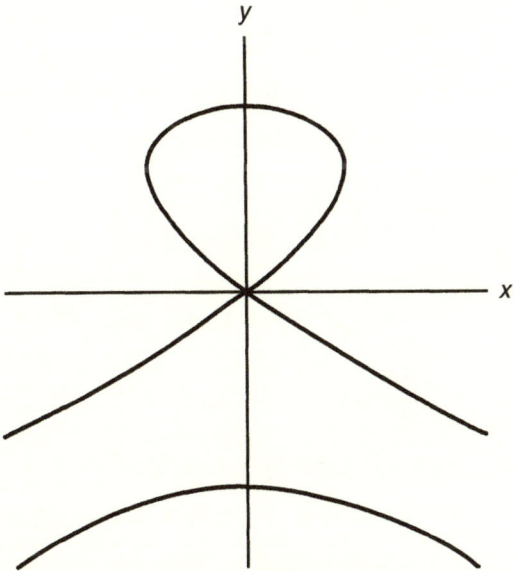

Figure 5.6. Two cubics, each with a double point at the origin.

when interpreting results we obtain in the complex projective plane and when drawing sketches of these curves. This very anomaly results in a need for a more detailed analysis and is of crucial importance when determining the visual properties of rational curves such as parametric splines. We will consider the visual properties of a wide variety of curves in Chapter 6. Let us now continue with our investigation of tangents and asymptotes.

Definition 5.2 *An **asymptote** is a line, other than the line at infinity, that is tangent to a curve at a point on the line at infinity.*

Two results follow immediately from this definition. Firstly, the concept of asymptotes is definitely not a projective idea; for in the projective plane the ideal line has no special significance and is not a projective invariant. Secondly, we see that a curve of order n has, at most, n asymptotes.

Example 5.8 *Determine the asymptotes of the following curves:*

(a) $x^2y + xy^2 + xy + y^2 + x = 0$,

(b) $x^2 + 4y^2 - 1 = 0$,

(c) $y - x^3 = 0$.

The first and third examples can have at most three asymptotes each, while the conic can have at most two.

(a): Writing the curve in its projective form

$$x^2y + xy^2 + xyz + y^2z + xz^2 = 0,$$

we see that $z = 0$ meets the curve where $xy(x + y) = 0$. The three points of intersection are therefore $(0, 1, 0)$, $(1, 0, 0)$, and $(1, -1, 0)$. Furthermore,

$$\frac{\partial f}{\partial x} = 2xy + y^2 + yz + z^2,$$

$$\frac{\partial f}{\partial y} = x^2 + 2xy + xz + 2yz,$$

$$\frac{\partial f}{\partial z} = xy + y^2 + 2xz.$$

We use (5.1) and find that the three asymptotes are

$$x(1) + y(0) + z(1) = 0, \qquad x(0) + y(1) + z(0) = 0,$$

$$x(-1) + y(-1) + z(0) = 0,$$

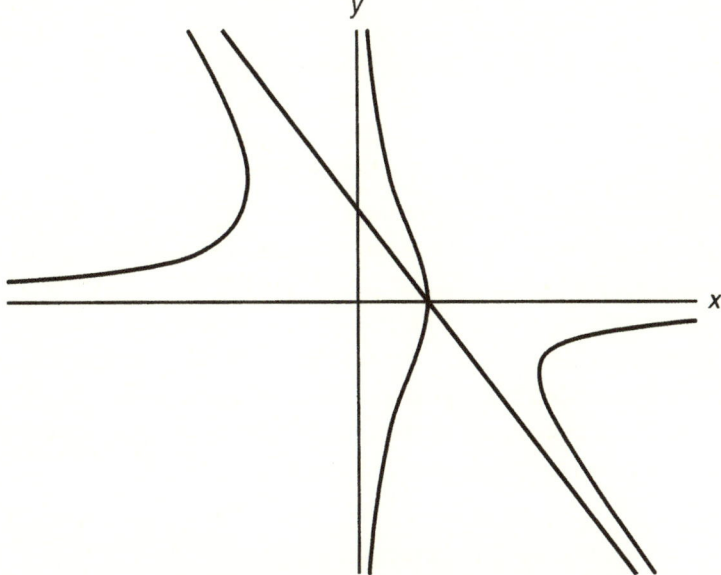

Figure 5.7. Cubic with three asymptotes.

or, in affine coordinates,

$$x + 1 = 0, \qquad y = 0, \quad \text{and} \quad x + y = 0.$$

The curve and the asymptotes are shown in Figure 5.7.

(b): In this case $z = 0$ meets the curve at $(2i, 1, 0)$ and $(-2i, 1, 0)$, resulting in the complex asymptotes $x = \pm 2iy$.

(c): The line $z = 0$ meets the curve three times at $(0, 1, 0)$. Since this point cannot be a triple point of the curve, the line $z = 0$ is the tangent at this point. The case in question does not, therefore, have any asymptote.

The principle of duality enables us to write equations of curves in either point or line coordinates. We have done this already in the case of the conic. From (5.1) we know that the tangent to a curve at the point $\underline{x}_1 = (x_1, y_1, z_1)$ is given by

$$x \frac{\partial f}{\partial x}\bigg|_{\underline{x}_1} + y \frac{\partial f}{\partial y}\bigg|_{\underline{x}_1} + z \frac{\partial f}{\partial z}\bigg|_{\underline{x}_1} = 0.$$

Therefore, if the curve is tangent to the line $\alpha x + \beta y + \gamma z = 0$, we know that, in vector notation, the line coordinates (α, β, γ) are parallel to $\nabla f(\underline{x}_1)$, and we must have

$$\frac{1}{\alpha}\frac{\partial f}{\partial x}\bigg|_{\underline{x}_1} = \frac{1}{\beta}\frac{\partial f}{\partial y}\bigg|_{\underline{x}_1} = \frac{1}{\gamma}\frac{\partial f}{\partial z}\bigg|_{\underline{x}_1},$$

and

$$\alpha x_1 + \beta y_1 + \gamma z_1 = 0.$$

Since the point (x_1, y_1, z_1) is an arbitrary point of the curve, we can eliminate the ratios x_1/z_1 and y_1/z_1, and the resulting equation, which involves only α, β, and γ, must be the condition that the line $\alpha x + \beta y + \gamma z = 0$ is tangent to the curve. This equation is therefore the line equation or tangential equation of the curve. Its order is called the *class* of the curve and gives the number of tangents to the curve that can be drawn through an arbitrary point. In the case of conics the line equation is also of second order but, in general, the class is greater than the order. We return to this question in Section 5.8.

5.5 Elimination

The process of eliminating a variable among linear equations can be effected by a simple substitution. The variable to be eliminated is solved for from one of the equations, and the corresponding expression in terms of the remaining variables is substituted for this variable in all the remaining equations. We are now interested in the points of intersection of curves more general than lines, and, since their equations are nonlinear, the procedure of solving for one variable followed by the substitution in the other equations is not, in general, viable.

The technique we will use is known as dialytic elimination and is due to James Joseph Sylvester, an outstanding mathematician who, like his great friend Arthur Cayley, made many contributions to mathematics, particularly in the field of algebra.

Consider the two polynomial equations $f = 0$ and $g = 0$ given by

$$f = \sum_{i=0}^{n} a_i x^i, \quad g = \sum_{i=0}^{m} b_i x^i, \quad a_n b_m \neq 0, \quad n, m \geq 1.$$

The coefficients a_i and b_i may themselves be polynomials in other variables. Multiplying each of these equations by x provides two additional equations. The four equations are

$$
\begin{aligned}
a_n x^{n+1} + a_{n-1} x^n + \cdots + a_0 x \quad &= 0, \\
a_n x^n + a_{n-1} x^{n-1} + \cdots + a_0 &= 0, \\
b_m x^{m+1} + b_{m-1} x^m + \cdots + b_0 x \quad &= 0, \\
b_m x^m + b_{m-1} x^{m-1} + \cdots + b_0 &= 0.
\end{aligned}
$$

If we continue constructing new equations from $f = 0$ by successive multiplication by x until we have done this $m - 1$ times, and similarly derived new equations from $g = 0$ until we have multiplied by x, $n - 1$ times, we will have a total of $m + n$ equations of the form

$$a_n x^{m+n-1} + a_{n-1} x^{m+n-2} + \cdots + a_0 x^{m-1} = 0,$$

$$a_n x^{m+n-2} + a_{n-1} x^{m+n-3} + \cdots + a_0 x^{m-2} = 0,$$

$$\vdots$$

$$a_n x^n + a_{n-1} x^{n-1} + \cdots + a_0 = 0,$$

$$b_m x^{m+n-1} + b_{m-1} x^{m+n-2} + \cdots + b_0 x^{n-1} = 0,$$

$$b_m x^{m+n-2} + b_{m-1} x^{m+n-3} + \cdots + b_0 x^{n-2} = 0,$$

$$\vdots$$

$$b_m x^m + b_{m-1} x + \cdots + b_0 = 0.$$

These equations can be written in matrix form as

$$A\underline{x} = \underline{0}, \quad \text{where} \quad A = \begin{pmatrix} A_1 \\ A_2 \end{pmatrix}$$

The submatrix A_1 has dimensions $m \times (n + m)$ and the form

$$A_1 = \begin{pmatrix} a_n & a_{n-1} & \cdots & \cdots & a_1 & a_0 & 0 & 0 & \cdots & 0 \\ 0 & a_n & a_{n-1} & \cdots & \cdots & a_1 & a_0 & 0 & \cdots & 0 \\ \vdots & \vdots & \ddots & \ddots & & & & \ddots & \ddots & \vdots \\ 0 & 0 & \cdots & \cdots & a_n & a_{n-1} & \cdots & \cdots & a_1 & a_0 \end{pmatrix},$$

whereas A_2 has dimensions $n \times (n + m)$ and the form

$$A_2 = \begin{pmatrix} b_m & b_{m-1} & \cdots & \cdots & \cdots & b_1 & b_0 & 0 & \cdots & 0 \\ 0 & b_m & b_{m-1} & \cdots & \cdots & \cdots & b_1 & b_0 & \cdots & 0 \\ \vdots & \ddots & \ddots & \ddots & & & & \ddots & \ddots & \vdots \\ 0 & 0 & \cdots & b_m & b_{m-1} & \cdots & \cdots & \cdots & b_1 & b_0 \end{pmatrix},$$

and $\underline{x} = (x^{m+n-1} \quad x^{m+n-2} \quad \cdots \quad x \quad 1)^T$.

This is a system of $m + n$ linear homogeneous equations, and hence, if they are to have a solution, the determinant of the matrix A must be zero. The determinant of this matrix is called the *resultant* R of the polynomials f and g with respect to x. Therefore,

$$R = \det A.$$

The equation $R = 0$ is the result of the elimination of x from the two polynomials. This process of elimination is called *Sylvester's dialytic elimination*.

The elimination of the ratio x/y from the equations $\sum_{i=0}^{n} a_i x^i y^{n-i} = 0$ and $\sum_{i=0}^{m} b_i x^i y^{m-i} = 0$ would yield precisely the same result as the elimination of x from the equations $\sum_{i=0}^{n} a_i x^i = 0$ and $\sum_{i=0}^{m} b_i x^i = 0$. This fact enables us to extend our application of the definition of resultants to the case where it is a ratio that is to be eliminated. This is useful when dealing with homogeneous line coordinates; for in that case it is the ratio of the two variables that is independent.

Example 5.9 *Find the resultants with respect to the ratios λ/μ and μ/λ of the polynomials $\lambda^3 - \mu^3$ and $\lambda + \mu$.*

The resultant with respect to the ratio λ/μ is

$$
\det \begin{pmatrix}
1 & 0 & 0 & -1 \\
1 & 1 & 0 & 0 \\
0 & 1 & 1 & 0 \\
0 & 0 & 1 & 1
\end{pmatrix} = 2.
$$

The resultant with respect to the ratio μ/λ is

$$
\det \begin{pmatrix}
-1 & 0 & 0 & 1 \\
1 & 1 & 0 & 0 \\
0 & 1 & 1 & 0 \\
0 & 0 & 1 & 1
\end{pmatrix} = -2.
$$

Example 5.10 *Use Sylvester's dialytic elimination to find the intersection points of the conics*

$$x^2 + 3xy + y^2 - xz - yz = 0,$$

$$x^2 + y^2 - xz - yz = 0.$$

Let us eliminate x. We then rewrite these equations as polynomials in x, namely

$$x^2 + (3y - z)x + y^2 - yz = 0,$$

$$x^2 - zx + y^2 - yz = 0.$$

Multiplying each of these equations by x, we get the system of four equations

$$
\begin{pmatrix}
1 & 3y - z & y^2 - yz & 0 \\
0 & 1 & 3y - z & y^2 - yz \\
1 & -z & y^2 - yz & 0 \\
0 & 1 & -z & y^2 - yz
\end{pmatrix}
\begin{pmatrix}
x^3 \\
x^2 \\
x \\
1
\end{pmatrix}
=
\begin{pmatrix}
0 \\
0 \\
0 \\
0
\end{pmatrix}.
$$

Then the determinant of the matrix is

$$R(y, z) = 9y^3(y - z),$$

and hence $R(y, z) = 0$ implies $y = 0$ or $y = z$. Upon substitution of these solutions back into the original equations, we obtain the three intersection points $(0, 0, 1)$, $(1, 0, 1)$, and $(0, 1, 1)$.

There seems to be some cause for concern here. For some time now we have been claiming that two conics meet in four points, whereas this example shows that two conics can meet in only three points. In this simple case we have already provided an explanation, for in Section 2.12 we discussed multiple-point contact of conics. Writing the two conics in the form

$$x^2 + 3xy + y^2 - xz - yz \equiv (x + y)(x + y - z) + xy = 0$$

and

$$x^2 + y^2 - xz - yz \equiv (x + y)(x + y - z) - 2xy = 0,$$

we see that both are members of the pencil of conics given by

$$(x + y)(x + y - z) + \alpha xy = 0,$$

which is the pencil of conics through the points $(0, 0, 1)$, $(1, 0, 1)$, and $(0, 1, 1)$ and tangent to the line $x + y = 0$ at the point $(0, 0, 1)$ (see Section 2.11). These two conics are then considered to have *double* contact at the point $(0, 0, 1)$, and simple contact at the points $(1, 0, 1)$ and $(0, 1, 1)$ (Figure 5.8).

The two conics do, therefore, meet in four points—provided that we define the multiplicities of the intersections in an appropriate way.

5.6 More about Rational Curves

The use of elimination also gives us a simple way of proving the following theorem.

Theorem 5.8 *A plane rational curve is a plane algebraic curve.*

Proof. Let $\tilde{x} = \tilde{X}(s, t)$, $\tilde{y} = \tilde{Y}(s, t)$, $\tilde{z} = \tilde{Z}(s, t)$ be the parametrization of a plane rational curve, where s and t are projective line coordinates and $\tilde{x}$, $\tilde{y}$, and $\tilde{z}$ are projective coordinates for the plane. Since the curve is actually a function of the single parameter s/t, it is this ratio that must be eliminated. Similarly, it is the ratios $\tilde{x}/\tilde{z}$ and $\tilde{y}/\tilde{z}$ that are the independent variables in the plane. We rewrite the parametrization in its affine form, namely

$$\frac{\tilde{x}}{\tilde{z}} \equiv x = \frac{X(u)}{Z(u)}, \qquad \frac{\tilde{y}}{\tilde{z}} \equiv y = \frac{Y(u)}{Z(u)}, \tag{5.2}$$

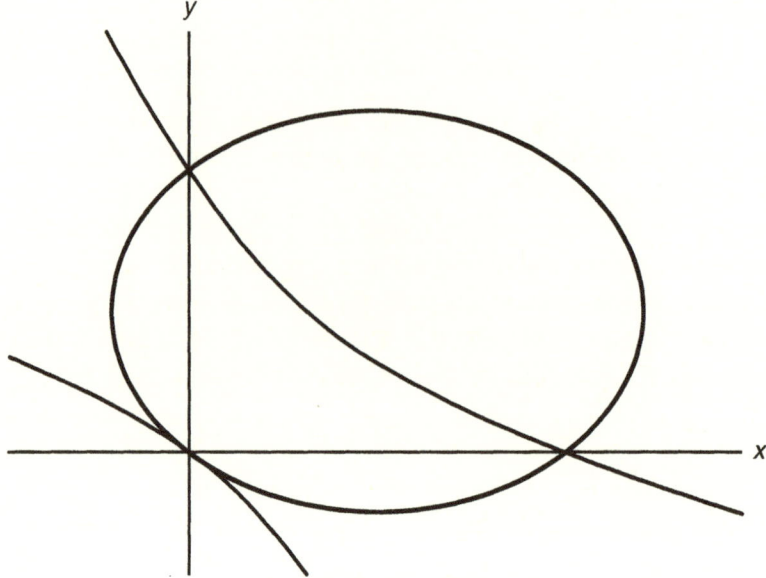

Figure 5.8. Intersections of two conics.

where $u = s/t$ is the affine line parameter, and $X(u) = \tilde{X}(u, 1)$, $Y(u) = \tilde{Y}(u, 1)$, and $Z(u) = \tilde{Z}(u, 1)$. The two equations (5.2) can be written as

$$X(u) - xZ(u) = 0, \qquad Y(u) - yZ(u) = 0,$$

or

$$\sum_{i=0}^{n} x_i u^i = 0, \qquad \sum_{i=0}^{n} y_i u^i = 0,$$

where the coefficients x_i are linear in x, the coefficients y_i are linear in y, and all of them depend upon the coefficients in the original parametrization.

Since both equations hold simultaneously, the resultant $R(x, y)$ of these two equations with respect to u must be zero. Since the resultant is a determinant and since a determinant is an algebraic function of its entries, the equation $R(x, y) = 0$ must be an algebraic equation and must, therefore, represent an algebraic curve, at least one component of which is satisfied by the rational parametrization. That component is therefore the plane algebraic curve corresponding to the plane rational curve. ∎

Example 5.11 *Use Sylvester's dialytic elimination to determine the algebraic curve corresponding to the rational curve*

$$\tilde{x} = s^2 t, \qquad \tilde{y} = s^3, \qquad \tilde{z} = s^3 + t^3.$$

The corresponding affine parametrizations are

$$x = \frac{u^2}{u^3 + 1}, \qquad y = \frac{u^3}{u^3 + 1},$$

or

$$xu^3 - u^2 + x = 0 \quad \text{and} \quad (y - 1)u^3 + y = 0.$$

Their resultant with respect to u is given by

$$R(x, y) = \det \begin{pmatrix} x & -1 & 0 & x & 0 & 0 \\ 0 & x & -1 & 0 & x & 0 \\ 0 & 0 & x & -1 & 0 & x \\ y - 1 & 0 & 0 & y & 0 & 0 \\ 0 & y - 1 & 0 & 0 & y & 0 \\ 0 & 0 & y - 1 & 0 & 0 & y \end{pmatrix} = x^3 + y^3 - y^2.$$

The algebraic equation of the given curve is therefore $x^3 + y^3 - y^2 = 0$ or, in homogeneous form, $\tilde{x}^3 + \tilde{y}^3 - \tilde{y}^2\tilde{z} = 0$.

Theorems 5.8 and 5.3 provide the following result.

Theorem 5.9 *A rational curve with parametrization of degree n is an algebraic curve of order at most n.*

Proof. Theorem 5.8 shows us that the rational curve is an algebraic curve. We must show only that its order is at most n. From Theorem 5.3 we know that the order will be the number of times a line cuts the curve. Let the curve have parametrization

$$x = X_n(s, t), \qquad y = Y_n(s, t), \qquad z = Z_n(s, t),$$

where X_n, Y_n, and Z_n are homogeneous polynomials in s and t of degree less than or equal to n, and where at least one of these polynomials is of exact degree n. Let $\alpha x + \beta y + \gamma z = 0$ be an arbitrary line. We do not know the algebraic equation of the curve. However, this line meets the curve where

$$\alpha X_n(s, t) + \beta Y_n(s, t) + \gamma Z_n(s, t) = 0,$$

which is a polynomial equation of degree n in the ratio s/t. The curve is therefore of order at most n. ∎

In Theorem 5.9 we may be tempted to say that the order of the curve is precisely n. However, consider the case

$$x = at^4, \qquad y = 2at^2.$$

Parametrizations of a given algebraic curve are not unique, and the algebraic curve corresponding to this rational curve is the same as the curve implied by putting $u = t^2$, in which case we have

$$x = au^2, \qquad y = 2au,$$

which is a well-known parametrization of the parabola $y^2 = 4ax$. The curve is therefore of second order, and not fourth order as the original parametrization might have led us to believe. However, as the parameters range over all real values, the points given by the second parametrization of the parabola sweep out the entire parabola, whereas those given by the first parametrization give us only that part which corresponds to points where x and y have the same sign. This is an important fact, for sometimes a suitable redefinition of the parameter of a rational curve, while not changing the curve, may provide a much more usable parametrization of the curve.

5.7 Resultants, Common Factors, and Intersections

Resultants play a very important part in the analysis of curve intersections and in the theory of rational curves. We will use resultants to find out whether two polynomials have common factors and therefore intersect in infinitely many points. We require the following lemma.

Lemma 5.2 *If f and g are polynomials of degrees precisely n and m, respectively, in a variable x, then they have a common nonconstant factor involving x if and only if there exist polynomials p and q such that the degree of p is less than m, the degree of q is less than n, and $pf = qg$.*

Proof. If f and g have a common nonconstant factor h, then $f = hq$ and $g = hp$ for some polynomials p and q, and $pf = phq = qg$.

Conversely, if $pf = qg$, let us factor g into its irreducible components, that is, let

$$g = \prod_{i=1}^{k} g_i,$$

where the polynomials g_i are nonconstant and irreducible. Then

$$pf = q \prod_{i=1}^{k} g_i.$$

Since the degree of p is less than m, not all of the factors g_i can be factors of p. Therefore, at least one of the g_i, say g_j, is a factor of f, and hence f and g have a common nonconstant factor. ∎

Theorem 5.10 *If f and g are polynomials in a variable x, then f and g have a common nonconstant factor involving x if and only if the resultant of f and g with respect to x is zero.*

Proof. Let f and g be of precise degree n and m, respectively, in x, with $m \leq n$. Then, by Lemma 5.2, f and g have a common nonconstant factor involving x if and only if there exist polynomials p and q with the degree of p less than m and the degree of q less than n such that $pf = qg$. Let

$$f = \sum_{i=0}^{n} f_i x^i, \quad g = \sum_{i=0}^{m} g_i x^i, \quad f_n g_m \neq 0,$$

and let

$$p = \sum_{i=0}^{m-1} p_i x^i, \quad q = -\sum_{i=0}^{n-1} q_i x^i, \quad pq \neq 0.$$

Then f and g have a common nonconstant factor involving x if and only if

$$\left(p_{m-1}x^{m-1} + p_{m-2}x^{m-2} + \cdots + p_0\right)\left(f_n x^n + f_{n-1}x^{n-1} + \cdots + f_0\right)$$
$$= -\left(q_{n-1}x^{n-1} + q_{n-2}x^{n-2} + \cdots + q_0\right)$$
$$\times \left(g_m x^m + g_{m-1}x^{m-1} + \cdots + g_0\right),$$

that is, if and only if

$$f_n p_{m-1} + g_m q_{n-1} = 0,$$
$$f_{n-1}p_{m-1} + f_n p_{m-2} + g_{m-1}q_{n-1} + g_m q_{n-2} = 0,$$
$$\vdots$$
$$f_{n-m+1}p_{m-1} + f_{n-m+2}p_{m-2} + \cdots + f_n p_0 + g_1 q_{n-1} + \cdots + g_m q_{n-m} = 0,$$
$$\vdots$$
$$f_1 p_{m-1} + f_2 p_{m-2} + \cdots + f_m p_0 + g_0 q_m + g_1 q_{m-1} + \cdots + g_m q_0 = 0,$$
$$f_0 p_{m-1} + f_1 p_{m-2} + \cdots + f_{m-1}p_0 + g_0 q_{m-1} + \cdots + g_{m-1}q_0 = 0,$$
$$\vdots$$
$$f_0 p_1 + f_1 p_0 + g_0 q_1 + g_1 q_0 = 0,$$
$$f_0 p_0 + g_0 q_0 = 0.$$

These $m + n$ equations in the unknown coefficients p_i and q_i can be written in the matrix form $A\underline{p} = \underline{0}$, where the vector $\underline{p}$ contains all the p_i and q_i, $\underline{p} =$

$(p_{m-1} \; p_{m-2} \; \cdots \; p_0 \; q_{n-1} \; q_{n-2} \; \cdots \; q_1 \; q_0)^T$, and the matrix A has the form

$$\begin{pmatrix}
f_n & 0 & \cdots & 0 & g_m & 0 & \cdots & \cdots & \cdots & \cdots & 0 \\
f_{n-1} & f_n & \cdots & 0 & g_{m-1} & g_m & 0 & \cdots & \cdots & \cdots & 0 \\
\vdots & \vdots & \ddots & \vdots & \vdots & \vdots & & & & & \vdots \\
f_{n-m+1} & \vdots & & f_n & g_1 & g_2 & \cdots & \cdots & g_m & 0 & 0 \\
\vdots & \vdots & & \vdots & \vdots & & & & & \ddots & \vdots \\
f_1 & f_2 & \cdots & f_m & \cdots & 0 & g_0 & \cdots & \cdots & \cdots & g_m \\
f_0 & f_1 & \cdots & f_{m-1} & 0 & \vdots & 0 & g_0 & \cdots & \cdots & g_{m-1} \\
\vdots & & & \vdots & \vdots & \vdots & \vdots & & \ddots & \ddots & \vdots \\
0 & \cdots & 0 & f_0 & f_1 & 0 & \cdots & \cdots & 0 & g_0 & g_1 \\
0 & \cdots & \cdots & 0 & f_0 & 0 & \cdots & \cdots & \cdots & 0 & g_0
\end{pmatrix}.$$

This system of homogeneous equations has a solution if and only if the matrix A is singular. This matrix is the transpose of the matrix associated with the resultant, and hence f and g have a common, nonconstant factor involving x if and only if their resultant with respect to x is zero. ∎

Example 5.12 *Use resultants to show that the polynomials*

$$f = xy + xz + yz + z^2,$$
$$g = xy + xz - yz - z^2$$

do not have a common nonconstant factor involving x, but do have such a component involving y.

Writing f and g as polynomials in x, we have

$$f = (y+z)x + z(y+z),$$
$$g = (y+z)x - z(y+z).$$

Then

$$R(y,z) = \det \begin{pmatrix} y+z & z(y+z) \\ y+z & -z(y+z) \end{pmatrix} = -2z(y+z)^2 \neq 0.$$

Therefore, f and g do not have a common nonconstant factor involving x.

Writing f and g as polynomials in y, we have

$$f = (x+z)y + z(x+z),$$
$$g = (x-z)y + z(x-z).$$

Then

$$R(y, z) = \det \begin{pmatrix} x + z & z(x + z) \\ x - z & z(x - z) \end{pmatrix} = 0.$$

Therefore, f and g do have a common nonconstant factor involving y. In fact, $f = (x + z)(y + z)$ and $g = (x - z)(y + z)$.

The degree of the resultant of two polynomials is of crucial importance. We require another lemma.

Lemma 5.3 *If $f(x_1, x_2, \ldots, x_r)$ is a polynomial in the variables $x_1, x_2, \ldots, x_r$, then f is homogeneous of degree n if and only if*

$$f(tx_1, tx_2, \ldots, tx_r) = t^n f(x_1, x_2, \ldots, x_r).$$

Proof. The necessity of the condition follows immediately from the definition of a homogeneous polynomial given in Section 4.2.

For sufficiency, suppose that $f = \sum_{i=1}^m g_i$, where each g_i is a homogeneous polynomial of degree s_i, $i = 1, 2, \ldots, m$. Then

$$f(tx_1, tx_2, \ldots, tx_r) = \sum_{i=1}^m t^{s_i} g_i(x_1, x_2, \ldots, x_r).$$

However,

$$f(tx_1, tx_2, \ldots, tx_r) = t^n f(x_1, x_2, \ldots, x_r);$$

therefore, $s_i = n$, $i = 1, 2, \ldots, m$, and all the g_i are homogeneous of the same degree. Then

$$f(tx_1, tx_2, \ldots, tx_r) = t^n \sum_{i=1}^m g_i = t^n h, \qquad \text{where} \quad h = \sum_{i=1}^m g_i.$$

Upon substitution of 1 for the value of t, we get the required result.   ∎

Theorem 5.11 *If $f = \sum_{i=1}^n a_i x_r^i$ and $g = \sum_{i=1}^m b_i x_r^i$, where the a_i and b_i are homogeneous polynomials of degree $n - i$ and $m - i$, respectively, and $a_n b_m \neq 0$, then the resultant R of f and g with respect to x_r is either zero or a homogeneous polynomial of degree nm.*

Proof. If $R(x_1, x_2, \ldots, x_{r-1})$ is the resultant of f and g with respect to x_r, then $R(tx_1, tx_2, \ldots, tx_{r-1})$ is given by

$$\det \begin{pmatrix} a_n & ta_{n-1} & \cdots & \cdots & \cdots & \cdots & t^n a_0 & 0 & \cdots & 0 \\ 0 & a_n & ta_{n-1} & \cdots & \cdots & \cdots & t^{n-1}a_1 & t^n a_0 & & \vdots \\ \vdots & \vdots & \vdots & & & & & & & 0 \\ 0 & \vdots & 0 & a_n & \cdots & \cdots & \cdots & \cdots & \cdots & t^n a_0 \\ b_m & tb_{m-1} & \cdots & \cdots & t^m b_0 & 0 & \cdots & \cdots & \cdots & 0 \\ 0 & b_m & tb_{m-1} & \cdots & \cdots & t^m b_0 & & & & \vdots \\ \vdots & \vdots & \vdots & & \vdots & & & & & 0 \\ 0 & 0 & 0 & \cdots & 0 & b_m & tb_{m-1} & \cdots & \cdots & t^m b_0 \end{pmatrix}.$$

If we multiply the ith row of the rows involving the a's and the ith row of the rows involving the b's by t^i, then we have multiplied the determinant by t^K, where

$$K = \sum_{i=1}^{m} i + \sum_{i=1}^{n} i = \frac{1}{2}m(m+1) + \frac{1}{2}n(n+1).$$

Then $t^K R(tx_1, tx_2, \ldots, tx_{r-1})$ is given by the determinant of

$$\begin{pmatrix} ta_n & t^2 a_{n-1} & \cdots & \cdots & \cdots & \cdots & t^{n+1}a_0 & \cdots & 0 \\ 0 & t^2 a_n & t^3 a_{n-1} & & & & & & \vdots \\ \vdots & \ddots & \ddots & & & & & & 0 \\ 0 & \cdots & 0 & t^m a_n & \cdots & \cdots & \cdots & \cdots & t^{m+n}a_0 \\ tb_m & t^2 b_{m-1} & \cdots & t^m b_1 & t^{m+1}b_0 & 0 & \cdots & \cdots & 0 \\ 0 & t^2 b_m & & \vdots & \vdots & & & & \vdots \\ \vdots & & & \vdots & \vdots & & & & 0 \\ 0 & \cdots & \cdots & \cdots & 0 & t^n b_m & t^{n+1}b_{m-1} & \cdots & t^{m+n}b_0 \end{pmatrix}.$$

Column j of this last determinant has t^j as a factor. Therefore,

$$t^K R(tx_1, tx_2, \ldots, tx_{r-1}) = t^L R(x_1, x_2, \ldots, x_{r-1}),$$

where

$$L = \sum_{i=1}^{m+n} i = \frac{1}{2}(m+n)(m+n+1).$$

Therefore,

$$R(tx_1, tx_2, \ldots, tx_{r-1}) = t^{L-K} R(x_1, x_2, \ldots, x_{r-1}).$$

However,

$$L - K = \tfrac{1}{2}[(m+n)(m+n+1) - m(m+1) - n(n+1)] = mn,$$

and hence, using Lemma 5.3, we have the required result. ∎

Example 5.13 *Verify Theorem 5.11 in the case where* $f = x^2 + y^2$ *and* $g = x^2 + xy + yz$.

The resultant with respect to x is

$$R(y, z) = \det \begin{pmatrix} 1 & 0 & y^2 & 0 \\ 0 & 1 & 0 & y^2 \\ 1 & y & yz & 0 \\ 0 & 1 & y & yz \end{pmatrix} = 2y^4 - 2y^3z + y^2z^2,$$

which is homogeneous of degree four. Similarly, the resultant with respect to y is

$$R(x, z) = \det \begin{pmatrix} 1 & 0 & x^2 \\ x+z & x^2 & 0 \\ 0 & x+z & x^2 \end{pmatrix} = x^2(2x^2 + 2xz + z^2),$$

which is also homogeneous of degree four. Since f is of degree two in y and g is only of degree one in y, the theorem predicts that $R(x, z)$ should be of degree two. However, in this case the theorem does not apply, for if

$$f = a_2 y^2 + a_1 y + a_0 \quad \text{and} \quad g = b_1 y + b_0,$$

then the theorem requires a_2 and b_1 to be of degree zero, which is not the case. This also explains why the first result in Example 5.12 was not linear.

Note that $R(x, y)$ is not defined, since f, considered as a polynomial in z, is not of degree at least one.

If (a, b, c) is a point common to two curves f and g, then

$$R_x(b, c) = R_y(a, c) = R_z(a, b) = 0,$$

where R_x, R_y, and R_z denote the resultants of f and g with respect to x, y, and z, respectively.

Example 5.14 *The curves* $x^2 + y^2 - z^2 = 0$ *and* $x - y - z = 0$ *meet at* $(0, -1, 1)$ *and* $(1, 0, 1)$. *Examine the zeros of the resultants with respect to* x, y, *and* z, *respectively.*

$R_x(y, z) = 2y(y + z)$, so that $R_x(-1, 1) = R_x(0, 1) = 0$, showing that the y- and z-coordinates of the common points of the two curves are zeros of the resultant with respect to x. Similarly, $R_y(y, z) = 2x(x - z)$, so that $R_y(0, 1) = R_y(1, 1) = 0$, and $R_z(x, y) = 2xy$, so that $R_z(0, -1) = R_z(1, 0) = 0$.

We must be careful when using resultants to find the common points of curves, because if two points are collinear with a vertex of the triangle of reference, the ratio of two of their coordinates will be the same. Hence, a multiple zero of the resultant does not necessarily imply a multiple intersection point of the two curves.

Example 5.15 *Relate the zeros of $R(y, z)$ to the common zeros of the curves* $x^2 + y^2 - 5z^2 = 0$ *and* $x^2 + 2y^2 - 6z^2 = 0$.

We know that $R(y, z) = (y^2 - z^2)^2$, so that this resultant has two double roots at $(1, 1)$ as well as at $(-1, 1)$. However, the two curves do not have multiple intersections, for substitution into the equations of the curves shows that the intersections are $(2, 1, 1)$, $(-2, 1, 1)$, $(2, -1, 1)$, and $(-2, -1, 1)$.

5.8 Equations for a Finite Point Set

The principle of duality is one of the most important features of the projective plane. From it we learned the dual interpretations of the equation $\alpha x + \beta y + \gamma z = 0$ as the equation of a line, the line coordinates (α, β, γ) being given, or as the equation of a point, the point coordinates (x, y, z) being given.

Using this concept, we could, without ambiguity, consider the equation

$$(\alpha x_1 + \beta y_1 + \gamma z_1)(\alpha x_2 + \beta y_2 + \gamma z_2) = 0$$

as the equation of the points (x_1, y_1, z_1) and (x_2, y_2, z_2). The equation

$$(\alpha x_1 + \beta y_1 + \gamma z_1)^2(\alpha x_2 + \beta y_2 + \gamma z_2)^3 = 0$$

is a representation of the point (x_1, y_1, z_1) taken twice and the point (x_2, y_2, z_2) taken three times. In general, if P_i is the point with coordinates (x_i, y_i, z_i), then the finite point set comprising the points P_i, $i = 1, 2, \ldots, n$, each taken r_i times, would have the equation

$$S(\alpha, \beta, \gamma) \equiv \prod_{i=1}^{n} (\alpha x_i + \beta y_i + \gamma z_i)^{r_i} = 0.$$

Alternatively, we could represent the point (x_1, y_1, z_1) by the equation

$$\det \begin{pmatrix} x_1 & y_1 & z_1 \\ p & q & r \\ u & v & w \end{pmatrix} = 0,$$

which, as we have seen in Section 4.2, is the condition for the points (p, q, r) and (u, v, w) to be collinear with the given point (x_1, y_1, z_1). Once more, since (x_1, y_1, z_1) is given and the (p, q, r) and (u, v, w) are variables, this equation can unambiguously be considered as the equation of the point (x_1, y_1, z_1).

In the cause of simplicity let us make a few minor changes in notation. Let $\underline{a}$ denote the triple (a_1, a_2, a_3), $\underline{x}$ and $\underline{y}$ the triples (x_1, x_2, x_3) and (y_1, y_2, y_3), and so on. The determinant point equation given above is unaltered if we consider the transpose of the matrix and reorder the columns, so it can be written in the form

$$\det \begin{pmatrix} p & u & x_1 \\ q & v & y_1 \\ r & w & z_1 \end{pmatrix} = 0.$$

Let the notation $|\underline{x}\ \underline{y}\ \underline{a}|$ denote

$$\det \begin{pmatrix} x_1 & y_1 & a_1 \\ x_2 & y_2 & a_2 \\ x_3 & y_3 & a_3 \end{pmatrix}.$$

Then the equation of the point $\underline{a}$ can be written in the form

$$|\underline{x}\ \underline{y}\ \underline{a}| = 0.$$

Generalizing this idea to produce the equation of the finite point set of points P_i, $i = 1, 2, \ldots, n$, each taken r_i times, we let the point P_i have coordinates $\underline{a}^{[i]} = (a_1^{[i]}, a_2^{[i]}, a_3^{[i]})$, and the corresponding equation is

$$\prod_{i=1}^{n} |\underline{x}\ \underline{y}\ \underline{a}^{[i]}|^{r_i} = 0.$$

Our first representation of the equation of a point involves the coordinates of a variable line. The second representation involves the coordinates of two points. There are, therefore, six variables in the equation $|\underline{x}\ \underline{y}\ \underline{a}| = 0$, and it is certainly a more complicated equation than $\alpha a_1 + \beta a_2 + \gamma a_3 = 0$. It is, however, the form of the equation of a point that we shall find most useful in proving the Maclaurin–Bézout theorem, which is perhaps the most fundamentally important theorem in our study of plane curves.

The equation $|\underline{x}\ \underline{y}\ \underline{a}| = 0$ is an example of what is called a two-form. A *two-point* is an ordered pair of triples and is denoted by

$$(\underline{x}; \underline{y}) \equiv (x_1, x_2, x_3; y_1, y_2, y_3),$$

where $\underline{x}$ and $\underline{y}$ are points in the projective plane. Since, for $\lambda \mu \neq 0$, $\lambda \underline{x}$ and $\mu \underline{y}$ are the same points as $\underline{x}$ and $\underline{y}$, respectively, we regard $(\lambda \underline{x}; \mu \underline{y})$ as the same two-point as $(\underline{x}; \underline{y})$.

A *two-form* $f(\underline{x}; \underline{y})$ of degree $(l; m)$ is a polynomial in the six variables $x_1, x_2, x_3, y_1, y_2,$ and y_3 that is homogeneous of degree l in the three variables $x_1, x_2,$ and x_3 and homogeneous of degree m in the three variables $y_1, y_2,$ and y_3. The expression $f(\underline{x}; \underline{y}) = |\underline{x} \ \underline{y} \ \underline{a}|$ is therefore a two-form of degree $(1; 1)$. A two-point $(\underline{a}; \underline{b})$ is said to be a zero of the two-form $f(\underline{x}; \underline{y})$ if and only if $f(\underline{a}; \underline{b}) = 0$.

Since two-forms can be thought of as ordinary polynomials in six variables, the unique factorization results that hold in this setting must also hold for two-forms, and hence, if $f(\underline{x}; \underline{y})$ and $g(\underline{x}; \underline{y})$ are two two-forms that have the same zeros, they must have the same irreducible factors. These factors may appear with different exponents.

Lemma 5.4 *If $f = 0$ and $g = 0$ are algebraic curves of precise orders m and n, respectively, with no common components, then the resultant of $f(\lambda \underline{x} + \mu \underline{y})$ and $g(\lambda \underline{x} + \mu \underline{y})$ with respect to the ratio λ/μ (or μ/λ) is a two-form of degree $(mn; mn)$.*

Proof. The expression $f(\lambda \underline{x} + \mu \underline{y})$ can be rewritten in the form

$$f(\lambda \underline{x} + \mu \underline{y}) = \sum_{i=0}^{m} a_i(\underline{x}; \underline{y}) \lambda^i \mu^{m-i},$$

where $a_i(\underline{x}; \underline{y})$ is a homogeneous polynomial of degree $m - i$ in the variables $y_1, y_2,$ and y_3. Similarly, $g(\lambda \underline{x} + \mu \underline{y})$ can be written in the form

$$g(\lambda \underline{x} + \mu \underline{y}) = \sum_{i=0}^{n} c_i(\underline{x}; \underline{y}) \lambda^i \mu^{n-i},$$

where $c_i(\underline{x}; \underline{y})$ is a homogeneous polynomial of degree $n - i$ in the variables $y_1, y_2,$ and y_3. Since f is of precise degree m, we know that $a_m(\underline{x}; \underline{y}) \neq 0$, and since g is of precise degree n, we know that $c_n(\underline{x}; \underline{y}) \neq 0$. Therefore, from Theorem 5.10, the resultant of $f(\lambda \underline{x} + \mu \underline{y})$ and $g(\lambda \underline{x} + \mu \underline{y})$ with respect to the ratio λ/μ is a homogeneous polynomial of degree mn in the variables $y_1, y_2,$ and y_3. However, we can also write

$$f(\lambda \underline{x} + \mu \underline{y}) = \sum_{i=0}^{m} b_i(\underline{x}; \underline{y}) \mu^i \lambda^{m-i},$$

where $b_i(\underline{x}; \underline{y}) \ [= a_{m-i}(\underline{x}; \underline{y})]$ is a homogeneous polynomial of degree $m - i$ in the variables $x_1, x_2,$ and x_3, and

$$g(\lambda \underline{x} + \mu \underline{y}) = \sum_{i=0}^{n} d_i(\underline{x}; \underline{y}) \mu^i \lambda^{n-i},$$

where $d_i(\underline{x}; \underline{y})\ [= c_{n-i}(\underline{x}; \underline{y})]$ is a homogeneous polynomial of degree $n - i$ in the variables x_1, x_2, and x_3. Again, the precise orders of the curves $f = 0$ and $g = 0$ being m and n, respectively, enable us to apply Theorem 5.10 to show that the resultant of $f(\lambda\underline{x} + \mu\underline{y})$ and $g(\lambda\underline{x} + \mu\underline{y})$ with respect to the ratio μ/λ is a homogeneous polynomial of degree mn in the variables x_1, x_2, and x_3. However, the resultant of these two polynomials with respect to the ratio λ/μ can differ from their resultant with respect to the ratio μ/λ by at most a factor -1. Therefore, this resultant must be homogeneous of degree mn in both the triple of coordinates (x_1, x_2, x_3) and the triple (y_1, y_2, y_3). It is, therefore, a two-form of degree $(mn; mn)$ ∎

Example 5.16 *For* $f = x_1^2 + x_2^2 - 2x_2x_3$ *and* $g = x_2 - x_3$, *find the resultant of* $f(\lambda\underline{x} + \mu\underline{y})$ *and* $g(\lambda\underline{x} + \mu\underline{y})$ *with respect to the ratio* λ/μ.

We have $f(\lambda\underline{x} + \mu\underline{y}) = A\lambda^2 + 2B\lambda\mu + C\mu^2$, where

$$A = x_1^2 + x_2^2 - 2x_2x_3,$$

$$B = x_1y_1 + x_2y_2 - x_3y_2 - x_2y_3,$$

$$C = y_1^2 + y_2^2 - 2y_2y_3,$$

and $g(\lambda\underline{x} + \mu\underline{y}) = (x_2 - x_3)\lambda + (y_2 - y_3)\mu$. Therefore,

$$R(\underline{x}; \underline{y}) = \det \begin{pmatrix} A & 2B & C \\ x_2 - x_3 & y_2 - y_3 & 0 \\ 0 & x_2 - x_3 & y_2 - y_3 \end{pmatrix}$$

$$= [(x_2y_3 - x_3y_2) - (x_1y_3 - x_3y_1) + (x_1y_2 - x_2y_1)]$$

$$\times [-(x_2y_3 - x_3y_2) - (x_1y_3 - x_3y_1) + (x_1y_2 - x_2y_1)]$$

$$= |\underline{x}\ \underline{y}\ \underline{a}|\ |\underline{x}\ \underline{y}\ \underline{b}|,$$

where $\underline{a}$ is the point $(1, 1, 1)$ and $\underline{b}$ is the point $(-1, 1, 1)$. This is a two-form of degree $(2; 2)$. The points $\underline{a}$ and $\underline{b}$ are the intersections of the curves $f = 0$ and $g = 0$.

5.9 The Maclaurin–Bézout Theorem

Colin Maclaurin was one of the most outstanding mathematicians of his time. He entered Glasgow University at the tender age of eleven and graduated in 1714, aged sixteen. Although he is most often remembered because of the series which bears his name, most of his work had a strong geometrical flavor. At the age of twenty-one he became a Fellow of the Royal Society, and

the following year published his *Geometria Organica*, which extended Newton's geometrical studies. In this book he did much to enhance the theory of
higher plane curves, and it was in connection with this theory that he postulated the result that two curves, of orders m and n, respectively, meet in
mn points. Twenty-six years later Etienne Bézout, one of a long line of great
French geometers, gave a somewhat incomplete proof, there still remaining
a confusion about multiple intersections. It remained thus for over a century
until, in 1873, another Frenchman, Georges-Henri Halphen, finally settled the
issue.

Theorem 5.12 (*Maclaurin–Bézout*) *Two plane algebraic curves of precise
orders m and n, respectively, and having no common nonconstant component,
have mn intersections.*

Proof. Let the curves be given by $f(\underline{x}) = 0$ and $g(\underline{x}) = 0$, and let $\underline{y}$ and
$\underline{z}$ denote two arbitrary points. Then, by Lemma 5.4, the resultant $R(\underline{y}; \underline{z})$ of
$f(\lambda \underline{y} + \mu \underline{z})$ and $g(\lambda \underline{y} + \mu \underline{z})$ with respect to the ratio λ/μ (or μ/λ) is a two-
form of degree $(mn; mn)$. This resultant is zero if and only if the polynomials
$f(\lambda \underline{y} + \mu \underline{z})$ and $g(\lambda \underline{y} + \mu \underline{z})$ in the ratio λ/μ (or μ/λ) have a common factor.
This will be the case if and only if the points $\underline{y}$ and $\underline{z}$ are collinear with a point
of intersection of the two given curves. Let the points $\underline{a}^{[i]}$, $i = 1, 2, \ldots, k$, be
the points common to the curves $f(\underline{x}) = 0$ and $g(\underline{x}) = 0$. Then $R(\underline{y}; \underline{z})$ has
the same zeros as

$$\prod_{i=1}^{k} \left| \underline{y} \; \underline{z} \; \underline{a}^{[i]} \right|.$$

Since $R(\underline{y}; \underline{z})$ is of degree $(mn; mn)$, k must be less than or equal to mn, and

$$R(\underline{y}; \underline{z}) = \alpha \prod_{i=1}^{k} \left| \underline{y} \; \underline{z} \; \underline{a}^{[i]} \right|^{r_i},$$

where α is a nonzero constant and $\sum_{i=1}^{k} r_i = mn$. The multiplicity of f and
g at the point $\underline{a}^{[i]}$ is defined to be the exponent r_i in the above expression for
$R(\underline{y}; \underline{z})$. With this interpretation of the multiplicity of intersections, the theorem
is proved.   ∎

We have finally justified our earlier hypothesis (Section 2.8) that two conics
meet in four points. We also note here the weaker statement that if two curves of
orders m and n, respectively, have more than mn intersections, then they must
have a common nonconstant component and infinitely many points in common.

The Maclaurin–Bézout theorem is the most fundamental result necessary for understanding algebraic curves. As an application, let us show that a cubic curve cannot have more than one double point, a basic result that will be used in Example 5.18 to find the class of a cubic curve. We use a *reductio ad absurdum* argument. If a cubic curve had two double points, the line joining them would meet the cubic in four points, hence violating the statement of the theorem.

Theorem 5.13 *The multiplicity of the intersections of two plane algebraic curves is a projective invariant.*

Proof. Let $\tilde{\underline{x}} = A\underline{x}$ denote the projective transformation. Then

$$(\det A)^{mn} R(\underline{y}; \underline{z}) = \prod_{i=1}^{k} \left| A\underline{y} A\underline{z} A\underline{a}^{[i]} \right|^{r_i} = \prod_{i=1}^{k} \left| \tilde{\underline{y}} \, \tilde{\underline{z}} \, \tilde{\underline{a}}^{[i]} \right|^{r_i},$$

and hence the multiplicities are unchanged. ∎

Although the method of proving the Maclaurin–Bézout theorem is very straightforward, calculating the multiplicities in this fashion would prove prohibitively cumbersome in all but the most special of cases. We could somewhat simplify the process by a suitable choice of coordinate system. For example, if the point $(1, 0, 0)$ does not lie on both curves, we could choose $\underline{y} = (1, 0, 0)$. With this choice $R(\underline{y}; \underline{z})$ is the resultant of $f(\lambda + \mu z_1, \mu z_2, \mu z_3)$ and $g(\lambda + \mu z_1, \mu z_2, \mu z_3)$ with respect to the ratio λ/μ and is of the form

$$R(\underline{y}; \underline{z}) = \prod_{i=1}^{k} \left[\det \begin{pmatrix} 1 & z_1 & a_1^{[i]} \\ 0 & z_2 & a_2^{[i]} \\ 0 & z_3 & a_3^{[i]} \end{pmatrix} \right]^{r_i}$$

$$= \alpha \prod_{i=1}^{k} \left(z_2 a_3^{[i]} - z_3 a_2^{[i]} \right)^{r_i},$$

which is simply the resultant of $f(z_1, z_2, z_3)$ and $g(z_1, z_2, z_3)$ with respect to z_1. Therefore we can find the intersection multiplicities by taking simple resultants without resorting to the concept of two-forms. We must be careful, however, to distinguish between two different points that are collinear with the reference point $(1, 0, 0)$, or whatever reference point is chosen. In that case intersections of multiplicity r_1 at $(a_1^{[1]}, a_2^{[1]}, a_3^{[1]})$ and r_2 at $(a_1^{[2]}, a_2^{[1]}, a_3^{[1]})$ would result only in the term $(z_2 a_3^{[1]} - z_3 a_2^{[1]})^{r_1 + r_2}$ and we must do further analysis to determine whether this corresponds to a single intersection point of multiplicity $r_1 + r_2$ or to several points of intersection, the sum of whose multiplicities is $r_1 + r_2$. The following theorem, which will be used in Section 5.11 to calculate intersection multiplicities at singular points, is stated here without proof.

Theorem 5.14 *If P is a point of multiplicity i on a curve f and of multiplicity j on a curve g, then the intersection multiplicity of f and g at P is at least ij. If the tangents to f at P are all distinct from the tangents to g at P, the multiplicity of the intersection of f and g at P is precisely ij.*

In addition to postulating the general result for the number of intersections of two algebraic curves, Maclaurin was the first to note the following paradox, which emphasizes the problems associated with interpolation in two dimensions predicted by Haar's theorem (Section 1.6). Since a cubic is determined by nine parameters (Section 5.2), interpolation at nine points, in general, determines a unique cubic. By Theorem 5.12, two different cubics meet in nine points. Clearly, then, the nine points of intersection do not determine a unique cubic. It was almost thirty years later that Euler and Cramer, independently, gave an explanation. The nine points of intersection of two cubics cannot be independent. Since a cubic is determined by nine points, there will be a pencil of cubics through eight given points in general position. Let this pencil be given by $\lambda f_1 + \mu f_2 = 0$, where f_1 and f_2 represent two different cubics through the eight points. Then f_1 and f_2 meet in the eight given points and one other point P. If f_3 is any other member of the pencil, then f_3 also passes through P. That is, the point P is determined by the eight given points. This result is often stated, in somewhat abstruse fashion, as "every cubic through eight points also passes through a ninth" (Figure 5.9). Much later Julius Plücker gave further generalizations of this paradox (see Exercises 5.21, 5.22).

A beautiful example of the power of the Maclaurin–Bézout theorem and the elegance of the concept of pencils of curves is given by the following theorem.

Theorem 5.15 *If mn of the n^2 intersections of two curves of order n lie on a curve of order m, then the remaining intersections are on a curve of order $n - m$.*

Proof. Let $f_1 = 0$ and $f_2 = 0$ be the two curves of order n, and let $g = 0$ be the curve of order m. Let P be any point of g other than any of the previously specified mn points that are common to f_1, f_2, and g. Then there exists a unique ratio λ/μ such that the curve $\lambda f_1 + \mu f_2 = 0$ passes through P. Therefore, the curve $\lambda f_1 + \mu f_2 = 0$ has $mn + 1$ intersections with g, and hence, by the Maclaurin–Bézout theorem, must have g as a component. The remaining component (or components) must be of order $n - m$ and contains the remaining $n^2 - mn$ intersection points of f_1 and f_2.   ∎

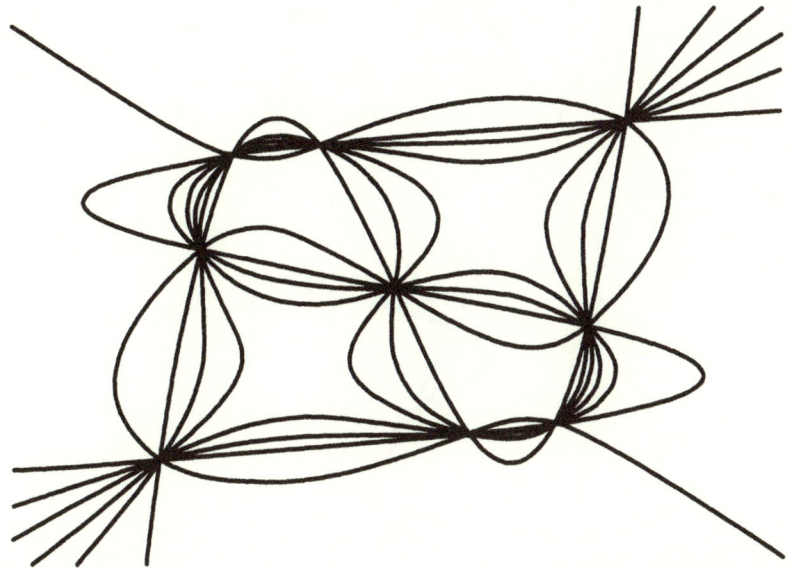

Figure 5.9. All cubics through eight points pass through a ninth.

For the case $n = 3$ (Figure 5.10) the theorem states that if three of the nine intersections of two cubics lie on a line, then the other six points lie on a conic. Once again, this is no contradiction to the general cases of a unique cubic through nine points and a unique conic through five points. We can, in fact, say that if we have eight points, three of which are collinear, then the pencil of cubics through the eight points determines a ninth point, and that this ninth point will also lie on the conic determined by the five points of the pencil that do not lie on the line. Pascal's elegant result on the hexagon inscribed in a conic, originally proved by synthetic means, can be shown to be a simple special case of Theorem 5.15 (see Exercise 5.23).

5.10 Polars, Class, and Inflections

Let $f(\underline{x})$ be a curve of order n, and let $\underline{a}$ and $\underline{b}$ be two distinct points. The line determined by these two points is $\underline{x} = \lambda\underline{a} + \mu\underline{b}$, and this line will meet the curve where

$$f(\lambda\underline{a} + \mu\underline{b}) = 0.$$

This equation is homogeneous of degree n in λ and μ and can be written in the form

$$\lambda^n f(\underline{a}) + \cdots + \frac{1}{r!}\lambda^{n-r}\mu^r (\underline{b} \cdot \nabla_{\underline{a}})^r f(\underline{a}) + \cdots + \frac{1}{n!}\mu^n (\underline{b} \cdot \nabla_{\underline{a}})^n f(\underline{a}) = 0,$$

$$(5.3)$$

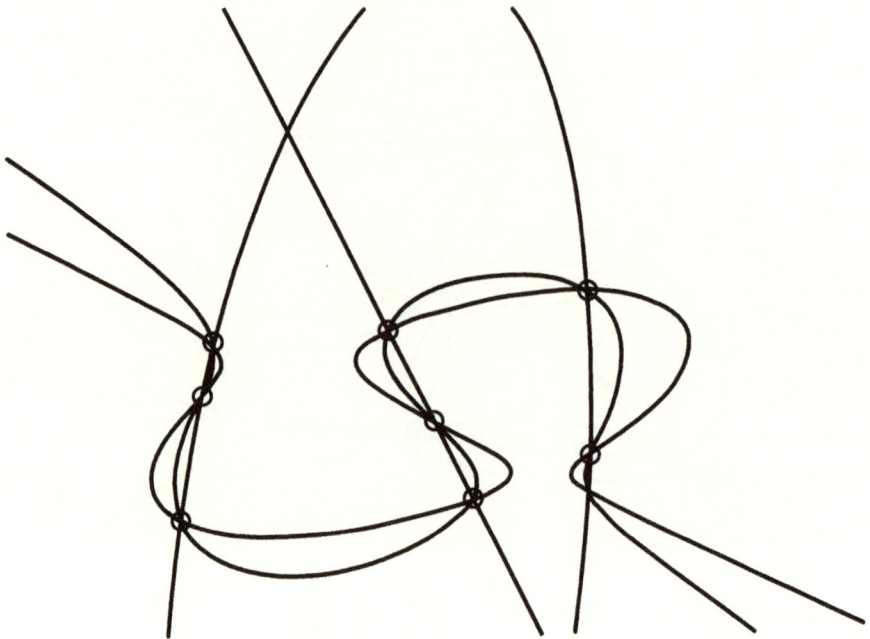

Figure 5.10. Two cubics through nine points, with three points collinear.

where

$$(\underline{b} \cdot \nabla_{\underline{a}}) f(\underline{a}) \equiv \left(b_1 \frac{\partial}{\partial a_1} + b_2 \frac{\partial}{\partial a_2} + b_3 \frac{\partial}{\partial a_3} \right) f(\underline{a})$$

and

$$(\underline{b} \cdot \nabla_{\underline{a}})^n \equiv (\underline{b} \cdot \nabla_{\underline{a}})(\underline{b} \cdot \nabla_{\underline{a}})^{n-1}.$$

From the symmetry between $\lambda \underline{a}$ and $\mu \underline{b}$ this can also be written in the form

$$\mu^n f(\underline{b}) + \cdots + \frac{1}{r!} \mu^{n-r} \lambda^r (\underline{a} \cdot \nabla_{\underline{b}})^r f(\underline{b}) + \cdots + \frac{1}{n!} \lambda^n (\underline{a} \cdot \nabla_{\underline{b}})^n f(\underline{b}) = 0,$$

and hence we have the *symmetry relation*

$$\frac{1}{(n-r)!} (\underline{b} \cdot \nabla_{\underline{a}})^{n-r} f(\underline{a}) = \frac{1}{r!} (\underline{a} \cdot \nabla_{\underline{b}})^r f(\underline{b}).$$

Given a curve $f(\underline{x}) = 0$ of order n, the curve given by $(\underline{a} \cdot \nabla_{\underline{x}})^r f(\underline{x}) = 0$ is of order $n - r$ and is called the rth *polar* of $\underline{a}$ with respect to $f(\underline{x})$ or, in the appropriate context, the rth polar of $f(\underline{x})$ with respect to $\underline{a}$. For a curve $f(\underline{x})$ of order n, the line $(\underline{a} \cdot \nabla_{\underline{x}})^{n-1} f(\underline{x})$ is called the *polar line* of $\underline{a}$ with respect to $f(\underline{x})$, and $(\underline{a} \cdot \nabla_{\underline{x}})^{n-2} f(\underline{x})$ is called the *polar conic*. If $\underline{a}$ lies on the curve, then, since $(\underline{a} \cdot \nabla_{\underline{x}})^{n-1} f(\underline{x})$ is simply a scalar multiple of $(\underline{x} \cdot \nabla_{\underline{a}}) f(\underline{a})$, the polar line with respect to $\underline{a}$ is the tangent to the curve at $\underline{a}$. Furthermore, the symmetry relationship tells us directly that if $\underline{b}$ lies on the rth polar of $\underline{a}$, then $\underline{a}$ lies on the $(n - r)$th polar of $\underline{b}$.

Example 5.17 *Verify the symmetry relationships for the curve*

$$x^3 + y^3 - 3x^2z + 3y^2z = 0.$$

We first calculate all the partial derivatives:

$$f_x = 3x^2 - 6xz, \qquad f_y = 3y^2 + 6yz, \qquad f_z = -3x^2 + 3y^2,$$

$$f_{xx} = 6x - 6z, \qquad f_{xz} = -6x, \qquad f_{yy} = 6y + 6z, \qquad f_{yz} = 6y,$$

$$f_{xxx} = f_{yyy} = f_{yyz} = -f_{xxz} = 6,$$

and all the other second and third partial derivatives are zero. Therefore,

$$\frac{1}{3!}(\underline{b} \cdot \nabla_{\underline{a}})^3 f(\underline{a}) = \frac{1}{3!}(6b_1^3 + 6b_2^3 - 18b_1^2b_3 + 18b_2^3b_3)$$

$$= b_1^3 + b_2^3 - 3b_1^2b_3 + 3b_2^3b_3$$

$$= f(\underline{b}),$$

$$\frac{1}{2!}(\underline{b} \cdot \nabla_{\underline{a}})^2 f(\underline{a}) = \frac{1}{2!}\big[b_1^2(6a_1 - 6a_3) + 2b_1b_3(-6a_1)$$

$$+ 2b_2b_3(6a_2) + b_2^2(6a_2 + 6a_3)\big]$$

$$= a_1(3b_1^2 - 6b_1b_3) + a_2(3b_2^2 + 6b_2b_3)$$

$$+ a_3(-3b_1^2 + 3b_2^2)$$

$$= (\underline{a} \cdot \nabla_{\underline{b}})f(\underline{b}),$$

$$(\underline{b} \cdot \nabla_{\underline{a}})f(\underline{a}) = b_1(3a_1^2 - 6a_1a_3) + b_2(3a_2^2 + 6a_2a_3)$$

$$+ b_3(-3a_1^2 + 3a_2^2)$$

$$= \frac{1}{2!}\big[a_1^2(6b_1 - 6b_3) + 2a_1a_3(-6b_1)$$

$$+ 2a_2a_3(6b_2) + a_2^2(6b_2 + 6b_3)\big]$$

$$= \frac{1}{2!}(\underline{a} \cdot \nabla_{\underline{b}})^2 f(\underline{b}),$$

and finally,

$$f(\underline{a}) = a_1^3 + a_2^3 - 3a_1^2a_3 + 3a_2^3a_3$$

$$= \frac{1}{3!}(6a_1^3 + 6a_2^3 - 18a_1^2a_3 + 18a_2^3a_3)$$

$$= \frac{1}{3!}(\underline{a} \cdot \nabla_{\underline{b}})^3 f(\underline{b}).$$

Clearly, a point of multiplicity s of a curve f is a point of multiplicity at least $s - r$ of its rth polar. This fact, together with the Maclaurin–Bézout theorem, shows us that an algebraic curve has a finite number of multiple points (see Exercise 5.25).

Theorem 5.16 *The first polar of a point P with respect to a curve f meets the curve at its multiple points and at the points where the tangents to the curve through P meet the curve.*

Proof. The first part of the statement follows directly from the fact that multiple points of a curve are at least simple points of its first polar and hence common to both curves.

For the second part, let P have coordinates $\underline{a}$, and let $\underline{b}$ be a point of intersection of the curve and its first polar that is a simple point of the curve. Then

$$(\underline{a} \cdot \nabla_{\underline{b}}) f(\underline{b}) = 0,$$

that is, $\underline{a}$ lies on the line

$$(\underline{x} \cdot \nabla_{\underline{b}}) f(\underline{b}) = 0,$$

which is tangent to f at $\underline{b}$. This is true for any simple point common to f and its first polar, and hence the theorem is proved.   ∎

In Section 4.2.2 we discussed the duality between point and line and expressed conics as point equations and also as line equations. The lines of the line equation of a conic are tangential to the conic, so that the line equation represents the envelope of the given conic (see Example 4.5). The order of the point equation gives us, via the Maclaurin–Bézout theorem, the number of points of an arbitrary line that are incident to the curve. The dual of this statement is that the order of the line equation will give us the number of lines through an arbitrary point that are tangent to the curve. The order of this line equation is then the dual of the order of the point equation (in the case of conics, both are second-order), and is called the *class* of the curve. The class m is then the number of tangents that can be drawn to the curve through an arbitrary point. Theorem 5.16 gives us an immediate upper bound of $n(n - 1)$ for the class of an nth-order curve (see Exercise 5.30).

If P is an r-fold point of a curve f of order n, then, by a change of coordinate system, we can always select P to have coordinates $(0, 0, 1)$, in which case f

can be written in the form

$$\sum_{i=r}^{n} g_i(x, y)z^{n-i} = 0,$$

where each coefficient function g_i, $(i = r, r + 1, \ldots, n)$, is a homogeneous polynomial of degree i. The first polar of a point $\underline{a}$ with respect to f has the equation

$$a_1 \left(\sum_{i=r}^{n} \frac{\partial g_i}{\partial x} z^{n-i} \right) + a_2 \left(\sum_{i=r}^{n} \frac{\partial g_i}{\partial y} z^{n-i} \right) + a_3 \left(\sum_{i=r}^{n-1} (n-i)g_i z^{n-i-1} \right) = 0.$$

The tangents to f at P are given by the factors of $g_r(x, y)$, and those of the first polar of P are the factors of (from Theorem 5.7)

$$a_1 \frac{\partial g_r}{\partial x} + a_2 \frac{\partial g_r}{\partial y}. \tag{5.4}$$

If P is an ordinary singular point of f (the tangents to f at P are distinct), then there are no repeated factors in $g_r(x, y)$, and, since $\underline{a}$ is a general point, no factor of $g_r(x, y)$ will also be a factor of (5.4). Therefore, by Theorem 5.13, the intersection multiplicity of f and its first polar at P is precisely $r(r-1)$. Hence, if f is a curve of order n with k ordinary singular points of multiplicities r_i, $i = 1, 2, \ldots, k$, then the class m of f is given by

$$m = n(n-1) - \sum_{i=1}^{k} r_i(r_i - 1).$$

For nonordinary multiple points we restrict our discussion to those of multiplicity two, that is, cusps. Assume that the curve f has a cusp at $P(0, 0, 1)$. A rotation of axes enables us to choose $y = 0$ as the equation of the double tangent. With this coordinate choice f can be written in the form

$$y^2 z^{n-2} + \sum_{i=3}^{n} g_i(x, y)z^{n-i} = 0,$$

where each coefficient g_i, $i = r, r + 1, \ldots, n$, is a homogeneous polynomial of degree i. Then the first polar of $\underline{a}$ with respect to f is

$$a_1 \left(\sum_{i=3}^{n} \frac{\partial g_i}{\partial x} z^{n-i} \right) + a_2 \left(2yz^{n-2} + \sum_{i=3}^{n} \frac{\partial g_i}{\partial y} z^{n-i} \right)$$
$$+ a_3 \left((n-2)y^2 z^{n-3} + \sum_{i=3}^{n-1} (n-i)g_i z^{n-i-1} \right) = 0,$$

from which expression we see that $(0, 0, 1)$ is a simple point of the first polar and that $y = 0$ is the equation of the tangent at that point. Since f and the first polar share a tangent at $(0, 0, 1)$, Theorem 5.14 tells us that their intersection multiplicity there is at least two. We state without proof that, in this case, the intersection multiplicity is three, and that the following theorem gives the class of a curve with only ordinary singular points and cusps.

Theorem 5.17 *The class m of an irreducible curve of order n that has δ ordinary singular points of multiplicities r_i, $i = 1, 2, \ldots, \delta$, and κ cusps is given by*

$$m = n(n - 1) - \sum_{i=1}^{\delta} r_i(r_i - 1) - 3\kappa.$$

Example 5.18 *Show that the class of the curve $2(x^3 + y^3) - 9xy = 0$ is four, and find the tangents through the point $(6, 6)$ to the curve.*

The curve is a cubic that has a double point with distinct tangents at the origin, that is, the origin is an ordinary double point. Being a cubic, the curve can have no more than a single double point (see Section 5.9). Hence, by Theorem 5.17, the class of the curve is given by

$$m = (3)(2) - (2)(1) = 4.$$

Using homogeneous coordinates, the curve becomes $2(x^3 + y^3) - 9xyz = 0$, and the point of interest has coordinates $(6, 6, 1)$. The equation of the first polar with respect to $(6, 6, 1)$ is

$$6(6x^2 - 9yz) + 6(6y^2 - 9xz) + 1(-9xy) = 0.$$

In affine coordinates this can be written as

$$4y^2 - y(6 + x) + (4x^2 - 6x) = 0.$$

Upon elimination of y we find that the first polar meets the cubic when $x = 0, 1, 2$, or $(-\frac{3}{2} \pm i\sqrt{15}/2)$. The solution $x = 0$ corresponds to the first polar meeting the cubic at the double point. Therefore, the points of tangency will come from the other values of x. Substituting $x = 1$ in the equation of the first polar, we find that $y = 2$ or $y = -\frac{1}{4}$. The point $(1, 2)$ lies on the cubic, but the point $(1, -\frac{1}{4})$ does not. Similarly, the point $(2, 1)$ lies on the cubic, as do the points $(-\frac{3}{2} \pm i\sqrt{15}/2, -\frac{3}{2} \mp i\sqrt{15}/2)$. The cubic, its first polar with respect to the point $(6, 6)$, and the real tangents are shown in Figure 5.11.

Another important qualitative concept is that of an *inflection*. Inflections at multiple points can be defined, but we shall not consider these, and for our

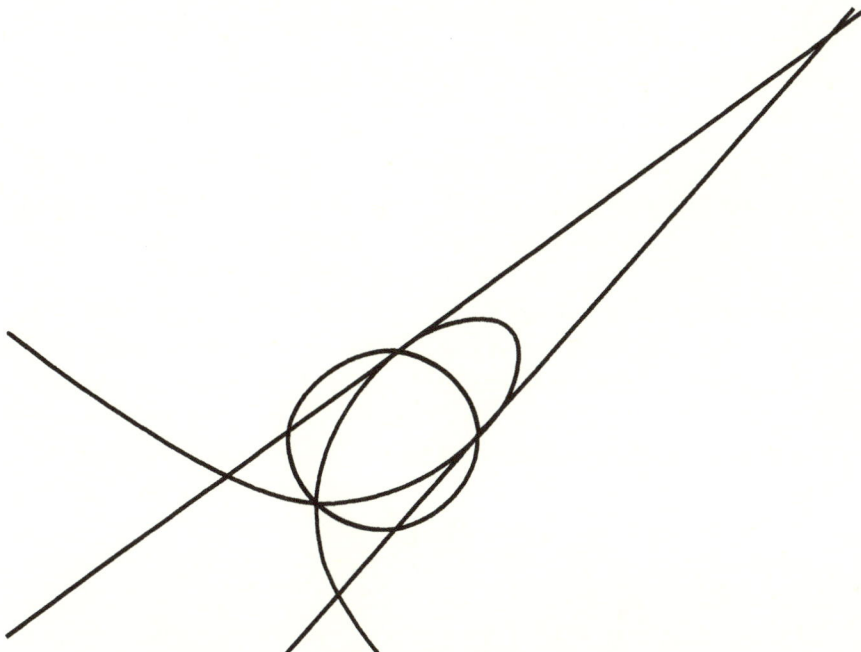

Figure 5.11. A cubic, with a first polar and tangent lines.

purposes we define an inflection of a curve as a simple point of the curve at which the tangent meets the curve in an odd number of coincident points. Since a tangent meets the curve in at least two coincident points, this odd number must be at least three. We can think of an inflection point as a simple point where the curve touches the tangent line, possibly more than once, and then crosses over it.

Letting $\underline{a}$ be a point of the curve f and rewriting the arbitrary point $\underline{b}$ as $\underline{x}$, we see, from the expansion (5.3) at the beginning of the section, that the line determined by $\underline{a}$ and $\underline{x}$ will meet the curve f where

$$\lambda^n f(\underline{a}) + \lambda^{n-1}\mu(\underline{x} \cdot \nabla_{\underline{a}})f(\underline{a}) + \frac{1}{2!}\mu^{n-2}\lambda^2(\underline{x} \cdot \nabla_{\underline{a}})^2 f(\underline{a}) + \cdots$$

$$+ \frac{1}{r!}\mu^{n-r}\lambda^r(\underline{x} \cdot \nabla_{\underline{a}})^r f(\underline{a}) + \cdots + \frac{1}{n!}\mu^n(\underline{x} \cdot \nabla_{\underline{a}})^n f(\underline{a}) = 0.$$

Since $\underline{a}$ is on the curve, we have $f(\underline{a}) = 0$, and we interpret this equation (see Section 5.2) as saying that every line through $\underline{a}$ meets the curve at least once there. The tangent is defined by the equation $(\underline{x} \cdot \nabla_{\underline{a}})f(\underline{a}) = 0$, and this line will meet the curve at least twice at $\underline{a}$. However, if $\underline{a}$ is an inflection, we know that the tangent meets the curve at least three times at $\underline{a}$, and hence the quantity

$(\underline{x} \cdot \nabla_{\underline{a}})^2 f(\underline{a})$ must also be zero. That is, at an inflection point $\underline{a}$ we have

$$(\underline{x} \cdot \nabla_{\underline{a}}) f(\underline{a}) = 0 \quad \text{and} \quad (\underline{x} \cdot \nabla_{\underline{a}})^2 f(\underline{a}) = 0.$$

Since both equations must be satisfied and there are infinitely many zeros of the first equation that will also satisfy the second equation, the Maclaurin–Bézout theorem tells us that $(\underline{x} \cdot \nabla_{\underline{a}}) f(\underline{a})$ must be a factor of $(\underline{x} \cdot \nabla_{\underline{a}})^2 f(\underline{a})$. The curve $(\underline{x} \cdot \nabla_{\underline{a}})^2 f(\underline{a}) = 0$ is the polar conic of $\underline{a}$ with respect to f [since $(\underline{x} \cdot \nabla_{\underline{a}}) f(\underline{a})$ is simply a scalar multiple of $(\underline{a} \cdot \nabla_{\underline{x}})^{n-2} f(\underline{x})$], and hence a simple point P of a curve is an inflection if and only if its polar conic is reducible, the components being the tangent at P and another line. This other line cannot pass through P (see Exercise 5.31).

The polar conic can be written in the form

$$(\underline{x} \cdot \nabla_{\underline{a}})^2 f(\underline{a}) = \underline{x}^T \begin{pmatrix} \dfrac{\partial^2 f}{\partial a_1^2} & \dfrac{\partial^2 f}{\partial a_1 \partial a_2} & \dfrac{\partial^2 f}{\partial a_1 \partial a_3} \\[2ex] \dfrac{\partial^2 f}{\partial a_2 \partial a_1} & \dfrac{\partial^2 f}{\partial a_2^2} & \dfrac{\partial^2 f}{\partial a_2 \partial a_3} \\[2ex] \dfrac{\partial^2 f}{\partial a_3 \partial a_1} & \dfrac{\partial^2 f}{\partial a_3 \partial a_2} & \dfrac{\partial^2 f}{\partial a_3^2} \end{pmatrix} \underline{x}$$

$$= \underline{x}^T H_{\underline{a}} \underline{x} = 0.$$

We know from our discussion of conics that the conic $\underline{x}^T H_{\underline{a}} \underline{x} = 0$ is a line pair if and only if the matrix $H_{\underline{a}}$ is singular.

For a curve $f(x, y, z) = 0$ the equation

$$\det \begin{pmatrix} f_{xx} & f_{xy} & f_{xz} \\ f_{yx} & f_{yy} & f_{yz} \\ f_{zx} & f_{zy} & f_{zz} \end{pmatrix} = 0 \tag{5.5}$$

is called the *Hessian* of the curve f. Clearly the inflection points of a curve lie on the curve and on its Hessian. We restate this as a theorem.

Theorem 5.18 *The inflections of a curve are its simple points that also lie on its Hessian.*

Example 5.19 *Use Theorem 5.18 to determine the inflection points of the curve* $x^3 - yz^2 = 0$.

The Hessian of this curve is

$$H = \begin{pmatrix} 6x & 0 & 0 \\ 0 & 0 & -2z \\ 0 & -2z & -2y \end{pmatrix} = -24xz^2 = 0.$$

The nine intersections of the cubic curve with its Hessian are one at $(0, 0, 1)$ and eight at $(0, 1, 0)$. The point $(0, 1, 0)$ is a cusp of the given curve, and hence $(0, 0, 1)$ is the only inflection point.

Theorem 5.18 gives us an upper bound on the number of inflections of a curve of order n. Since the Hessian is of order $3(n - 2)$, the maximum number of inflections is $3n(n - 2)$. As in the case of the number of tangents to a curve, we can expect the existence of multiple points to affect the number of inflections. It can be shown that if a curve has an ordinary singular point of multiplicity r then it meets its Hessian $3r(r - 2)$ times there and eight times at any cusp of the curve. This agrees with the result of the last example.

Theorem 5.19 *The number i of inflection points of an irreducible curve of order n which has δ ordinary singular points of multiplicities r_j, $j = 1, 2, \ldots, \delta$, respectively, and κ cusps, is given by*

$$i = 3n(n - 2) - \sum_{j=1}^{\delta} 3r_j(r_j - 1) - 8\kappa.$$

In the specific case where the ordinary singular points are nodes, the formulae of Theorems 5.17 and 5.19 reduce to

$$m = n(n - 1) - 2\delta - 3\kappa \quad \text{and} \quad i = 3n(n - 2) - 6\delta - 8\kappa.$$

The dual statement to "a point of a curve at which there are two distinct tangents" is "a line that is tangent to the curve at two distinct points." Such a line is called a *bitangent*. A bitangent is therefore the dual of a node. Just as a node becomes a cusp by the tangents coalescing, so a bitangent becomes an *inflectional tangent* as the points of contact coalesce. Hence, the dual of a cusp is an inflectional tangent, and therefore each of the two equations above has its dual. Writing τ for the number of bitangents, we obtain

$$n = m(m - 1) - 2\tau - 3i \quad \text{as the dual of} \quad m = n(n - 1) - 2\delta - 3\kappa,$$

and

$$\kappa = 3m(m - 2) - 6\tau - 8i \quad \text{as the dual of} \quad i = 3n(n - 2) - 6\delta - 8\kappa.$$

These four equations, only three of which are independent (see Exercise 5.32), are known as *Plücker's equations*.

5.11 Deficiency and Rational Curves

In Section 5.2 we saw that, in general, a plane algebraic curve of order n can be made to pass through $\frac{1}{2}n(n+3)$ points in general position in the plane. This number was referred to as the number of degrees of freedom. In the same section we introduced and defined multiple points. From Definition 5.1 it is a simple matter to count the linear conditions that must be satisfied in order for a point to be of multiplicity r. This number is $\sum_{i=1}^{r} i = \frac{1}{2}r(r+1)$. If a curve has a given number s of prescribed points, the remaining number of degrees of freedom will be less than $\frac{1}{2}n(n+3)$. If the s points are prescribed to have multiplicities $r_i, i = 1, 2, \ldots, s$, then the remaining number of degrees of freedom will be

$$k = \frac{1}{2}n(n+3) - \frac{1}{2}\sum_{i=1}^{s} r_i(r_i + 1).$$

We call this number k the *freedom* of the curve.

We have already seen hints of a connection between multiple points and rationality of a curve. In a sense a curve will be rational if it has its maximum number of double points. It is therefore important to know how many fewer than this maximum possible number is the number of multiple points of a given curve. This difference is called the *deficiency* D of the curve. It is defined by

$$D = \frac{1}{2}(n-1)(n-2) - \frac{1}{2}\sum_{i=1}^{s} r_i(r_i + 1),$$

where the curve has order n and s multiple points of multiplicities $r_i, i = 1, 2, \ldots, s$, respectively.

Theorem 5.20 *The deficiency of an irreducible curve is positive or zero.*

Proof. Using subscripts to denote the order of the curve, let f_n be an irreducible curve of order n with s multiple points of multiplicities $r_i, i = 1, 2, \ldots, s$. Let f_{n-1} be a curve of order $n-1$ that has points of multiplicity $r_i - 1$ at each of the corresponding multiple points of f_n. Such a curve certainly exists, since any first polar of f_n would satisfy the requirements. Then the freedom k of f_{n-1} is given by

$$k = \frac{1}{2}(n-1)(n+2) - \frac{1}{2}\sum_{i=1}^{s}(r_i - 1)r_i$$

$$= \frac{1}{2}(n-1)(n-2) - \frac{1}{2}\sum_{i=1}^{s}(r_i - 1)r_i + 2(n-1)$$

$$= D + 2(n-1).$$

The freedom k is the number of points that still remain to be chosen in order for f_{n-1} to be determined. If we choose these points to lie on f_n, we see immediately that this number cannot exceed the number of intersections f_{n-1} has with f_n, except for those at multiple points of f_n. From Theorem 5.14 we see that the intersection multiplicity of f_{n-1} with f_n at a multiple point is at least $r_i(r_i - 1)$. Therefore, the number of intersections at points other than the multiple points of f_n is at most

$$n(n - 1) - \sum_{i=1}^{s} r_i(r_i - 1) = 2D + 2(n - 1).$$

Hence, the freedom k of f_{n-1} cannot exceed this number. In other words, $D + 2(n - 1) \le 2D + 2(n - 1)$, or $D \ge 0$. ∎

Example 5.20 *Show that an irreducible curve of order n cannot have more than $\frac{1}{2}(n - 1)(n - 2)$ double points.*

From Theorem 5.20 we have

$$\sum_{i=1}^{s} r_i(r_i - 1) \le (n - 1)(n - 2).$$

If all the multiple points are double points, then $r_i = 2$, $i = 1, 2, \ldots, s$. Then

$$\sum_{i=1}^{s} 2 \le (n - 1)(n - 2),$$

which implies that

$$S \le \tfrac{1}{2}(n - 1)(n - 2).$$

We see from this example that, if a curve has only simple and double points, then its deficiency is the difference between the maximum possible number of double points the curve could have and the actual number of such points it has. Theorem 5.5 showed us that an irreducible curve of order n with a point of multiplicity $n - 1$ is rational. The proof of this result used the concept of a pencil of lines through the multiple point and the fact that any member of this pencil could meet the curve in only one variable point (the other points being prescribed). The method of proof actually provided a way of producing the parametrization. The proof of the following theorem is a slight extension of the same ideas.

Theorem 5.21 *If the deficiency of an irreducible curve is zero, then the curve is rational.*

Proof. Let the given curve f_n be of order n, with s multiple points of multiplicities r_i, $i = 1, 2, \ldots, s$. We will consider a curve of order $n - 2$ which has points of multiplicity $r_i - 1$ at the multiple points of f_n. To show that such a curve exists we note that the freedom of a general curve of order $n - 2$ is $\frac{1}{2}(n - 2)(n + 1)$. The number of constraints imposed by specifying s points of multiplicities $r_i - 1$, $i = 1, 2, \ldots, s$, is $\frac{1}{2}\sum_{i=1}^{s} r_i(r_i - 1)$. This number must be less than or equal to $\frac{1}{2}(n - 2)(n + 1)$, or no such curve can exist. However, we know that

$$\frac{1}{2}\sum_{i=1}^{s} r_i(r_i - 1) \leq \frac{1}{2}(n - 1)(n - 2)$$

$$= \frac{1}{2}(n - 2)(n + 1) - n + 2$$

$$\leq \frac{1}{2}(n - 2)(n + 1) \qquad \text{for} \quad n \geq 2.$$

Lines are rational and have deficiency zero. Hence, there is no loss of generality in assuming that $n \geq 2$, in which case the required curve exists. Its freedom is

$$\frac{1}{2}(n - 2)(n + 1) - \frac{1}{2}\sum_{i=1}^{s} r_i(r_i - 1) = D + n - 2.$$

Hence, for $D = 0$ the freedom is $n - 2$.

Select $n - 3$ points on f_n. There is a pencil of curves of order $n - 2$ that have points of multiplicity $r_i - 1$, $i = 1, 2, \ldots, s$, at the s multiple points of f_n, and simple points at the $n - 3$ additional prescribed points of f_n. This pencil can be written in the form $\alpha g + \beta h = 0$, where g and h are homogeneous polynomials of degree $n - 2$, and α and β are the homogeneous parameters of the pencil. Any member of this pencil meets f_n at all the prescribed points (by construction) and precisely one other point. This variable point is therefore in rational correspondence to the ratio α/β (or β/α), and hence the curve f_n is a rational curve. ∎

Example 5.21

(a) *Show that the curve*

$$\sum_{\substack{i,j \geq 0 \\ i+j=3}}^{4} a_{ij} x^i y^j = 0$$

    is rational.

(b) *Produce a parametrization of the curve* $x^4 + xy^3 - y^4 - x^2 y = 0$ *using* (i) *the technique of Theorem 5.5,* (ii) *the technique of Theorem 5.21.*

(a): The given curve is a quartic with a triple point at the origin. By Theorem 5.5 the curve must be rational. The deficiency is given by

$$D = \frac{1}{2}(3)(2) - \frac{1}{2}(3)(2) = 0.$$

Hence, Theorem 5.21 also indicates that the curve is rational.

(b)(i): Using the technique of Theorem 5.5, the pencil of lines through the origin can be written as $y - mx = 0$. This pencil meets the given curve three times at the point $(0, 0)$ and once at the point given by

$$x = \frac{m}{m^4 - m^3 - 1}, \qquad y = \frac{m^2}{m^4 - m^3 - 1}. \tag{5.6}$$

(ii): Following Theorem 5.21, we look for a pencil of conics that have a double point at the origin and pass through some other prescribed point of the quartic. Let us take this last point to be $(1, 1)$. Then the pencil of conics can be written in the form $(y - mx)(y - x) = 0$. This pencil of conics meets the quartic six times at the origin (see Theorem 5.14), once at $(1, 1)$, and once at the point given, as in (i) above, by (5.6).

Example 5.22 *Show that the pencil of quartics given by*

$$\alpha f(x, y, z) + \beta g(x, y, z) = 0,$$

where

$$f(x, y, z) = (x - y + z)(x + y + z)(x^2 + y^2 - z^2),$$
$$g(x, y, z) = (x^2 + 2y^2 - xz - 2z^2)(x^2 + 2y^2 - 2z^2 + xy - xz),$$

is rational. Select a value of the ratio α/β such that the resulting curve passes through the point $(3, 0, 1)$. Produce a parametrization of the resulting curve.

Sketching the two representative members of the pencil given by $\alpha = 0$ and $\beta = 0$ indicates that all the members of the pencil have double points at $(-1, 0, 1)$, $(0, 1, 1)$, and $(0, -1, 1)$. Example 5.20 shows us that there can be no more double points, and a simple application of the Maclaurin–Bézout theorem precludes the possibility of any additional points of greater multiplicity. We have therefore found all the multiple points of the pencil. The deficiency is given by

$$D = \frac{1}{2}(3)(2) - \frac{1}{2}\sum_{i=1}^{3}(2)(1) = 0.$$

Hence, by Theorem 5.21, the pencil is rational.

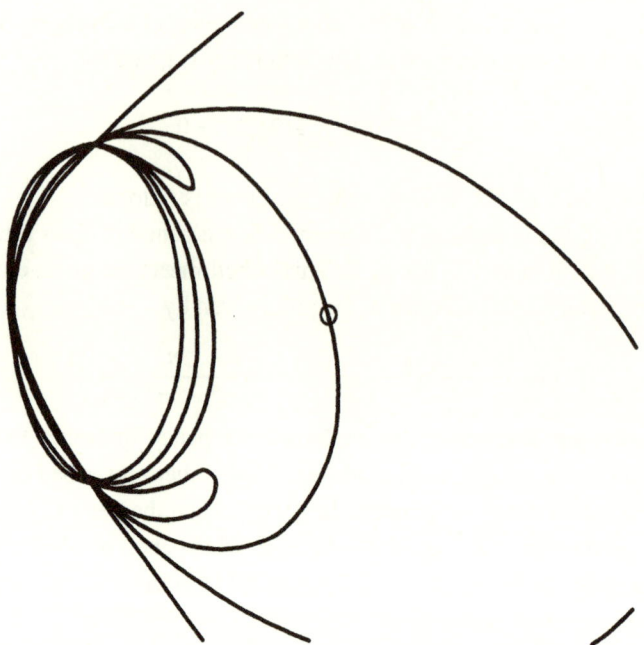

Figure 5.12. A pencil of quartics with three double points.

Substituting the coordinates of the additional point in the equation of the pencil shows that $\alpha = 1$ and $\beta = -8$ are suitable values to force the corresponding member of the pencil to pass through the point $(3, 0, 1)$. This quartic, together with three other members of the pencil corresponding to the parameter pairs (α, β) equal to $(1, -2)$, $(1, 24)$, and $(2, -1)$, respectively, are shown in Figure 5.12.

Theorem 5.5 is of little use to us in producing the parametrization. Following the technique used in the proof of Theorem 5.21, we seek a pencil of conics through $(-1, 0, 1)$, $(0, 1, 1)$, and $(0, -1, 1)$ and one other point of the given quartic. Taking this last point to be $(3, 0, 1)$, a suitable pencil of conics is

$$s(x^2 + 3y^2 - 3z^2 - 2xz) + txy = 0.$$

This pencil will have eight intersections with the quartic. There will, by construction, be two intersections at each of the double points of the quartic (Theorem 5.14) and one intersection (again by construction) at $(3, 0, 1)$. Since seven of the eight intersections are known and prescribed, the single remaining intersection must be in rational correspondence with the parameters of the pencil. Eliminating y between the two equations and ignoring the factors corresponding to the prescribed intersections gives $A(s, t)x - B(s, t)z = 0$,

where

$$A(s, t) = 64s^4 - 160s^3 t + 228s^2 t^2 - 200st^3 + 77t^4,$$

$$B(s, t) = -64s^4 + 160s^3 t + 156s^2 t^2 - 352st^3 + 121t^4.$$

Similarly, elimination of x leads to $A(s, t)y - C(s, t)z = 0$, where $A(s, t)$ is as given, and

$$C(s, t) = 12st(16s^2 - 32st + 15t^2).$$

Therefore, a suitable parametrization is

$$x = -64s^4 + 160s^3 t + 156s^2 t^2 - 352st^3 + 121t^4,$$

$$y = 12st(16s^2 - 32st + 15t^2),$$

$$z = 64s^4 - 160s^3 t + 228s^2 t^2 - 200st^3 + 77t^4.$$

In the proof of Theorem 5.21 we only needed to show that the points of f_n were in rational correspondence to the ratio α/β (or β/α). The proof, however, also shows that the members of the pencil $\alpha g + \beta h = 0$ are in rational correspondence to the points of f_n, for to a variable point of f_n there corresponds a unique member of the pencil and hence a unique ratio α/β (or β/α). A curve of deficiency zero is in birational correspondence with some pencil of curves. Any pencil of curves can be placed in birational correspondence with a line, and hence a curve of deficiency zero is in birational correspondence with a line. Working directly with parametrizations of a rational curve and a line, it is an even simpler matter to show that a rational curve is birationally equivalent to a line. We state the theorem without proof.

Theorem 5.22 *A rational curve is birationally equivalent to a line.*

Example 5.23 *Show, by computation of the transformations, that the curve* $x^3 + y^3 - xyz = 0$ *is birationally equivalent to the line* $x + y - z = 0.$

The cubic has a double point at $(0, 0, 1)$ and hence is rational. Taking the pencil of lines $\alpha x - \beta y = 0$ through the double point, we obtain a parametrization of the cubic in the form

$$x = \alpha\beta^2, \qquad y = \alpha^2\beta, \qquad z = \alpha^3 + \beta^3.$$

The line $x + y - z = 0$ is determined by any two of its points, for example, the points $(1, 0, 1)$ and $(0, 1, 1)$, and can be thought of as the pencil of points given by $\underline{x} = \mu(1, 0, 1) + \nu(0, 1, 1)$, that is,

$$x = \mu, \qquad y = \nu, \qquad z = \mu + \nu.$$

We set up a correspondence between the two pencils by introducing the relation $\alpha/\beta = \mu/\nu$. A point (x_L, y_L, z_L) on the line $x + y - z = 0$ is associated with a unique ratio μ/ν, namely

$$\frac{\mu}{\nu} = \frac{x_L}{y_L}.$$

This is associated with a point on the cubic by setting

$$\frac{\alpha}{\beta} = \frac{\mu}{\nu} = \frac{x_L}{y_L}$$

to give the rational correspondence

$$x = x_L y_L^2, \qquad y = x_L^2 y_L, \qquad z = x_L^3 + y_L^3$$

from the line to the cubic.

Let (x_c, y_c, z_c) be a point on the cubic. The member of the pencil $\alpha x - \beta y = 0$ that passes through this point satisfies

$$\alpha x_c - \beta y_c = 0, \qquad \text{or} \qquad \frac{\alpha}{\beta} = \frac{y_c}{x_c}.$$

To obtain the correspondence between the cubic and the line we can identify

$$\frac{\mu}{\nu} = \frac{\alpha}{\beta} = \frac{y_c}{x_c}$$

to obtain

$$x = y_c, \qquad y = x_c, \qquad z = y_c + x_c$$

as a simple correspondence between the cubic and the line. Hence, the two curves are in birational correspondence.

Now let us consider the rational curve

$$x = s^2 t^2, \qquad y = t^4, \qquad z = s^4 - st^3.$$

Elimination of the ratio s/t yields the quartic $(x^2 - yz)^2 - xy^3 = 0$. This quartic has a single double point at the origin (see Exercise 5.35) and no other multiple points. Therefore, its deficiency is two. We have, by example, produced a rational curve whose deficiency is greater than zero. Clearly, Theorem 5.21 is only a partial answer to the question of when a curve is rational.

5.12 Rational Transformations

The tools we have developed so far are not sophisticated enough to draw any distinction between the cusp of the quartic $(x^2 - yz)^2 - xy^3 = 0$ and the cusp of, for example, the cubic $x^3 + y^3 + y^2z = 0$. We will use transformations to discover the difference. In Chapters 3 and 4 we discussed projective transformations. These transformations were linear and could be represented as matrices. One interpretation of such a transformation was to consider it as defining a linear mapping of the plane onto itself. Thus a point with coordinates (x, y, z) is mapped to a point with coordinates (x', y', z'), where

$$\rho x' = \phi_1(x, y, z), \qquad \rho y' = \phi_2(x, y, z), \qquad \rho z' = \phi_3(x, y, z),$$

and the functions ϕ_1, ϕ_2, and ϕ_3 are linear in x, y, and z. We know from Theorem 5.6 that the multiplicity of a point of an algebraic curve is a projective invariant. We now find it necessary to relax the condition that the functions ϕ_1, ϕ_2, and ϕ_3 are linear.

Let ϕ_1, ϕ_2, and ϕ_3 be linearly independent homogeneous polynomials of degree n. Then the equation

$$\alpha\phi_1 + \beta\phi_2 + \gamma\phi_3 = 0$$

represents a net of curves (see Section 2.10) of order n. Call this net Φ. Associated with this net of curves there is the net L of lines given by the equation

$$\alpha x' + \beta y' + \gamma z' = 0.$$

As a point $\underline{x}$ traces out a member of the net Φ, its transform $\underline{x}'$ traces out the corresponding member of the net of lines L. To a general point in the plane there correspond members of both nets that pass through the chosen point. Therefore the transformation is a rational mapping of the plane onto itself.

There may be a finite number of exceptional points. If ϕ_1, ϕ_2, and ϕ_3 share common points, then the transformation would yield $\underline{x}' = \underline{0}$, which is a nonexistent point of the complex projective plane. We call such points *base points* of the net. Since the only triple of numbers that does not correspond to a point of the complex projective plane is the triple $(0, 0, 0)$, we see that the base points are the only points at which the transformation is not defined.

5.12.1 Cremona Transformations

Let Φ_1 and Φ_2 be any two members (curves) of the net Φ. These two curves meet in n^2 points, and among these points must certainly be counted all the base points of the net Φ. Let S denote the subset of these n^2 points we obtain by exclusion of the base points. The number of points in S is called the *grade* N of the net. The two members Φ_1 and Φ_2 form a pencil $\mu\Phi_1 + \nu\Phi_2 = 0$, and to this corresponds the pencil of lines

$$\mu L_1 + \nu L_2 = 0, \tag{5.7}$$

where, if

$$\Phi_1 = \alpha_1\phi_1 + \beta_1\phi_2 + \gamma_1\phi_3 \quad \text{and} \quad \Phi_2 = \alpha_2\phi_1 + \beta_2\phi_2 + \gamma_2\phi_3,$$

then

$$L_1 = \alpha_1 x' + \beta_1 y' + \gamma_1 z' \quad \text{and} \quad L_2 = \alpha_2 x' + \beta_2 y' + \gamma_2 z'.$$

This latter pencil (5.7), being a pencil of lines, represents all lines through some point Q. This point is defined by $L_1 = 0$ and $L_2 = 0$ and hence is the image of any of the N points of S under the rational transformation. Conversely, given a point Q, we can associate with it a pencil of lines through Q and from this a pencil of curves of the net. In this way, to each point of the plane, except for the base points of the net (a finite number), there correspond N other points, each of which is an inverse image of the given point. Algebraically, given a point (x', y', z'), we obtain the inverse image by solving the equations

$$x'\phi_3(x, y, z) = z'\phi_1(x, y, z), \qquad y'\phi_3(x, y, z) = z'\phi_2(x, y, z).$$

These two equations represent nth-order curves in the plane and hence will have n^2 common points. These n^2 solutions include the base points of the net, but we discount these because they do not have a valid transform. There are therefore precisely N valid solutions to this system. Of course, the inverse transformation is unique if and only if $N = 1$. In this case it is obvious that, since we know $n^2 - 1$ of the solutions to this system of two equations regardless of the values of (x', y', z'), the remaining solution is the only one that depends on (x', y', z') and must be a rational function of these values. That is, when $N = 1$, the rational transformation of the plane onto itself given by

$$\rho x' = \phi_1(x, y, z), \qquad \rho y' = \phi_2(x, y, z), \qquad \rho z' = \phi_3(x, y, z)$$

is one-to-one (base points excluded) and has a rational inverse of the form

$$\rho x = \psi_1(x', y', z'), \qquad \rho y = \psi_2(x', y', z'), \qquad \rho z = \psi_3(x', y', z').$$

Such a transformation is called a *Cremona transformation*.

Notice the subtle difference between the requirement of grade one and the requirement that the separate curves of the net each share $n^2 - 1$ points. Certainly the latter condition implies the former, but the converse is not true, as the following example shows.

Example 5.24 *Produce a cubic Cremona transformation and its inverse.*

Since we need a net of cubics of grade one, we must find three cubics, each pair of which has eight contact points in common. If we consider cubics through

$3^2 - 1$ points, then there can be only a pencil of cubics, not a net. There do not exist three linearly independent cubics each sharing the same eight points. By choosing cubics sharing a double point and four other simple points, we will prescribe the equivalent of seven linear constraints and hence have enough freedom left to define a net.

A net of cubics through seven simple points would have grade two, but a net based on a double point and four simple points has grade one. Let us construct three cubics, each of which has a double point at the origin. Such cubics must be of the form

$$ax^3 + bx^2y + cxy^2 + dy^3 + ex^2 + fxy + gy^2 = 0.$$

Since two curves with a point of multiplicity two at a common point must have at least a fourfold intersection, there we need only force our cubics through four (rather than seven) additional points. Choose, for example, the points $(0, 1)$, $(1, 0)$, $(1, 1)$, and $(1, -1)$. Imposing these constraints leads to the equation

$$ax^2(x - 1) + by(x^2 - xy - y^2 + y) - fy(y + x)(y - 1) = 0,$$

as the equation representing cubics with a double point at the origin and passing through the four additional points. Representative members of this net are

$$
\begin{array}{lll}
a = b = 0, \quad f = 1: & y(y + x)(y - 1) & = 0, \\
a = f = 1, \quad b = 0: & x^3 - xy^2 - y^3 - x^2 + xy + y^2 = 0, \\
a = f = 0, \quad b = 1: & y(x^2 - xy - y^2 + y) & = 0
\end{array}
$$

Hence, the homogeneous transformation

$$\rho x' = y(y + x)(y - z),$$
$$\rho y' = x^3 - xy^2 - y^3 - x^2z + xyz + y^2z,$$
$$\rho z' = y(y^2 + xy - x^2 - yz),$$

must be a Cremona transformation.

The inverse is found by solving the equations for x, y, and z, finally obtaining

$$\rho x = (x' + y')(x'^2 - 3x'z' - y'z' + z'^2),$$
$$\rho y = (x' - z')(x'^2 - 3x'z' - y'z' + z'^2),$$
$$\rho z = x'^3 - x'^2y' - x'y'^2 - 5x'^2z' - 2x'y'z' + 4x'z'^2 + y'z'^2 - z'^3.$$

We notice that the order of the inverse transformation is the same as that of the original transformation. This is no coincidence; for, noting that we cannot create nor destroy ordinary intersections of two curves with a Cremona transformation, we have the following argument. Consider, in the (x, y, z) plane, a line and some member of the net of curves that specifies the transformation. The order of the transformation is n, which is the number of the intersections of the line with a member of the net. Now apply the transformation. The line maps to a member of the net corresponding to the inverse transformation, and, as we have already seen, any member of the original net maps to a line. Since the number of intersections must remain unchanged, there must be n intersections of this line with a member of the net that specifies the inverse transformation. Hence the inverse transformation must also be of order n.

5.12.2 Quadratic Transformations

The Cremona transformation of most interest to us is the *quadratic transformation*. We have, in fact, already seen a quadratic transformation, for in Example 4.7 the points of a circle were placed in rational correspondence to those of the projective line. Extending this idea to spheres, we get the stereographic map projections, which were probably the earliest significant application of quadratic transformations.

In the quadratic case we seek a net of conics of grade one. Equivalently, this is the net of conics through three points. Without loss of generality we may take these points to be the vertices of the triangle of reference, in which case the net Φ is given by

$$\Phi \equiv \alpha yz + \beta zx + \gamma xy = 0.$$

Any set of three linearly independent members of this net could be used as the ϕ_1, ϕ_2, and ϕ_3 to define the quadratic transformation. In the cause of simplicity and symmetry let us choose

$$\phi_1(x, y, z) = yz, \qquad \phi_2(x, y, z) = zx, \qquad \phi_3(x, y, z) = xy;$$

hence, the transformation is

$$\rho x' = yz, \qquad \rho y' = zx, \qquad \rho z' = xy.$$

A simple calculation will show that this transformation is its own inverse, that is,

$$\rho x = y'z', \qquad \rho y = z'x', \qquad \rho z = x'y'.$$

This particular Cremona transformation is called the *standard quadratic transformation*.

5.12.3 Proper Transforms and Fundamental Lines

Given a curve $f(x, y, z) = 0$, we can associate with it a curve $g(x', y', z') = 0$, where we have used the standard quadratic transformation to map from (x, y, z)

coordinates to (x', y', z') coordinates. Denote this transformation by T. We shall see in what follows that even when f is irreducible, its transform may (and often does) have factors corresponding to lines through pairs of base points. Ignoring such factors, we refer to the remaining curve as the *proper transform* of f, denoted by $T(f)$.

Any point on a line joining any two base points has a zero in the coordinate position corresponding to the third base point. For example, any point on the line determined by the base points $(1, 0, 0)$ and $(0, 0, 1)$ has coordinates $(a, 0, b)$. Substitution into the transformation shows immediately that all such points are mapped onto the third base point. In our example all points of the form $(a, 0, b)$ are mapped to the point $(0, 1, 0)$ [except the base points themselves, which have no transform, since the point $(0, 0, 0)$ does not exist in the complex projective plane]. Thus each line of the triangle formed by the three base points maps onto the base point not contained in that line. These lines are called the *fundamental lines* of the transformation. It is evident that any quadratic Cremona transformation will have three such lines. In our situation, with the particular choice of coordinate system we have made, these lines have equations $x = 0$, $y = 0$, and $z = 0$, respectively. Let us look at an arbitrary line l through a base point. Such a line has equation

$$\alpha x + \beta y + \gamma z = 0,$$

where *precisely one* of α, β, and γ is zero. For example, an arbitrary line through $(1, 0, 0)$ has equation

$$\beta y + \gamma z = 0.$$

This line is transformed to

$$\beta z' x' + \gamma x' y' = x'(\beta z' + \gamma y') = 0.$$

Ignoring the factor corresponding to the fundamental line $x' = 0$, the proper transform of the line $\beta y + \gamma z = 0$ is the line given by

$$\beta z' + \gamma y' = 0,$$

which is another line through the corresponding base point. Thus, ignoring factors corresponding to fundamental lines, we see that a line, other than a fundamental line, through a base point is mapped to a line through the same base point in the transformed coordinate plane. Consider a point P and its transform $Q = T(P)$. Through Q we can construct a pencil of lines $\alpha L_1 + \beta L_2 = 0$, and to this there corresponds a pencil of conics given by $\alpha \Phi_1 + \beta \Phi_2 = 0$. Hence, the constructed Φ_1 and Φ_2 are zero at the point P, which is the inverse transform of Q. However, Φ_1 and Φ_2 are also zero at the three base points. Hence both

Φ_1 and Φ_2 are zero at four points given by P and the original base points of the ϕ_1, ϕ_2, and ϕ_3 that defined the transformation [i.e., the points $(1, 0, 0)$, $(0, 1, 0)$, and $(0, 0, 1)$, with our choice of coordinates]. Let us recall the interpretation of a tangent to a curve, which we discussed in Section 2.9, as the limiting case of a chord as the two points defining the chord coalesce. Let us label the three points $(1, 0, 0)$, $(0, 1, 0)$, and $(0, 0, 1)$ as X, Y, and Z, respectively. Our pencil of conics given by $\alpha\Phi_1 + \beta\Phi_2 = 0$ then passes through the four points X, Y, Z, and P. Let P move towards one of the base points, X, say, along a line l. As P coalesces with X, the pencil of conics $\alpha\Phi_1 + \beta\Phi_2 = 0$ becomes the pencil through X, Y, and Z, and tangent to l at the point X. Although the transform of a base point does not exist, the transform of the pencil $\alpha\Phi_1 + \beta\Phi_2 = 0$ does exist as P coalesces with X, and the corresponding point Q will, in the limit, be the vertex of the pencil of lines which correspond to the limiting case of the pencil of conics through X, Y, and Z and tangent to l at the point X. Where is this point?

Assume that the line l has the equation $\beta y + \gamma z = 0$, with β/γ fixed. It must be of this form, since it passes through $(1, 0, 0)$, and the ratio β/γ is its slope. The pencil of conics through the three base points and tangential to this line at X is given by

$$\delta yz + \gamma zx + \beta xy = 0.$$

This is a pencil and not a net, for we have fixed the ratio β/γ. The image of this pencil of conics is the pencil of lines given by

$$\delta x' + \gamma y' + \beta z' = 0,$$

which is the pencil of lines through the point given by

$$\delta x' = 0 \quad \text{and} \quad \gamma y' + \beta z' = 0,$$

that is, through the point $(0, \beta, -\gamma)$. Thus, as P tends to X along a line given by $\beta y + \gamma z = 0$, its transform Q goes to the point on the line $Y'Z'$ whose coordinates are $(0, \beta, -\gamma)$. Therefore, to each direction at a base point there corresponds a point on the fundamental line that does not contain the base point in question. Consequently, if a curve passes through a base point and the tangent to the curve at that base point is not a fundamental line, then the proper transform of the curve passes through the corresponding point on the fundamental line not containing the base point in question. With reference to our specific example, if a curve passes through the base point $(1, 0, 0)$ and the tangent there is $\beta y + \gamma z = 0$, $\beta\gamma \neq 0$, then the proper transform of the curve passes through the point $(0, \beta, -\gamma)$ on the fundamental line $x' = 0$. Since the transform under consideration is its own inverse, the converse must also be true.

To recapitulate, let us list the properties of the standard quadratic transformation that we have deduced so far.

1. Since all Cremona transformations are one-to-one (apart from base points), the number of intersections of two curves, away from base points, is an invariant.
2. Any point on a fundamental line, apart from a base point, is mapped to the base point not on that line.
3. Any line through a base point, apart from a fundamental line, is mapped to another line through the same base point.
4. To each direction at a base point there corresponds a point on the fundamental line that does not pass through the base point.

We use the notation $\{n; i, j, k\}$ to indicate the set of curves of order n with the base points X, Y, and Z as points of multiplicities i, j, and k, respectively, and where no fundamental line is tangent to a curve at a base point.

By $N(fg)$ we denote the number of points, other than base points, that are common to f and g.

Let $f \in \{n; i, j, k\}$, and let $g = T(f)$ be the proper transform of f. Then $g \in \{n'; i', j', k'\}$ for some n', i', j', and k'. From the Maclaurin–Bézout theorem, n' is the number of times any line meets g. Let L be a line not through any base point. Then

$$
\begin{aligned}
n' &= N(Lg) \\
&= N(T(L)T(g)) \quad \text{(from the property 1)} \\
&= N(\Phi f) \\
&= 2n - i - j - k \quad \text{(by Theorem 5.14).}
\end{aligned}
$$

Likewise, let L_X be a line through the base point X. Then

$$
\begin{aligned}
i' &= n' - N(L_X g) \\
&= n' - N(T(L_X)T(g)) \\
&= n' - N(L'_X f), \\
&= 2n - i - j - k - (n - i) \\
&= n - j - k,
\end{aligned}
$$

where L'_X is some other line through X (see property 3). Similarly,

$$
j' = n - k - i \quad \text{and} \quad k' = n - i - j.
$$

We encapsulate these last deductions in the following theorem.

Theorem 5.23 *If $f \in \{n; i, j, k\}$, and g is the proper transform of f, then*

$$g \in \{2n - i - j - k; n - j - k, n - k - i, n - i - j\}.$$

Example 5.25 *Show* (a) *synthetically,* (b) *algebraically that the proper transform of a conic inscribed in the triangle of reference is a quartic with cusps at the base points. Show, also, that the converse statement is true.*

(a): The conic having, as it does in this case, double contact with the fundamental lines at the points of tangency cannot, being only a second-order curve, have any further intersections with the fundamental lines. Therefore, the conic cannot pass through a base point, and hence its proper transform is indeed a quartic. By the property 4 we see that points on the fundamental line correspond to tangent directions at a base point, and, since the conic meets each fundamental line at two coincident points, the proper transform must have two coincident tangents at each base point, that is, a cusp at each base point. Conversely, assume that a quartic has a cusp at each of the three base points. Then, once again from property 4, since tangents at a base point correspond to points on a fundamental line, a cusp, having two coincident tangents, must correspond to two coincident points on a fundamental line. Hence, the proper transform is a conic that has double contact with each of the fundamental lines and must, therefore, be inscribed in the triangle of reference.

(b): A conic inscribed in the triangle of reference has an equation of the form

$$a^2 x^2 + b^2 y^2 + c^2 z^2 + 2abxy + 2bcyz + 2cazx = 0.$$

The proper transform is

$$a^2 y'^2 z'^2 + b^2 z'^2 x'^2 + c^2 x'^2 y'^2 + 2x'y'z'(abz' + bcx' + cay') = 0,$$

which is a quartic. The lowest-order terms, after putting $z' = 1$, are

$$(ay' + bx')^2,$$

which indicates the existence of a double point at $(0, 0, 1)$ with coincident tangents given by $ay' + bx' = 0$, that is, the quartic has a cusp at $(0, 0, 1)$. Similarly, there are cusps at $(0, 1, 0)$ and $(1, 0, 0)$. Putting $x' = yz$, $y' = zx$, and $z' = xy$ in the quartic, we get

$$x^2 y^2 z^2 (a^2 x^2 + b^2 y^2 + c^2 z^2 + 2abxy + 2bcyz + 2cazx) = 0,$$

and hence the proper transform is

$$a^2 x^2 + b^2 y^2 + c^2 z^2 + 2abxy + 2bcyz + 2cazx = 0.$$

5.13 Resolution of Singularities

We now use these quadratic transformations to perform a more sophisticated analysis of singular points, an analysis that will enable us to resolve differences between singular points that hitherto have seemed identical.

We will resolve the singularities of a curve f of order n one at a time, by choosing a coordinate system where the singular point P in question is at one of the base points of the standard quadratic transformation, where the other two base points do not lie on the curve, where the fundamental lines through the singularity in question meet the curve elsewhere only in simple points, and where the third fundamental line meets the curve in n distinct points. It is always possible to construct such a coordinate system. Hence, if the singular point under consideration is one of multiplicity s, then $f \in \{n; s, 0, 0\}$, where we have placed the singular point at X.

Hence, by Theorem 5.23, the proper transform $g \in \{2n - s; n, n - s, n - s\}$. The choice of our coordinate system also ensures that the base points are ordinary multiple points of g. The multiple point of f that is under consideration [which was placed at the point $X = (1, 0, 0)$ by our choice of coordinate system] now corresponds to s points of g (none of which are base points) on the fundamental line YZ. Only if all of these points are simple points of g do we say that we have *resolved* the corresponding singularity of f. If this is not the case, our resolution of the original singularity is not yet complete. These points of g are said to constitute the *first neighborhood* of P, the originating s-fold point of f.

If we repeated the process of selecting a coordinate system and applying the standard quadratic transformation, choosing to resolve some t-fold point Q of the first neighborhood of P, we would produce points that would constitute the first neighborhood of Q. The set of first neighborhoods of all the points in the first neighborhood of P constitutes the *second neighborhood* of P. Continuing in this fashion, we can construct third, fourth, ... , mth neighborhoods of the originating point P. It is clear that the sum of the multiplicities of the points in the first neighborhood of an s-fold point cannot exceed s.

Example 5.26 *Examine the first neighborhoods of the double points of the curves*

(a) $x^3 + y^3 + y^2z = 0$,
(b) $(x^2 - yz)^2 - xy^3 = 0$.

(a): With the given coordinate system the double point is at $(0, 0, 1)$, with coincident tangents given by $y = 0$, which is one of the fundamental lines.

Therefore, we must change the coordinate system before performing the quadratic transformation. Although, in our discussions, we placed the multiple point under consideration at $(1, 0, 0)$, one fundamental point is as easy to deal with as the next, and hence we will leave the double point at $(0, 0, 1)$ but perform a simple rotation so that the tangent at the double point is no longer a fundamental line. This can be accomplished by changing y to $x + y$. The curve now becomes

$$f \equiv x^3 + (x + y)^3 + (x + y)^2 z = 0 \qquad \text{with} \quad f \in \{3; 0, 0, 2\}.$$

Performing the standard quadratic transformation and dropping the primes, we obtain the proper transform

$$g \equiv x^3 y + 2x^2 y^2 + xy^3 + x^3 z + 3x^2 yz + 3xy^2 z + 2y^3 z = 0.$$

As expected (from Theorem 5.23), $g \in \{4; 1, 1, 3\}$. The line $z = 0$ meets g at the base points and twice at $(1, -1, 0)$. However, this point is a simple point of g. Hence, the first neighborhood of the simple cusp of f comprises a single simple point.

(b): In this case we again have a double point at $(0, 0, 1)$ with coincident tangents given by $y = 0$. Also, the base point $(0, 1, 0)$ lies on the curve. We must determine a coordinate system that results in the double point in question being the only fundamental point of the curve and the tangent there being other than a fundamental line. The transformation

$$x = \tilde{x}, \qquad y = \tilde{x} + \tilde{y}, \qquad z = \tilde{x} + \tilde{y} + \tilde{z}$$

satisfies these purposes. For simplicity, we will always drop the tildes when we write the transformed equation. The curve then becomes

$$\begin{aligned} f \equiv\ & x^4 + 3x^3 y - x^2 y^2 - 3xy^3 - y^4 \\ & - 4x^2 yz - 6xy^2 z - 2y^3 z - x^2 z^2 - 2xyz^2 - y^2 z^2 = 0, \end{aligned}$$

with $f \in \{4; 0, 0, 2\}$. The proper transform g is given by

$$\begin{aligned} g \equiv\ & x^4 y^2 + 2x^3 y^3 + x^2 y^4 + 2x^4 yz + 6x^3 y^2 z \\ & + 4x^2 y^3 z + x^4 z^2 + 3x^3 yz^2 + x^2 y^2 z^2 - 3xy^3 z^2 - y^4 z^2 = 0, \end{aligned}$$

with $g \in \{6; 2, 2, 4\}$. In this case g also has double contact with $z = 0$ at the point $(1, -1, 0)$. However, and unlike the situation in (a), this point is a double point of g.

To determine more about this double point we make another change of coordinates,

$$x = \tilde{x}, \qquad y = \tilde{y} - \tilde{x}, \qquad z = \tilde{z}.$$

This coordinate change moves the double point to $(1, 0, 0)$. Putting $x = 1$, we get the corresponding affine equation

$$y^4 z^2 - y^3 z^2 - 4y^2 z^2 - 4y^3 z - y^4$$
$$+ 4yz^2 + 6y^2 z + 2y^3 - (y + z)^2 = 0. \tag{5.8}$$

The lowest-order terms are (as we would expect when the origin of the affine coordinate system is a double point) of second order. These terms give us the tangents at the double point. In this case, the tangents are coincident, showing us that the double point is a cusp. Hence the original curve has a cusp, and another cusp in its first neighborhood.

Although the curves in (a) and (b) both have cusps, the resolution of the cusps produced only simple points in the case of the first curve but produced another cusp in the second case.

In considering the curves of Example 5.26 from the viewpoint of rationality, we will show that the cubic curve is of deficiency zero whereas the quartic has deficiency two. It is instructive to ask what the deficiencies of the transformed curves are. Recall (Section 5.11) that the deficiency D is given by

$$D = \frac{1}{2}(n - 1)(n - 2) - \frac{1}{2}\sum_i r_i(r_i - 1),$$

where the curve is of order n with points of multiplicity r_i.

The cubic in question, namely $x^3 + y^3 + y^2 z = 0$, belongs to $\{3; 0, 0, 2\}$, with, of course, no other multiple points than the one already accounted for in this classification. For such a curve

$$D = \tfrac{1}{2}(2)(1) - \tfrac{1}{2}(2)(1) = 0.$$

The transformed curve $g \in \{4; 1, 1, 3\}$ (see Example 5.26 above) with, of course, no other multiple points. For this curve

$$D = \tfrac{1}{2}(3)(2) - \tfrac{1}{2}(3)(2) = 0.$$

In this case the resolution of the singularity produced a curve whose deficiency was unchanged.

For the quartic of Example 5.26(b) the original curve belonged to $\{4; 0, 0, 2\}$ and hence the deficiency is

$$D = \tfrac{1}{2}(3)(2) - \tfrac{1}{2}(2)(1) = 2.$$

In this case the transformed curve belonged to $\{6; 2, 2, 4\}$ with, in addition to the multiple points at the base points, a double point (a cusp) at $(1, -1, 0)$. In this case the deficiency is given by

$$D = \tfrac{1}{2}(5)(4) - \tfrac{1}{2}[(2)(1) + (2)(1) + (2)(1) + (4)(3)] = 1.$$

In this case the deficiency has been reduced by the application of the standard quadratic transformation. Some insight into what happens in general is embodied in the proof of the following fundamental theorem, which is often referred to as Noether's theorem.

Theorem 5.24 *(Noether's Theorem) Any irreducible algebraic curve in a plane can be transformed, by a finite succession of standard quadratic transformations, into a curve whose multiple points are all ordinary.*

Proof. Let f be an irreducible plane algebraic curve of order n with a nonordinary multiple point P of multiplicity k. Let f have s additional multiple points of multiplicities r_i, $i = 1, 2, \ldots, s$. Then f has deficiency D_f given by

$$D_f = \frac{1}{2}(n-1)(n-2) - \frac{1}{2}\sum_{i=1}^{s} r_i(r_i - 1) - \frac{1}{2}k(k-1).$$

Let us consider the application of the standard quadratic transformation to the resolution of the singularity at P. Therefore, we must choose a coordinate system such that the point P is at one of the base points, that no other base point lies on the curve, and that no fundamental line is a tangent to the curve at P. Let us choose the coordinate system so that P has coordinates $(1, 0, 0)$. Then $f \in \{n; k, 0, 0\}$. Let g be the proper transform of f. Then, by Theorem 5.23, $g \in \{2n - k; n, n - k, n - k\}$. Furthermore, all the multiple points of f, other than the one at P, will be transformed to similar multiple points of g. However, the multiple point at P will have been resolved (partly or wholly) into points on the line YZ, other than the base points Y and Z themselves. Let there be u such points with corresponding multiplicities t_i, $i = 1, 2, \ldots, u$. Let D_g be the deficiency of g. Then

$$D_g = \frac{1}{2}(2n - k - 1)(2n - k - 2)$$
$$- \frac{1}{2}[n(n-1) + (n-k)(n-k-1) + (n-k)(n-k-1)]$$
$$- \frac{1}{2}\sum_{i=1}^{s} r_i(r_i - 1) - \frac{1}{2}\sum_{i=1}^{u} t_i(t_i - 1)$$
$$= \frac{1}{2}(n-1)(n-2) - \frac{1}{2}\sum_{i=1}^{s} r_i(r_i - 1) - \frac{1}{2}k(k-1) - \frac{1}{2}\sum_{i=1}^{u} t_i(t_i - 1)$$
$$= D_f - \frac{1}{2}\sum_{i=1}^{u} t_i(t_i - 1).$$

If the singularity has not been wholly resolved, then there is at least one t_i,

$i = 1, 2, \ldots, u$, that is greater than unity. Consequently,

$$\sum_{i=1}^{u} t_i(t_i - 1) > 0 \quad \text{and} \quad D_g < D_f.$$

However, the deficiency of a curve is finite and nonnegative; hence the process of repeated applications of the standard quadratic transformation must result, after a finite number of steps, in a curve such that

$$\sum_{i=1}^{u} t_i(t_i - 1) = 0,$$

that is, the corresponding points are simple. They may be the images of a multiple point that has distinct tangents, for such a point must resolve into simple points; or they may be the images of points with coincident tangents, as in Example 5.26(a). In this latter case the proper transform will be tangent to a fundamental line at the simple point in question. The corresponding multiple point must, therefore, be an ordinary multiple point.

Applying these arguments to each of the nonordinary multiple points of f in turn, we get the required result. ∎

Example 5.27 *Continue the investigation of Example 5.26(b) by performing one other standard quadratic transformation to further resolve the original cusp at $(0, 0, 1)$ of the quartic $(x^2 - yz)^2 - xy^3 = 0$.*

Continuing from Example 5.26(b), let us consider the proper transform g of f in the form

$$x^4(y + z)^2 - 2x^3(y^3 + 3y^2z + 2yz^2)$$
$$+ x^2(y^4 + 4y^3z + 4y^2z^2) + xy^3z^2 - y^4z^2 = 0,$$

where we have written (5.8) in homogeneous form. The cusp is at $(1, 0, 0)$, and the tangent there is $y + z = 0$, which is not a fundamental line. However, the fundamental points $(0, 1, 0)$ and $(0, 0, 1)$ both lie on the curve (points of multiplicity two and four, respectively), and so we must perform one other linear transformation of coordinates before we are ready to apply the standard quadratic transformation. The transformation

$$x = \tilde{x}, \qquad y = \tilde{y} + 2\tilde{z}, \qquad z = \tilde{y} - \tilde{z},$$

is a simple one that satisfies our needs. After this transformation the curve is given by

$$x^2[x(2y + z) - 3y(y + 2z)]^2 - (y - z)^2(y + 2z)^3(y - x + 2z) = 0.$$

The only base point that lies on this curve is the one at $(1, 0, 0)$ and the tangent there is $2y + z = 0$. In this form, the curve belongs to $\{6; 2, 0, 0\}$. We are now ready to apply the standard quadratic transformation. The proper transform h is given, after simplification, by

$$x^3(2y + z)^3(y - z)^2[x(2y + z) - yz]$$
$$- y^2 z^4[3x(2y + z) - y(y + 2z)]^2 = 0,$$

which, as expected, belongs to $\{10; 6, 4, 4\}$. To determine where the fundamental line YZ meets this curve, we put $x = 0$, to obtain

$$y^4 z^4(y + 2z)^2 = 0,$$

showing us that the cusp at $(1, 0, 0)$ has been resolved into a double point at $(0, -2, 1)$. We now know that h has two double points and a point of multiplicity four, each corresponding to the multiple points of g that we were not resolving. In addition, from the resolution of the singularity in question, we see that h has a point of multiplicity six at $(1, 0, 0)$, two points of multiplicity four at $(0, 1, 0)$ and $(0, 0, 1)$, respectively, and a double point at $(0, -2, 1)$. Hence, the deficiency D_h of h is given by

$$D_h = \tfrac{1}{2}(9)(8) \qquad \left[\text{the } \tfrac{1}{2}(n - 1)(n - 2) \text{ term}\right]$$
$$- \tfrac{1}{2}[(2)(1) + (2)(1) + (4)(3)]$$

(from the other multiple points of h)

$$- \tfrac{1}{2}[(6)(5) + (4)(3) + (4)(3)] \qquad (\text{since } h \in \{10; 6, 4, 4\})$$
$$- \tfrac{1}{2}(2)(1) \qquad [\text{from the double point at } (0, -2, 1)]$$
$$= 0.$$

We see, therefore, that the application of the standard quadratic transformation to resolve the cusp has further reduced the deficiency. To examine the double point at $(0, -2, 1)$ let us move it to $(0, 0, 1)$ with the transformation

$$x = \tilde{x}, \qquad y = \tilde{y} - 2\tilde{z}, \qquad z = \tilde{z}.$$

Putting $z = 1$, the second-order terms are $-4(9x - 2y)^2$, showing us that the double point is another cusp. We could then describe the original quartic of Example 5.26(b), namely $(x^2 - yz)^2 - xy^3 = 0$, as a quartic with a cusp that has a cusp in its first neighborhood and a cusp in its second neighborhood. To resolve this last cusp we can make the change of coordinates

$$x = \tilde{x} + \tilde{y}, \qquad y = \tilde{x} - \tilde{y}, \qquad z = \tilde{z},$$

and then apply the standard quadratic transformation. The line $z = 0$ meets the proper transform twice at the point $(7, -11, 0)$, but examination of the curve

at this point shows, as expected, that it is a simple point of the proper transform. Why expected? Otherwise the proper transform would have a deficiency
less than that of h, which is already zero, and, deficiencies being nonnegative
integers, we would have a contradiction. Hence, there was no need to perform a
further quadratic transformation on h to resolve the cusp; for we knew, *a priori*,
that it would correspond to a simple point of the proper transform and that the
tangent there would be the fundamental line $z = 0$.

5.14 Genus of a Curve

Let us now make the important observation that we have, in effect, proved
that the curve investigated in Examples 5.26 and 5.27, namely $f \equiv (x^2 -
yz)^2 - xy^3 = 0$, is rational. Why is this the case? We have shown that
this quartic f has deficiency two (see Exercise 5.35, plus an application of
the formula for deficiency). We have also shown (in the discussion following Example 5.26) that the proper transform g of f has deficiency one, and
that the proper transform h of g has deficiency zero (Example 5.27). Hence,
by Theorem 5.21, h is rational. However, like all Cremona transformations,
the standard quadratic transformation has a rational inverse. Therefore, if h
is rational, g must be rational, and if g is rational, so is f. This remarkable
observation prompts us to extend the concept of deficiency to what is called
genus.

From Noether's theorem (Theorem 5.24) we see that each time we resolve a
singularity, the deficiency is reduced by an amount $\frac{1}{2} \sum r_i(r_i - 1)$, where the sum
extends over all the points in the first neighborhood of the singularity in question.
Hence, if D_0 is the deficiency of the original curve and D_R is the deficiency
of the curve with only ordinary multiple points, which we would obtain by a
complete resolution of all the nonordinary singularities of the original curve,
then

$$D_R = D_0 - \frac{1}{2} \sum r_i(r_i - 1),$$

where now the summation is over all the neighborhoods of all the singular points
of the original curve. (Although we say "all the neighborhoods," because of
the factor $r_i - 1$ there will be no contribution to this sum once a singularity
has been resolved. For the same reason, only neighborhoods of nonordinary
singular points will contribute to the sum.)

We have

$$D_0 = \frac{1}{2}(n - 1)(n - 2) - \frac{1}{2} \sum r_i(r_i - 1),$$

where the sum is only over points of the original curve. Hence,

$$D_R = \frac{1}{2}(n-1)(n-2) - \frac{1}{2}\sum r_i(r_i - 1),$$

where the sum is over points of the original curve and all the points in all the neighborhoods of all the singular points of the original curve. This is a number that can be associated with the original curve and is called the *genus*. We will usually use the letter p for this number. Hence, the genus of an irreducible plane algebraic curve of order n is defined by

$$p = \frac{1}{2}(n-1)(n-2) - \frac{1}{2}\sum r_i(r_i - 1),$$

the summation extending over all singular points and over all the neighborhoods of these points.

For the genus to be a property of the curve only, and not simply a property of the curve together with a specific collection of quadratic transformations, the number p must be independent of the particular transformations involved. This, indeed, is the case, although we state the result without proof.

Theorem 5.25 *The genus of an irreducible plane algebraic curve is a well-defined integer and is independent of any particular collection of quadratic transformations used in the calculation of its value.*

This result, together with the fact that curves with zero deficiency are rational (Theorem 5.21), gives us the following theorem.

Theorem 5.26 *If an irreducible plane algebraic curve has genus zero, then it is rational.*

Finally, since the genus is invariant we have, from Theorem 5.22, that a rational curve must have genus zero. Therefore, we have the final theorem of this chapter:

Theorem 5.27 *An irreducible plane algebraic curve is rational if and only if it has genus zero.*

5.15 Bibliographical Notes

Our attention has been focused on curves, and we have developed only a small part of algebraic geometry. Hence most books on algebraic geometry cover far more than what we have discussed here. Most of the modern books on algebraic geometry take an abstract approach, and this may be quite difficult for the student who wants to relate the subject to curves that may be used in applications. As a result, there are few books that we can recommend for further reading. Parts of Robert Walker's book, *Algebraic Curves*, published by Dover (1962), and a few pages of Semple and Roth, *Introduction to Algebraic Geometry*, published by Clarendon Press, Oxford (1949), may be useful.

Exercises

5.1. Sketch the projective completions of $y = x^2$ and $y = x^3$ in the real projective plane.

5.2. Show that the curve $x^4 - 2x^3y - x^2 + xy + y = 0$ is rational, and find a parametrization of the curve.

5.3. Show that $(0, 1, 0)$ is a simple point of $x^4 - x^2yz + y^3z - 3z^4 = 0$, and find the equation of the tangent at that point.

5.4. Prove that the multiplicity of a point of an algebraic curve is a projective invariant.

5.5. Show that the curve $x^3y + 2x^3z + y^3z + 3y^2z^2 + 3yz^3 + z^4 = 0$ has a triple point at $(0, -1, 1)$.

5.6. Prove that if $f(x, y) = 0$ has a point of multiplicity r at the origin, then the components of the curve obtained from f by setting the terms of degree r equal to zero are the tangents to f at the origin.

5.7. Find the singular points of
(a) $x^2y^2 - 2z^3(x + y) + 3z^4 = 0$,
(b) $x^4 + y^4 + 2x^2z^2 + 2y^2z^2 = 0$.

5.8. Find the values of α that result in the following curves being singular:
(a) $x^3 + y^3 + z^3 + \alpha(x + y + z)^3 = 0$,
(b) $x^2y + y^2z + z^2x + zx^2 + xy^2 + yz^2 + \alpha xyz = 0$,
(c) $z^2x = y(y - x)(y - \alpha x)$,
(d) $(x + y + z)^3 + 6\alpha xyz = 0$,
(e) $x^3 + y^3 + z^3 + 6\alpha xyz = 0$.

5.9. Show that the curve $3x^2y^2 - 4y^3 - 4cx^3 + 6cxy - c^2 = 0$ has a cusp at $(c^{1/3}, c^{2/3})$.

5.10. Find the double points, and identify their type:
(a) $x^4 - 2\alpha y^3z - 3\alpha^2y^2z^2 - 2\alpha^2x^2z^2 + \alpha^4z^4 = 0$,
(b) $x^2y^2 + 36\alpha^3xz^3 + 24\alpha^3yz^3 + 108\alpha^4z^4 = 0$,
(c) $(x^2 + y^2 - \alpha^2z^2)^3 + 27\alpha^2x^2y^2z^2 = 0$.

5.11. Prove that a nonsingular curve is irreducible.

5.12. Prove that a conic is singular if and only if it is reducible.

5.13. Prove that all ellipses have complex asymptotes.

5.14. Find the asymptotes of the following curves:
(a) $x(x+y)^2 - x(x+y) + 2 = 0$,
(b) $x^3 + y^3 - 3xy = 0$.

5.15. Use Sylvester's dialytic elimination to find the intersections of the following pairs of conics:
(a) $2xy - xz - yz = 0$, $5xz - 4xy - yz = 0$,
(b) $2x^2 - 2xy + 3y^2 - 3xz - 4yz + z^2 = 0$, $3x^2 - 5xy - 5y^2 - 8yz + 3z^2 = 0$.

5.16. Use Sylvester's dialytic elimination to find the algebraic curves which correspond to the following plane rational curves:
(a) $x = \frac{2t}{1+t^2}$, $y = \frac{1-t^2}{1+t^2}$,
(b) $x = st^2$, $y = s^2t$, $z = 2s^3 - t^3$,
(c) $x = \frac{t}{1-t}$, $y = \frac{1-t}{1+t}$.

5.17. Use resultants to determine any common nonconstant factors of the following pairs of polynomials:
(a) $xy - y^2 + xz - yz$, $xy - y^2 + 2xz - 2yz$,
(b) $x^2z + x^2 + z + 1$, $x^2z + x^2 - z - 1$,
(c) $x^2y + x^2z - xz^2 + xy^2 - y^2z - yz^2$, $x^2y - x^2z - xz^2 + xy^2 + y^2z - yz^2$.

5.18. The points $(-3, 0, 1)$ and $(0, 2, 1)$ are common to the curves

$$4x^2 + 9y^2 - 36z^2 = 0 \quad \text{and} \quad 2x - 3y + 6z = 0.$$

Following Example 5.12, find the three resultants, and verify that the pairs $(0, 1)$, $(2, 1)$; $(-3, 1)$, $(0, 1)$; and $(-3, 0)$, $(0, 2)$ are zeros of the corresponding resultants.

5.19. Calculate the resultant with respect to the ratio λ/μ of $f(\lambda \underline{x} + \mu \underline{y})$ and $g(\lambda \underline{x} + \mu \underline{y})$, where

$$f(\underline{x}) = x_1^2 + 4x_2^2 - 4x_3^2 \quad \text{and} \quad g(\underline{x}) = x_2 - x_3.$$

5.20. Find the intersection points, with multiplicities, of the following pairs of curves; in the light of Theorem 5.13, pay particular attention to intersections at $(0, 0, 1)$.
(a) $y^3 - x^2 = 0$, $x^3 + y^3 - 3xy = 0$,
(b) $(y - x)^3 - (y + x)^2 = 0$, $y^3 - x^2 = 0$,
(c) $x^3 - y^2 - x = 0$, $x^3 + y^3 - 3xy = 0$,
(d) $x^3 - y^2 - x = 0$, $x^3 + y^3 - 3(x^2 - y^2) = 0$.

5.21. Show that any quartic through thirteen points passes through three more.

5.22. Show that all nth-order curves through $\frac{1}{2}(n^2 + 3n - 2)$ points also pass through $\frac{1}{2}(n - 1)(n - 2)$ other points.

5.23. Prove, using Theorem 5.15, that pairs of opposite sides of a hexagon inscribed in an irreducible conic meet in collinear points.

5.24. Verify the symmetry relation

$$\frac{1}{(n-r)!}(\underline{b} \cdot \nabla_{\underline{a}})^{n-r} f(\underline{a}) = \frac{1}{r!}(\underline{a} \cdot \nabla_{\underline{b}})^r f(\underline{b})$$

for the curve $y^3 - x^2 z + y^2 z = 0$.

5.25. Prove that an algebraic curve has a finite number of multiple points.

5.26. Prove that the first polar of a conic with respect to a point P is the chord defined by the points of contact with the conic of the tangents through P.

5.27. Show that the tangents to the circle $x^2 + y^2 - r^2 = 0$ through the point (a, b) are complex if $a^2 + b^2 < r^2$.

5.28. Consider the quartic

$$ax^2 y^2 + by^2 z^2 + cz^2 x^2 + 2xyz(fx + gy + hz) = 0.$$

(a) Show that this curve has double points at the vertices of the triangle of reference.

(b) Show, by an application of the Maclaurin–Bézout theorem and *reductio ad absurdum*, that the curve, if irreducible, cannot have a fourth multiple point.

(c) Hence show that, if $h^2 - bc \neq 0$, $g^2 - ab \neq 0$, and $f^2 - ac \neq 0$, six tangents to the curve can be drawn from an arbitrary point.

5.29. Find the tangents to $y^3 - x^2 z = 0$ through the point $(0, 1, 1)$.

5.30. Prove that the class of an nth-order curve is bounded from above by $n(n-1)$.

5.31. Prove that if the polar conic with respect to an inflection point P of a curve has a double point at P, then P is a double point of the original curve and hence the nontangent line of the polar conic with respect to an inflection point P cannot pass through P.

5.32. Show that any three of Plücker's equations are independent.

5.33. Determine the maximum number of triple points of curves of order four, five, six, seven, and eight.

5.34. Show that for a given curve of order n the freedom of a curve of order $n - 3$, that shares the same multiple points, but with multiplicity one less than for the given curve, is $D - 1$, and such a curve can meet the given curve in $2(D - 1)$ additional points, where D is the deficiency of the given curve. Similarly, show that for such a curve of order $n - 4$ these numbers are $D - n + 1$ and $2D - n - 2$.

5.35. Show that the quartic $(x^2 - yz)^2 - xy^3 = 0$ has only one multiple point that is a cusp at the origin.

5.36. Show that of the following quartics each has three double points and each passes through $(1, 1, 1)$. Determine the nature of the double points, and produce parametrizations of the curves.

(a) $x^2y^2 + y^2z^2 - 2z^2x^2 = 0$,

(b) $(x^2 + y^2 - yz)^2 - z^2(x^2 + y^2) = 0$.

5.37. Repeat the process of Example 5.23 when

(a) the parametrization of the line is based upon points other than $(1, 0, 1)$ and $(0, 1, 1)$, and

(b) the double point of the cubic is not at a base point (fundamental point).

5.38. Find (i) synthetically, (ii) algebraically the proper transforms of the following curves:

(a) the general conic,

(b) a cubic through the base points with a node at one and a cusp at another,

(c) a cubic through two base points,

(d) a quintic with nodes at each of the base points.

6

Examples and Applications

6.1 Introduction

Our aim throughout the first five chapters has been to develop some rudimentary understanding of the geometrical foundations of geometry. Before going on with applications and a discussion of surfaces, let us assess what we know about curves.

In Chapter 1 we discussed interpolation. We mentioned the limited usefulness of curves that are represented as functions of the form $y = f(x)$. We pointed out that curves given in parametric form $x = X(t), y = Y(t)$ would not suffer from as many restrictions, but we did not pursue the discussion of such curves. We saw that not all interpolation problems are of the standard finite linear interpolation type, and hence we must be careful in setting up our curve definition procedure before we can begin to ask questions about the qualitative nature of the curve. Much of our discussion concerned polynomial interpolation. We restricted ourselves to two-variable polynomial interpolation and ended Chapter 1 with a discussion of the connection between this topic and algebraic curves.

In Chapter 2 we moved away from curves of the form $y = f(x)$ and discussed the conics, the simplest nontrivial algebraic curves. Such curves are of major importance in engineering design and in many of the current software graphics applications. We saw rather quickly that several of the fairly obvious ways of attempting to specify such curves did not lead to finite linear interpolation problems. For an understanding of what was happening we developed synthetic and algebraic methods. The synthetic methods gave insight into many problems, particularly into the construction of certain pencils of conics. Our algebraic methods were also powerful, and we developed many useful equations and analyzed many important properties by algebraic methods. However, our algebraic arguments included the assumption that a line always meets a conic twice and that two conics always meet in four points.

The path to the justification of this assumption would provide information useful to problems of much wider scope than the original question. We spent some time (in Chapter 3) on central projections, perspective and projective transformations, and the role of synthetic arguments in geometry. In Chapter 4 we introduced homogeneous coordinates in order to deal with such concepts in an algebraic fashion. In algebraic form a plane projective transformation is nothing more than a 3×3 nonsingular matrix. This discussion had started with the desire to justify our assumptions about the intersections of lines and conics, but led us to a discussion of projective transformations. We classified different geometrical properties in terms of the groups under which they were invariant, these groups being nothing more than special 3×3 nonsingular matrices. The affine transformations, the similarity transformations, rotations, reflections, and rigid motions became trivial special cases of the projective transformations.

A simple question about two conics intersecting in four points resulted in our gaining an understanding of a hierarchy of geometries, of the complex projective plane and homogeneous coordinates, and of the most important coordinate transformations. The Maclaurin–Bézout theorem finally put to rest our fears about two conics meeting in four points and was crucial in attaining much of the qualitative information we now know about curves. Theorem 5.8 told us that every rational curve is an algebraic curve. Therefore, all our information on algebraic curves applies also to rational curves. However, it was still not clear whether algebraic and rational curves are simply the same set of objects given in two different forms, namely the implicit equation and the explicit parametrization. In the case of conics this is certainly the case, since the general second-order parametrization represents a conic, and all conics have rational parametrizations. This is a special case of Theorem 5.5 which tells us that any curve of order n is rational if it contains a point of multiplicity $n - 1$. This theorem hints at some connection between rational curves and multiple points, but gives us little insight into this connection. Algebraic curves are not all rational, nor do they have to have a point of multiplicity $n - 1$ in order to be so. The rational curves form a proper subset of the algebraic curves and have special properties not shared by algebraic curves in general. Polynomial curves (those with polynomial, rather than rational, parametrization) form a distinct proper subset of the rational curves. It seemed, therefore, a rather minimal requirement that we know necessary and sufficient conditions for a curve to be rational or polynomial, and the qualitative properties of such curves in the projective and affine planes.

We investigated some of these aspects in Sections 5.11 to 5.14. If the deficiency of a curve is zero, then it is rational (Theorem 5.21), but this only gives us a sufficient condition for rationality. We had to use quadratic transformations to resolve all singularties of the curve in order to investigate the neighborhoods of singular points before we could come to the final conclusion: an irreducible algebraic curve is rational if and only if its genus is zero (Theorem 5.26). The algebraic analysis we performed in Chapter 5 provides the tools for the study of a rich variety of applications and examples of algebraic curves.

We begin these applications and examples with an investigation of the affine classification of rational cubics, and progress to the construction of higher-order algebraic curves by using our acquired knowledge of, among other things, multiple points. We consider different ways of constructing parametrizations of curves with zero deficiency, and construct direct parametrizations that satisfy certain constraints. We extend the finite interpolation property to include Hermite interpolation in the plane, and use multiple points to forge the links with geometry. We return once more to that old favorite, the parabola, and discuss a variety of interpolation and approximation methods based on its use. This leads to a general discussion of the closeness of curves as a geometrical property, and the use of special curves in approximation problems. The chapter is concluded with a short section on the role of geometrical interpolation in the derivation of difference methods for the solution of differential equations.

6.2 The Affine Classification of Rational Cubics

Cubic curves have come to play a very special role in design and approximation. We have used them in Section 4.8 of Chapter 1 and made heavy use of them in Section 5 of the same chapter where we discussed splines. All of the methods discussed in these sections can be used both when the curve is of the form $y = f_3(x)$ and when it is of the form $x = X_3(t)$, $y = Y_3(t)$, and extensions can be used in cases where the curve is of the form $x = X_3(t)/Z_3(t)$, $y = Y_3(t)/Z_3(t)$. In all of these cases, when $f_3(x)$, $X_3(t)$, $Y_3(t)$, and $Z_3(t)$ are polynomials of degree three, the curves are rational cubics. In some instances, when using parametric cubics to interpolate data, undesirable results can occur. It is for this reason that we now discuss the rational cubics more carefully.

We now know (from Theorem 5.27) that a cubic is rational if and only if it has a double point. We also know, from the application of Bézout's theorem, that an irreducible cubic cannot have more than one double point. We still need to do a little more work before we know what such curves may look like.

Let us first consider those cubics with a double point. Since a double point can be either a cusp (the tangents coincide) or a node (distinct tangents), there can be only two types of rational cubics. This is nice and simple, but it is true only in the context of projective geometry over the complex field, which is not the context of most applications of geometry. For example, in which of these

Figure 6.1. Acnode (two affine inflection points).

two classes does the simple cubic $y = x^3$ lie? We are used to thinking of this latter curve in the real affine plane, but our general result about the rational cubic was in the context of the complex projective plane. Therefore, in order to achieve a real affine classification of the cubic, we must consider both the projective and the complex aspects of our discussions so far.

Let us first deal with the issue of the complex plane. At a node the tangents may be real or complex. If we are working in the complex plane, there will be no difference between these two situations, but if we are working in the real plane there will be a difference, namely, that in one case there are tangents (in which case we call the node a *crunode* – see Section 5.4) and in the other case there are none (in which case we call the node an *acnode*). In the real projective plane the case of the node splits into two cases, corresponding to tangents or no tangents. These two situations look very different. The case with tangents looks like a normal crossing of the curve with itself. Figure 6.1 shows the case without tangents.

Let us also draw a distinction between the affine and the projective context. In most application environments we shall be working in affine space, so let us use this space for our classification. We must draw the distinction between the double point being on the line at infinity and its being in the affine plane. If we examine the curve $y = x^3$ (Figure 6.2) and write it in its homogeneous form $yz^2 = x^3$, we notice that it has a simple point at $(0, 0, 1)$ and a double point at $(0, 1, 0)$. Furthermore, the double point is a cusp.

These last observations give us four more real affine classes: one for the acnode (a node with complex tangents), and three more corresponding to the cusp, crunode, or acnode being on the line at infinity.

This has expanded the number of classes, but we have not yet finished. The curve $x^2y = 1$ (or $x^2y = z^3$ in homogeneous form) does not look like any of the ones we have identified so far (Figure 6.3). Recall that in Section 5.4 we defined an asymptote as a line, other than the line at infinity, that is tangent to the curve at a point on the ideal line. The curve $yz^2 = x^3$ (Figure 6.2). has a cusp at infinity, and the tangent there is the ideal line. The ideal line has three-point contact with the curve at the cusp and cannot meet the curve at any other point. This curve, therefore, has no asymptotes. However, if a cubic had a cusp at

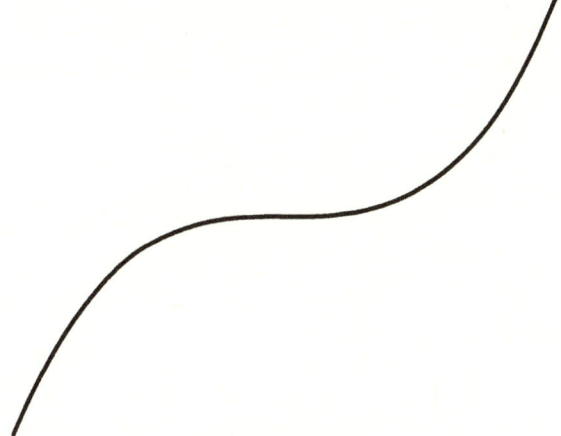

Figure 6.2. One affine inflection point.

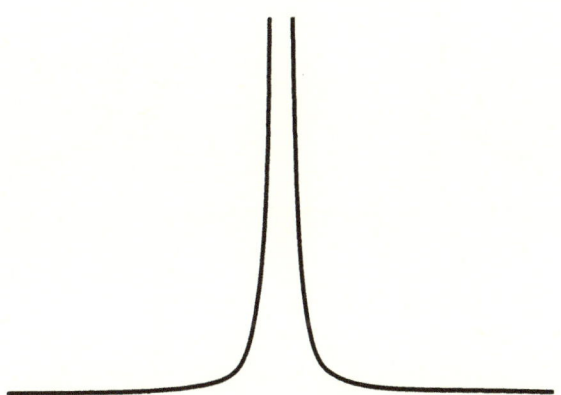

Figure 6.3. Cusp on ideal line and two asymptotes.

infinity and the tangent there were not the ideal line, then this tangent would be an asymptote. The ideal line must meet the curve at one other point and the tangent at this point will be another asymptote. The curve $x^2 y = z^3$ is such a curve. The cusp is at $(0, 1, 0)$ with tangent $x = 0$, and the ideal line also meets the curve at $(1, 0, 0)$, where the tangent is $y = 0$.

The crunode can be affine or on the ideal line. When it is on the ideal line, there are still two cases, for the ideal line may or may not be a tangent. If the ideal line is a tangent, then, as we have just seen, there can be no other point of the curve on this line, and the only asymptote is the other tangent at the crunode. If the ideal line is not a tangent at the crunode, then the curve will have three asymptotes, namely, the two lines corresponding to the tangents at the crunode and the third one corresponding to the tangent at the point, other than the crunode,

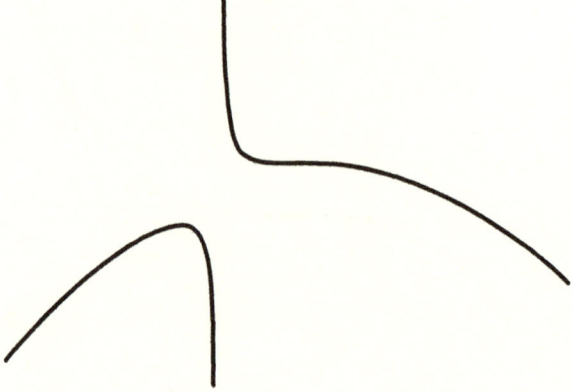

Figure 6.4. Crunode on ideal line and one asymptote.

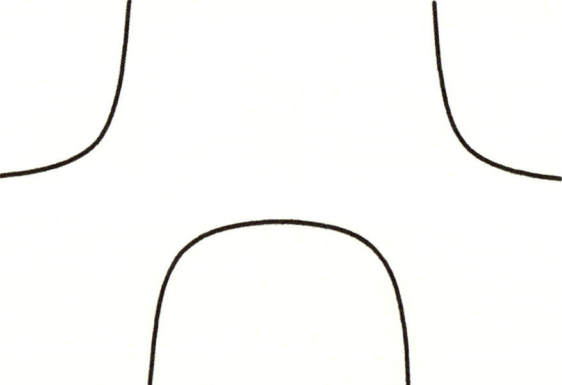

Figure 6.5. Crunode on ideal line and three asymptotes.

where the ideal line meets the curve. Figure 6.4 shows the situation where there is a crunode on the ideal line such that the ideal line is a tangent at the crunode. There is only one asymptote, which must correspond to the other tangent at the crunode. Figure 6.5 shows the case where the crunode is on the ideal line but the ideal line is not a tangent at the point. Hence both tangents at the crunode will be asymptotes. The ideal line must meet the curve at one other point, and the tangent there will also be an asymptote. Hence this curve has three asymptotes.

The acnode can also be on the ideal line, in which case, since the ideal line is real and the tangents at the acnode are complex, the ideal line cannot be a tangent at the acnode. Therefore, the ideal line will meet the curve at one other point, and the tangent there will be the only asymptote. Figure 6.6 shows an example of this case.

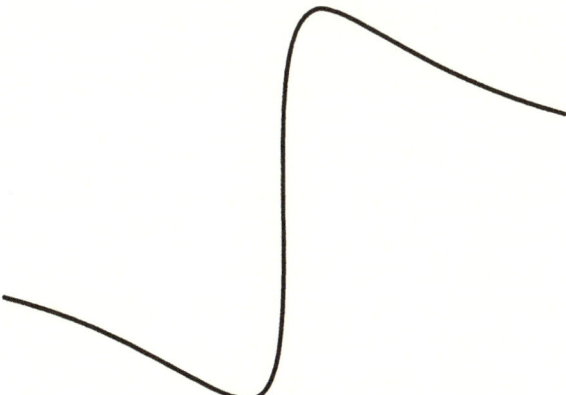

Figure 6.6. Acnode on ideal line and one asymptote.

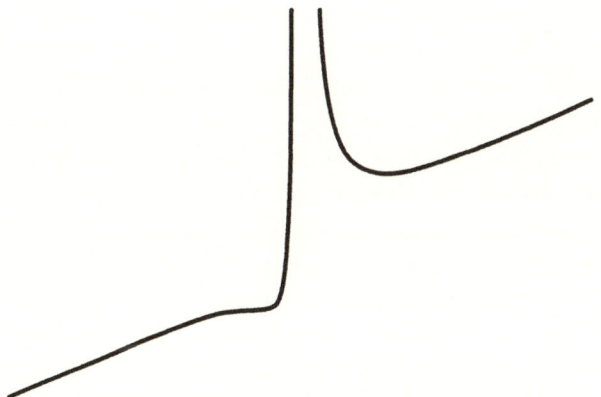

Figure 6.7. Cusp, two asymptotes, one inflection point.

Consider the function $yx^2 - (x + z)^3 = 0$. This curve has a cusp at $(0, 1, 0)$, where the ideal line is not a tangent. The ideal line also meets the curve at $(1, 1, 0)$. Hence this curve has a cusp on the ideal line and two asymptotes. This is exactly the same situation as for the curve drawn in Figure 6.3. However, the new curve (Figure 6.7) looks quite different. Each curve has two asymptotes, but in the present case the curve crosses one of the asymptotes at an affine point. A curve can only cross an asymptote and then curve back towards the asymptote if it passes through an inflection point. The last curve has an inflection point at $(-1, 0, 1)$. We must, therefore, include inflection points in the classification, for their existence and location will affect the affine appearance of the curve.

In Section 5.10 we showed that the inflection points of a curve are the non-singular points of the curve that also lie on its Hessian curve. This Hessian curve is of order $3(n-2)$, which, in the case of the cubic, will also be a cubic curve. We know that two cubics meet in nine points, so a cubic can have as many as nine inflection points. We also stated (though we did not prove) that an ordinary singular point of multiplicity r is a point of multiplicity $3r(r-1)$ of the intersection cycle of the curve and its Hessian. Hence, in the case of the rational cubic (which must have precisely one double point) the maximum number of inflection points is reduced to three. We further stated in Section 5.10 that if the multiple point is a cusp (a nonordinary multiple point), this cusp accounts for eight of the possible nine intersections of the curve with its Hessian. Therefore, a rational cubic with a cusp will have exactly one inflection point. This one inflection point can be affine or on the ideal line. This is precisely the difference between the two cubics of Figures 6.3 and 6.7. The first has its inflection point on the ideal line at $(1, 0, 0)$, and the second has an affine inflection point at $(-1, 0, 1)$.

We now leave it as an exercise to the reader (Exercise 6.2) to complete the classification for the cubic with a cusp. There are four cases in addition to the four we have discussed in this section. There are also many additional cases for the cubic with a crunode or acnode. We will give only a few examples.

Like the curve shown in Figure 6.4, the curve in Figure 6.8 has one point of inflection and one asymptote, but instead of being on the ideal line the crunode has become visible. The curve represents the equation $xy(x+z) - (x-y)^3 = 0$, which meets the ideal line at precisely one real point [at approximately $(2.74, 1, 0)$, which gives us the slope of the only asymptote], has a crunode at $(0, 0, 1)$, and has an inflection point at $(-1, -1, 1)$. We make a similar change to the situation illustrated in Figure 6.5 by moving both the crunode and the inflection point to visible positions. The curve equation is $y^3 - (x^2 - y^2)(x - y - z) = 0$, and the curve meets the ideal line in three real points, resulting

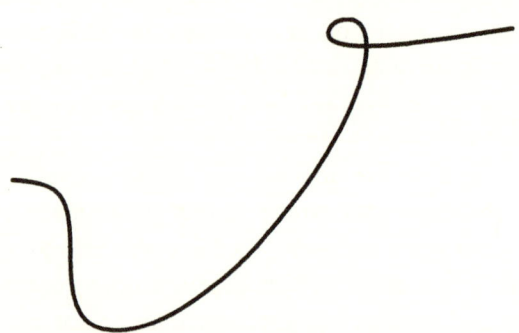

Figure 6.8. Crunode with asymptote and inflection point.

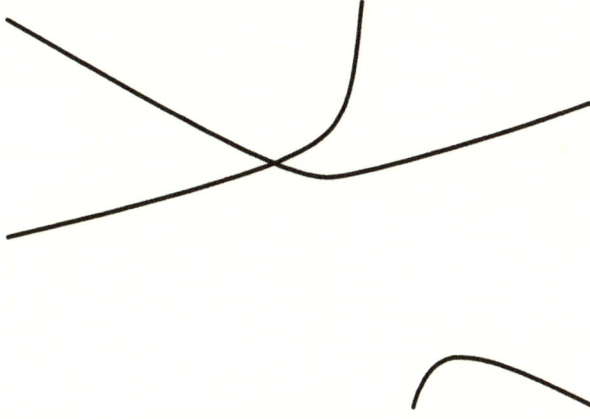

Figure 6.9. Crunode, three asymptotes, and inflection point.

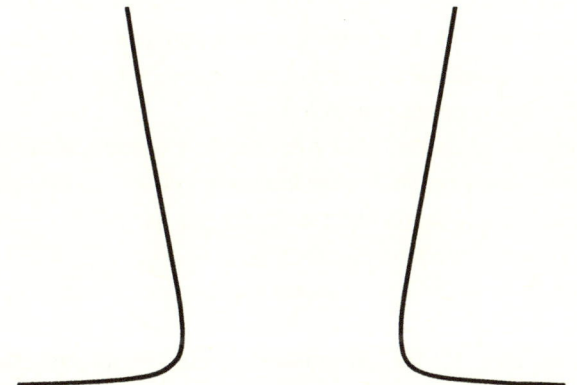

Figure 6.10. Acnode, two inflection points, and an asymptote.

in three asymptotes (Figure 6.9). The crunode is at $(0, 0, 1)$, and the inflection point at $(1, 0, 1)$.

Let us consider two more examples of curves with affine acnodes and show the effect of the presence of inflection points and/or asymptotes. As in Figure 6.1, the acnode of the curve in Figure 6.10 is affine and there are two affine inflection points, but we have introduced one asymptote. The curve equation is $x^2y - z(x^2 + y^2) = 0$, the acnode is at $(0, 0, 1)$, and, since it occurs for complex values of the parameter, it cannot be shown by using a real parametrization. The curve meets the ideal line once at $(1, 0, 0)$, causing the occurrence of an asymptote. The ideal line is tangent to the curve at $(0, 1, 0)$, so that there is no asymptote at this point. One inflection point is at the simple point on the ideal line, and the other two inflection points are at $(\pm 4, 4\sqrt{3}, \sqrt{3})$. The affine

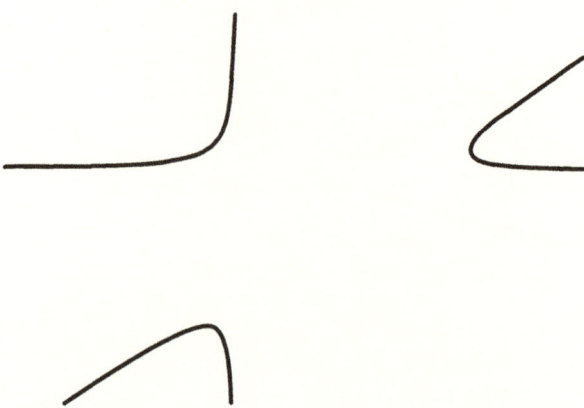

Figure 6.11. Acnode and three asymptotes.

inflections lie on a horizontal line, and this line meets the ideal line at the third inflection point, so that the three inflection points are collinear. This property of the three inflection points of a cubic with an acnode (or crunode) is examined in more detail in Exercises 6.5 and 6.6.

The curve shown in Figure 6.11 also has an affine acnode, but all its inflection points are on the ideal line, so that the three asymptotes are the tangents at the inflection points. The equation represented by the curve is $z(x^2 - xy + y^2) - xy(x - y) = 0$, with inflection points at $(0, 1, 0)$, $(1, 0, 0)$, and $(1, 1, 0)$. The acnode is at $(0, 0, 1)$, and the asymptotes are the lines $x + z = 0$, $y - z = 0$, and $x - y - z = 0$.

The equations used for the illustrations in this section, together with the parametrizations used for sketching the curves, are summarized in Table 6.1.

The eleven rational cubics we have shown are by no means all the different-looking ones that exist in the real affine plane. We have used them to emphasize the great complexity of such curves. This complexity resulted from two requirements we placed on ourselves: that results should be in the affine plane and that they should be real, not complex. The first of these resulted in the appearance of asymptotes that in the projective plane are simply tangents. The requirement that our points be real created new special cases where there were two inflection points and an isolated point. There is nothing special about this case in the complex plane, for it is simply a common node with distinct tangents. We should also bear in mind that we have restricted ourselves to the rational cubics. If we consider the cubics of genus one, we shall have many more cases – and then there are still the higher-order curves. We will make a few brief comments on quartics in Section 6.3.

Table 6.1. *Summary of illustrations.*

Figure	Equation	Parametrization
6.1	$y(y+z)^2 - x^2z = 0$	$x = t(t^2+1),\, y = t^2,\, z = 1$
6.2	$x^3 - yz^2 = 0$	$x = t,\, y = t^3,\, z = 1$
6.3	$x^2y - z^3 = 0$	$x = t^3,\, y = 1,\, z = t^2$
6.4	$xyz + (x-z)^3 = 0$	$x = t^2,\, y = (1-t)^3,\, z = t$
6.5	$z^3 - (x^2 - z^2)y = 0$	$x = t(t^2-1),\, y = 1,\, z = t^2 - 1$
6.6	$y(y+x)^2 - xz^2 = 0$	$x = t^3,\, y = t,\, z = t^2 + 1$
6.7	$yx^2 - (x+z)^3 = 0$	$x = t^3,\, y = (t+1)^3,\, z = t^2$
6.8	$xy(x+z) - (x-y)^3 = 0$	$x = 4t^2(t-1),\, y = 2t(t-1)^2,$ $z = -3t^3 + 7t^2 + 3t + 1$
6.9	$y^3 - (x^2 - y^2)(x - y - z) = 0$	$x = t^2 - 1,\, y = t(t^2 - 1),$ $z = t^2 + t - 1$
6.10	$x^2y - z(x^2 + y^2) = 0$	$x = t^2 + 1,\, y = t(t^2+1),\, z = t$
6.11	$z(x^2 - xy + y^2) - xy(x-y) = 0$	$x = t(t^2 - t + 1),\, y = t^2 - t + 1,$ $z = t(t-1)$

6.3 Construction of Algebraic Curves

We now turn our attention to a powerful and pleasing way of constructing algebraic curves. The tools we have developed are useful for both the analysis and the construction of algebraic curves. The main thrust of Chapter 5 was the analysis of curves to determine when they were rational. In Section 6.2, we analyzed a variety of rational cubic curves. Way back in Sections 1.2 and 2.8, we looked at the construction of a curve. We saw that although requiring a curve to pass through a point is equivalent to imposing a linear condition on the coefficients of the polynomial that defines the curve, it is not usually efficient to set up a large system of equations and then to solve for the coefficients. It was far more efficient to first construct a pencil of conics through four points. We now have the experience to apply this approach to the construction of a variety of curves more general than, but including, conics. We will illustrate the method by working through a series of examples and will suggest several others for the reader. We begin with our well-worn example of constructing a conic through five points in general position.

Example 6.1 *Construct a conic through five points.*

We construct the conic from lower-order curves. The only lower-order curves are lines, and a line is determined by two points. The product of two linear forms is quadratic, so taking the product of two lines (the product of the linear

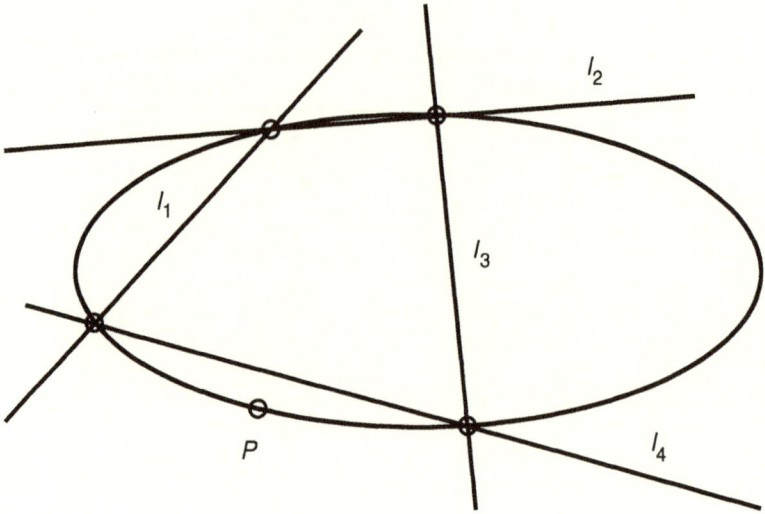

Figure 6.12. The conic through five points.

forms associated with the lines) will give us a conic. Working with line pairs, we can only force our conic to pass through four points. Temporarily, therefore, we ignore one of the given five points, the point P in Figure 6.12. We now construct two line pairs, each pair passing through the four points, and consider the pencil of conics given by

$$\alpha l_1 l_3 + \beta l_2 l_4 = 0.$$

We determine the value of the ratio α/β that will make the conic pass through the fifth point. This involves the solution of one equation in one unknown ratio, not five equations in five unknown ratios.

Another aspect of the construction is quite important. What would happen if we also included the third line pair defined by the four points and considered the equation

$$\alpha l_1 l_3 + \beta l_2 l_4 + \gamma l_5 l_6 = 0,$$

and then tried to make this conic pass through two additional points? This cannot be done, because we would then have passed a conic through six arbitrary points, and this is impossible. Therefore, the three line pairs are linearly dependent.

Example 6.2 *Construct the cubic through nine points.*

Once more we will construct the cubic from lower-order curves. We could use triple products of lines, or the product of a line and a conic. Which should

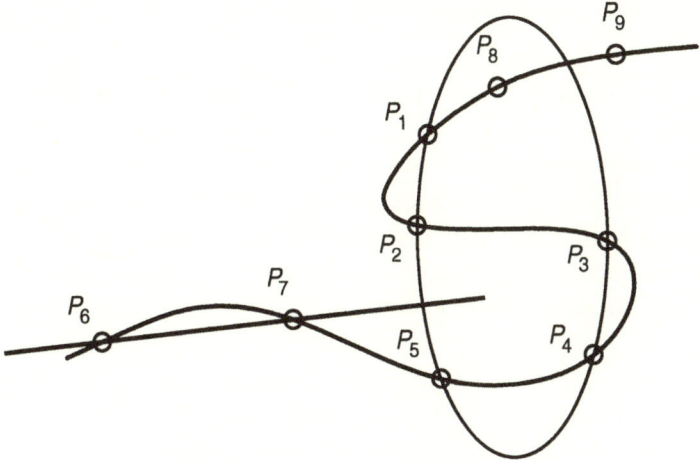

Figure 6.13. The cubic through nine points.

we use? If we use triples of lines, we shall be dealing with at most six of the nine given points. We shall therefore have to take four different triples of three lines through the six points and construct the web of cubics through these six points. Then we will solve for the three ratios that force the cubic to pass through the remaining three points. However, if we use a line and a conic, then, since the line is determined by two points but the conic by five, we shall be dealing with a collection of seven of the nine given points and only have to construct a net of cubics through the seven points, having only two linear equations to solve for the two ratios that make the cubic pass through the two remaining points. This is the more efficient way to proceed.

We ignore two of the nine points, P_8 and P_9 (Figure 6.13), and construct three cubics, each one being the product of a line and a conic. For example, one product could be the product of the line determined by the points P_6 and P_7 and the conic determined by the points P_1, P_2, P_3, P_4, and P_5, as shown. Let these curves be associated with the polynomials l_1 and c_1, respectively. We construct two more products in a similar fashion, each based on the same seven points. We construct the net of cubics

$$\alpha l_1 c_1 + \beta l_2 c_2 + \gamma l_3 c_3 = 0.$$

This net, by construction, passes through the seven points for any values of the parameters. We finally solve for the parameter ratios that force the cubic to pass through the remaining two points.

This procedure can also be applied in the case where there are singular points, as the following example shows.

Example 6.3 *Construct the cubic with a double point and six other points.*

Again, let us first satisfy ourselves that this is a meaningful question. A cubic will, in general, be determined by nine points. Passing through a double point will impose three conditions. There are six remaining conditions, which could be incidence with six points. The question is, therefore, well posed.

Arguing as before, we could use triples of lines or products of a line and a conic. We could also consider curves that have the double point as one of their own double points. This cannot be the case here, since neither a line nor a nondegenerate conic can have a double point. Again, triples of lines would be constructed from a set of six points, and products of a line and a conic would use seven points. It is more efficient to use the latter (ignoring any work involved in the construction of the conics). Since neither the line nor the conic can have a double point, we have to create a double point of the cubic by making the line and the conic pass through the point that is required to be the double point of the cubic. In this sense, the double point and five simple points can be thought of as seven points.

Assume that the double point is at P_1. We ignore one point (P_7, Figure 6.14). We now construct two cubics, each being the product of a line and a conic. For example, we may use the line determined by P_1 and P_2 and the conic determined by P_1, P_3, P_4, P_5, and P_6. We construct another line–conic pair, both passing

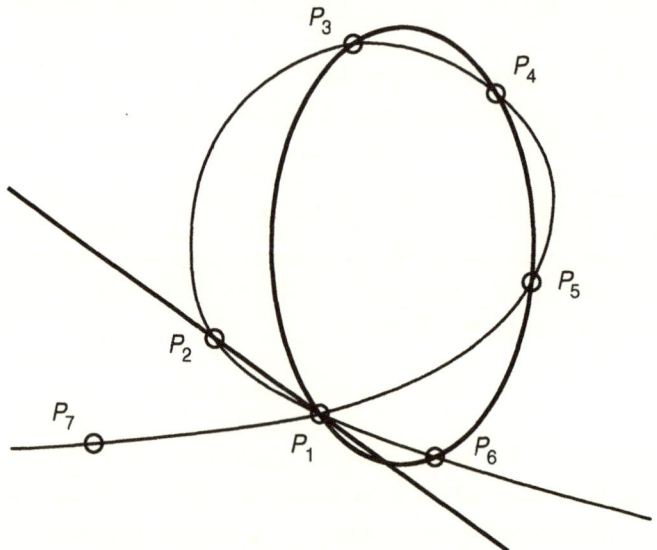

Figure 6.14. Cubic through a double point and six simple points.

through P_1, and form the pencil of cubics given by

$$\alpha l_1 c_1 + \beta l_2 c_2 = 0.$$

Then we find the value of the ratio α / β that makes the cubic pass through the point P_7.

Example 6.4 *Construct a quartic with two double points and passing through eight additional points in general position.*

Let us first check to make sure that the question is well posed. A quartic has fourteen degrees of freedom, and a double point imposes three conditions, so that two double points together with eight additional points will impose fourteen conditions.

We will use products of conics. To construct the two double points, both conics must pass through the two points that are required to be double points of the final quartic. Ignore two of the eight simple points. Construct three pairs of conics, each conic of a pair having the double points in common with the other member of the pair and passing through three of the other six points in such a way that the product of the pair is a quartic with the required two double points and simple points at the remaining six points. An example of such a pair of conics is shown in Figure 6.15. We partition the six points, in three different ways, into two disjoint sets of three points. In this way we have three independent conic pairs. We form the net of quartics given by

$$\alpha c_1 c_2 + \beta c_3 c_4 + \gamma c_5 c_6 = 0,$$

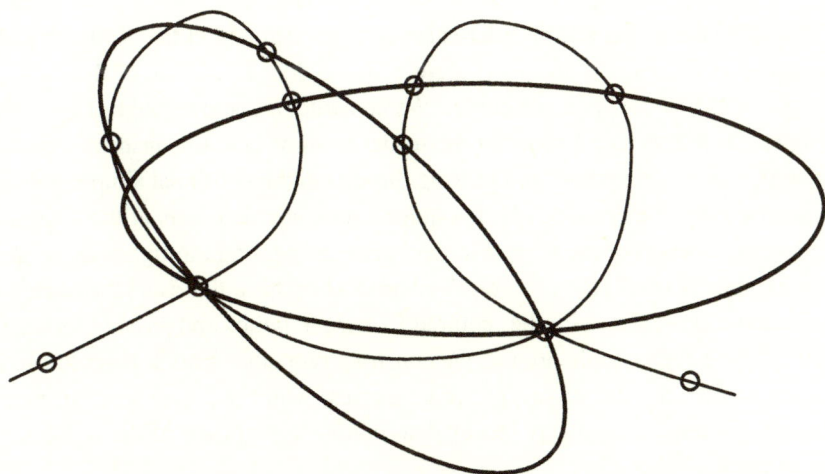

Figure 6.15. Quartic through two double points and eight simple points.

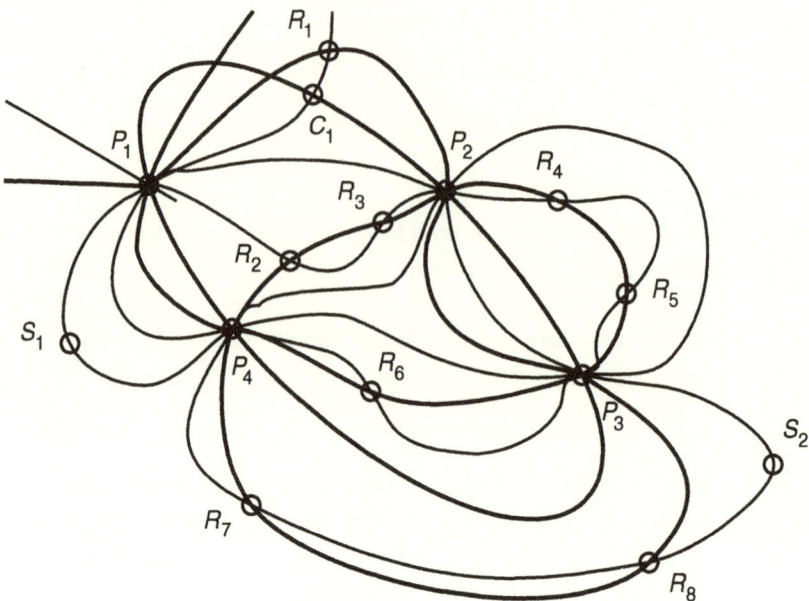

Figure 6.16. Seventh-order curve with four triple points.

and determine the parameters to make this quartic pass through the remaining two simple points.

Example 6.5 *Construct a seventh-order curve with four triple points and passing through an additional eleven simple points.*

We use a conic and a fifth-order curve as constructors. Let the desired triple points be at P_1, P_2, P_3, and P_4 (Figure 6.16), and let C_1 be one of the eleven simple points of the required curve. We construct a conic through these five points. It is left as an exercise for the reader to show how to construct a quintic with four double points and passing through eight additional simple points (Exercise 6.8). We construct such a quintic with double points at the four required triple points of the seventh-order curve and passing through eight of the remaining ten points $(R_1, \ldots, R_8)$. The product of this quintic and the conic is a seventh-order curve with the required four triple points and passing through nine of the eleven simple points. We construct two other similar pairs of such curves and form a net of seventh-order curves, which we then force to pass through the remaining two points by determining appropriate values of the net parameters.

We conclude this section with a numerical example.

Example 6.6 *Construct the quartic with three double points at* $P_1 = (0, 1, 1)$, $P_2 = (0, 0, 1)$, *and* $P_3 = (1, 0, 1)$, *and passing through the five points* $P_4 = (1, 1, 1)$, $P_5 = (-1, 1, 1)$, $P_6 = (-1, -1, 1)$, $P_7 = (1, -1, 1)$, *and the ideal point* $P_8 = (-2, 1, 0)$.

The three double points represent nine of the fourteen conditions on the quartic, and since it is also required to pass through five other points, the problem is well posed, provided that the positions of the given points do not violate any of our construction rules. For example, there cannot be more than the equivalent of four points on any line unless the quartic is degenerate.

We ignore the point at infinity and use products of conics to construct two independent quartics with the required double points and passing through the remaining four affine points. Each of the conics must contain the three points P_1, P_2, and P_3, so that their product will have double points at these positions. The following quadratic forms can be used in the construction:

$$\{P_1, P_2, P_3, P_4, P_5\}: \quad c_1 = y(y - z),$$

$$\{P_1, P_2, P_3, P_6, P_7\}: \quad c_2 = y(y + z) - 2(x - z)(x - y),$$

$$\{P_1, P_2, P_3, P_4, P_6\}: \quad c_3 = (x - y)(x + y - z),$$

$$\{P_1, P_2, P_3, P_5, P_7\}: \quad c_4 = (x + y)(x + y - z),$$

$$\{P_1, P_2, P_3, P_4, P_7\}: \quad c_5 = x(x - z),$$

$$\{P_1, P_2, P_3, P_5, P_6\}: \quad c_6 = x(x + z) - 2(y - z)(y - x).$$

The products $c_1 c_2$, $c_3 c_4$, and $c_5 c_6$ will all be quartics of the required form. We may use any two of these products to form a pencil of quartics. For example, we may use $\alpha c_1 c_2 + \beta c_5 c_6 = 0$ and require that this quartic pass through the remaining point $P_8 = (-2, 1, 0)$. This yields the condition $11\alpha + 8\beta = 0$, so that the required quartic has the form

$$(x + 2y)(11x^3 - 6xy^2 - 4y^3) - 2z(x^2 - y^2)(11x + 8y)$$
$$+ z^2(11x^2 - 6xy - 8y^2) = 0.$$

The deficiency of this quartic is zero. Therefore, by Theorem 5.21, the curve is rational and a parametrization can be found using the method described in that theorem.

6.4 Construction of Rational Curves

We have shown, in Section 6.2, that in the case of the cubic there is a great difference in complexity between the analysis in the complex projective plane and that in the real affine plane. From Theorem 5.27 we saw that the complete analysis, in the complex projective plane, of the rational cubic resulted in only two cases, yet the thirteen examples we gave in the real affine plane do not exhaust the possibilities. As the order of the curve increases, the situation becomes worse. In this section we will make a few comments on properties of rational curves in general and of quartics in particular, as well as on the construction of quartics.

From the fundamental theorem of Chapter 5 (Theorem 5.27) and the work on the resolution of singularities, we know that, without making any distinction between different kinds of points of a given multiplicity, there are four different types of rational quartics, corresponding to the four cases where the quartic has, respectively, a triple point, three double points, two double points and a double point in the first neighborhood of one of the other double points, and one double point with another double point in each of the first two neighborhoods. It would be quite bewildering to attempt to classify all the possibilities. Let us, instead, consider some of the questions relating to the parametrizations of such curves.

The proof of Theorem 5.27 relied on the fact that if a curve is birationally equivalent to a rational curve, then it, too, must be rational. The determination of whether or not a curve is birationally equivalent to a rational curve is a systematic process of applying quadratic transformations to resolve all the singularities. If the given curve is rational, then the resolution of singularities will eventually produce a proper transform that has deficiency zero. Not only does Theorem 5.21 show us that a curve with deficiency zero is rational, but the proof of this theorem gives a method for constructing the parametrization. We now have an algorithm for analyzing algebraic curves and for generating their parametrizations when they are rational. The algorithm is the following.

1. If the curve is of order one or two, it is rational and a parametrization can be produced as in Section 2.16. Go to step 5. If not, continue.
2. If the order of the curve is greater than two, determine whether or not there are any singular points. If there are no singular points, then the curve is not rational. End. If not, continue.
3. Check the deficiency of the curve. If it has deficiency zero, then it is rational. Go to step 5. If not, continue.
4. Resolve the singularities by means of the use of quadratic transformations to determine whether or not the curve has genus zero. If it does not, then the curve is not rational. End. If the genus is zero, then continue.
5. By using the method given in Theorem 5.21, produce the parametrization of the curve or of the final proper transform. If the original curve has deficiency zero, we have finished. If not, continue.

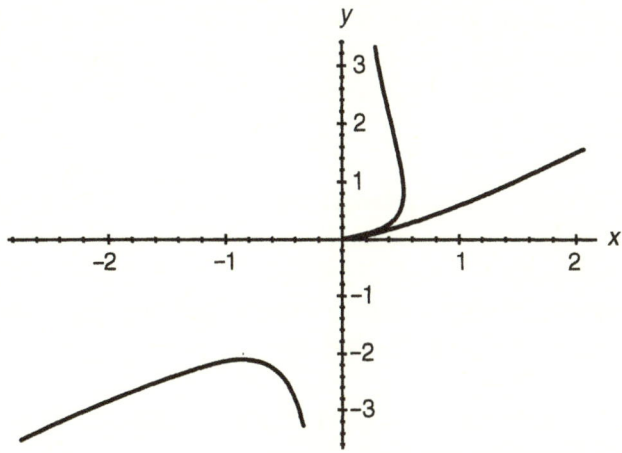

Figure 6.17. Quartic with double point in first two neighborhoods.

6. Apply the sequence of inverse quadratic transformations to produce the parametrization of the original curve.

This algorithm would work, but let us examine one case to see just how much work may be involved. In Examples 5.26 and 5.27 we examined the quartic $(x^2 - yz)^2 - xy^3 = 0$ (Figure 6.17). We showed that this curve has a cusp at $(0, 0, 1)$, a double point in its first neighborhood, and a double point in its second neighborhood. It is therefore rational, and we have completed the first four steps of the algorithm. The fourth step involved quite a bit of work, but let us look at step 5.

The final proper transform is a curve of order ten with one point of multiplicity six, three points of multiplicity four, and three double points. Its deficiency is zero, and it is rational. Applying the method of Theorem 5.21, we construct a curve of order eight that has these same points, but now of multiplicities, respectively, five, three, and one (simple points), and an additional seven simple points on the tenth-order curve. This will impose forty-three linear conditions on the eighth-order curve. Since an eighth-order curve has forty-four degrees of freedom, we have, as the theorem predicted, a one-parameter family of eighth-order curves. We now find the intersections of these two curves. There are eighty intersections, but we know, *a priori*, seventy-nine of them. The remaining one, the variable point, is given parametrically in terms of the parameter remaining in the pencil of eighth-order curves.

In the previous section we discussed how to construct algebraic curves with prescribed multiple and/or simple points. This method was certainly more interesting than solving large systems of equations. It is rare that we have to solve more than three equations at a time. However, the process builds up higher-order curves from lower-order ones, so there may be several sequences of smaller sets of equations. This approach, though tedious to do by hand, is quite practical with current computer technology, although still by no means trivial.

Is there, similarly, a more practical way to construct rational curves with specific required properties? One parametrization of the quartic $(x^2 - yz)^2 - xy^3 = 0$ was given in Section 5.11, namely, $x = t^2$, $y = t^4$, $z = 1 - t^3$. Let us consider some new examples to illustrate techniques that can be useful in the direct construction of parametrizations.

Example 6.7 *Construct a rational cubic with a cusp at* $(0, 0, 1)$, *given that the tangent at the cusp is the line* $x = 0$.

Since a cusp occurs at a point where two parameter values coalesce, let us decide that our affine line parameter t will be zero at this point. This means that the parametrization is of the form

$$x = t^2(a_1 t + b_1), \qquad y = t^2(a_2 t + b_2), \qquad z = a_3 t^3 + b_3 t^2 + c_3 t + c_4.$$

Since z cannot also be zero for $t = 0$, we normalize c_4 to have the value one. The slope of the affine tangent at the cusp is given by

$$\left. \frac{dy}{dx} \right|_{(0,0,1)} = \lim_{t \to 0} \frac{z \dfrac{dy}{dt} - y \dfrac{dz}{dt}}{z \dfrac{dx}{dt} - x \dfrac{dz}{dt}} = \lim_{t \to 0} \frac{3a_2 t^2 + 2b_2 t}{3a_1 t^2 + 2b_1 t} = \frac{b_2}{b_1},$$

which means that, for the tangent to be the y-axis, we must have $b_1 = 0$.

We have some remaining freedom in the construction, in particular with respect to the occurrence of asymptotes (the parameter values for which $z = 0$). We may, for example, require that the curve meet the ideal line exactly once when $t = 1$. This implies that $z = a_3(t - 1)^3$, with $a_3 \neq 0$. We normalize a_3 to the value one (using the fact that we only need the ratios x/z and y/z). Since the cubic cannot have another double point, this means that there will be an inflection point at $(a_1, a_2 + b_2, 0)$, with the ideal line as tangent. We now have

$$x = a_1 t^3, \qquad y = t^2(a_2 t + b_2), \qquad z = (t - 1)^3,$$

and there remain three undetermined coefficients that can affect the general shape of the curve, but not its geometrical classification (it has a cusp at the origin with tangent $x = 0$, an inflection point with the ideal line as tangent, and no asymptotes). Let us position the inflection point at $(1, -1, 0)$, so that $a_1 + a_2 + b_2 = 0$, and further require that the curve pass through $(1, 1, 1)$ when

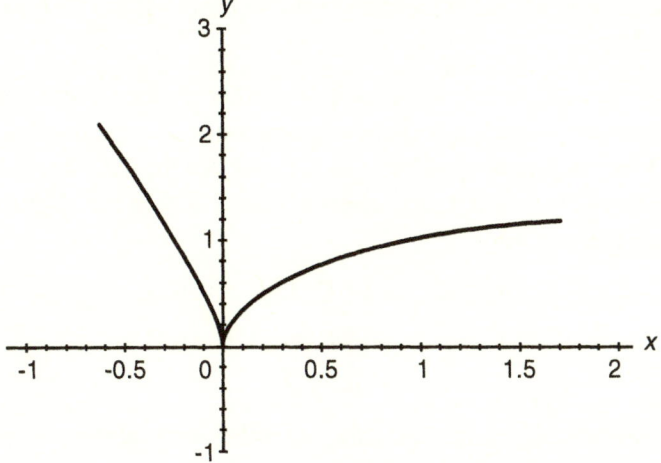

Figure 6.18. Parametrized cubic with prescribed properties.

$t = -1$, which implies that $a_1 = 8$, $a_2 = 0$, and $b_2 = -8$. This gives us the parametrization of the curve shown in Figure 6.18, namely

$$x = 8t^3, \qquad y = -8t^2, \qquad z = (t-1)^3.$$

Since the properties of the curve correspond to that of a polynomial cubic, the curve can be reparametrized to a polynomial form by setting $t/(t-1) = \tau$.

Example 6.8 *Construct a rational quartic with a crunode at $(0, 0, 1)$ with corresponding tangent lines $x = 0$ and $y = 0$, and that meets the ideal line for exactly two distinct values of the parameter.*

Since we are dealing with a fourth-order curve, the curve will intersect the ideal line in four points. To obtain exactly two points, we must ensure that one of the following is true:

1. Two of the points occur for complex values of the parameter, and the other two parameter values are distinct and correspond to simple points on the ideal line.
2. The curve intersects the ideal line twice in each of two distinct points, which means that such a point is either a point where the ideal line is tangent to the curve, or a double point, which must then be a cusp, since the parameter value corresponding to the point is unique.
3. The curve intersects the line three times for one parameter value, and the point must therefore be either a cusp or an inflection point with the ideal line

as tangent, while the other parameter value corresponds to a simple point on the ideal line.

We will illustrate the use the second possibility for the construction. The interested reader can investigate the other possibilities. Assume that the crunode is reached for the parameter values $t = \pm 1$, and that the ideal line is reached for two coalesced values of the parameter when $t = \pm 2$. At this stage the curve has the parametrized form

$$x = (t^2 - 1)(a_1 t^2 + b_1 t + c_1),$$
$$y = (t^2 - 1)(a_2 t^2 + b_2 t + c_2),$$
$$z = (t - 2)^2 (t + 2)^2,$$

where we have already used the fact that z cannot be identically zero to normalize its leading coefficient to one. As in the previous example, the slope of an affine tangent at $(0, 0, 1)$ is given by

$$\lim_{t \to \pm 1} \left(\frac{dy}{dt} \Big/ \frac{dx}{dt} \right) = \lim_{t \to \pm 1} \frac{2t(a_2 t^2 + b_2 t + c_2)}{2t(a_1 t^2 + b_1 t + c_1)} = \frac{a_2 \pm b_2 + c_2}{a_1 \pm b_1 + c_1}.$$

We select the tangent $y = 0$ to correspond to the parameter value $t = -1$, and the tangent $x = 0$ to correspond to the parameter value $t = 1$. Then we have to satisfy the conditions

$$a_2 - b_2 + c_2 = 0, \qquad a_1 + b_1 + c_1 = 0,$$

and the parametrization becomes

$$x = (t^2 - 1)(t - 1)(a_1 t + a_1 + b_1),$$
$$y = (t^2 - 1)(t + 1)(a_2 t - a_2 + b_2),$$
$$z = (t - 2)^2 (t + 2)^2.$$

In this form the curve has all the required properties, and we can use the remaining freedom in any way we want. We select the positions of the two points on the ideal line and decide that for $t = 2$ the point is at $(1, 1, 0)$, whereas for $t = -2$ it is at $(1, -1, 0)$. These positionings require that

$$3a_1 + b_1 = 3a_2 + 3b_2 \quad \text{and} \quad 3a_1 - 3b_1 = -3a_2 + b_2,$$

so that only two undetermined coefficients, for example a_1 and b_1, remain in the parametrization. If the closed loop of the crunode (there are no ideal points in the parameter interval $t \in [-1, 1]$) intercepts the y-axis at $(0, -5, 2)$ when $t = 0$, then

$$x = 12t(t^2 - 1)(t - 1),$$
$$y = -8(t^2 - 1)(t + 1)(2t - 5),$$
$$z = (t - 2)^2 (t + 2)^2.$$

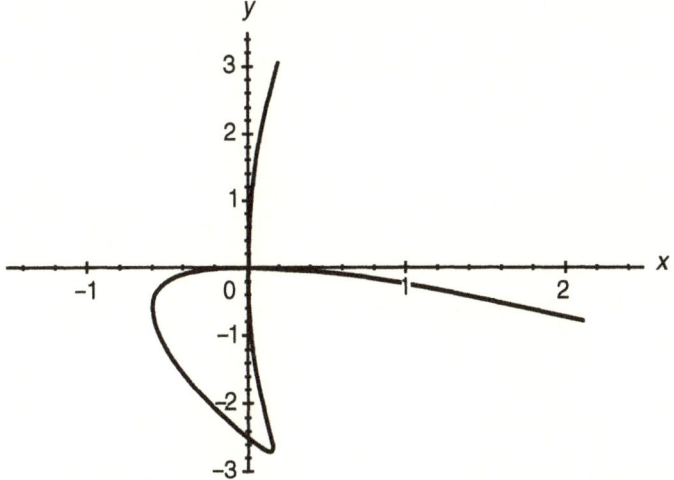

Figure 6.19. Parametrized quartic with a crunode at the origin.

A segment of this curve between the ideal points and containing the crunode is shown in Figure 6.19. The curve has another segment in the first and fourth quadrants, corresponding to the parameter interval ($t > 2, t < -2$), that is not shown. Further investigation will show that both ideal points are, in fact, points where the curve is tangent to the ideal line, and that the curve has no asymptotes.

It is possible to force one or both of the ideal points to correspond to cusps. One merely has to consider the direction of the tangent line at the point, and to require that the tangent be any line other than the ideal line.

Example 6.9 *Construct a rational quartic with a cusp at* $(0, 0, 1)$ *with tangent* $y = 0$, *and a crunode at* $(0, 1, 1)$ *with tangents* $y \pm x - z = 0$.

We use this example to illustrate the use of homogeneous line parameters. We assume that the cusp corresponds to the parameter value $t = 0$. We then have the general homogeneous parametrization ($z \neq 0$ for $t = 0$)

$$x = t^2 \left(a_1 t^2 + b_1 t s + c_1 s^2 \right),$$
$$y = t^2 \left(a_2 t^2 + b_2 t s + c_2 s^2 \right),$$
$$z = a_3 t^4 + b_3 t^3 s + c_3 t^2 s^2 + d_3 t s^3 + s^4.$$

The condition that $y = 0$ is the tangent at the cusp is, as before,

$$\lim_{t \to 0} \frac{2t \left(a_2 t^2 + b_2 t s + c_2 s^2 \right)}{2t \left(a_1 t^2 + b_1 t s + c_1 s^2 \right)} = \frac{c_2}{c_1} = 0,$$

so that $c_2 = 0$. Let us further assume that the crunode $(0, 1, 1)$ corresponds to parameter values $t \pm s = 0$, which implies the four conditions

$$a_1 + b_1 + c_1 = 0, \qquad a_2 + b_2 = a_3 + b_3 + c_3 + d_3 + 1,$$
$$a_1 - b_1 + c_1 = 0, \qquad a_2 - b_2 = a_3 - b_3 + c_3 - d_3 + 1.$$

Therefore,

$$a_1 = -c_1, \qquad b_1 = 0, \qquad c_3 = a_2 - a_3 - 1, \qquad d_3 = b_2 - b_3,$$

and the parametrization simplifies to

$$x = a_1 t^2 (t^2 - s^2),$$
$$y = t^3 (a_2 t + b_2 s),$$
$$z = t(a_3 t + b_3 s)(t^2 - s^2) + (a_2 - 1)t^2 s^2 + b_2 t s^3 + s^4.$$

We now have to consider the required tangents at the crunode. Since this is an affine point reached for affine parameter values, let us denote by $\dot{x}$, $\dot{y}$, and $\dot{z}$ the derivatives with respect to the affine parameter t/s. Then

$$\left. \frac{dy}{dx} \right|_{(0,1,1)} = \lim_{t \to \pm 1} \frac{z\dot{y} - y\dot{z}}{z\dot{x} - x\dot{z}} = \lim_{t \to \pm 1} \frac{\dot{y} - \dot{z}}{\dot{x}}.$$

If the tangent is the line $y - x - z = 0$ when $t = 1$, and the line $y + x - z = 0$ when $t = -1$, the conditions above become, after some algebraic manipulation,

$$a_2 + b_2 - a_3 - b_3 + 1 = a_1,$$
$$a_2 - b_2 - a_3 + b_3 + 1 = -a_1,$$

so that

$$a_3 = a_2 + 1, \qquad b_3 = b_2 - a_1,$$

and the parametrization becomes

$$x = a_1 t^2 (t^2 - s^2),$$
$$y = t^3 (a_2 t + b_2 s),$$
$$z = t^3 (a_2 t + b_2 s) - a_1 t s (t^2 - s^2) + (t^2 - s^2)^2.$$

There are still three remaining coefficients to play with. We choose $a_1 = a_2 = 1$ and $b_2 = 0$, so that the curve has four-point contact with the tangent at the cusp and for $s = 0$ the curve passes through $(1, 1, 2)$. Then the parametrization of the curve, shown in Figure 6.20, is

$$x = t^2 (t^2 - s^2), \qquad y = t^4, \qquad z = 2t^4 - t^3 s - 2t^2 s^2 + t s^3 + s^4.$$

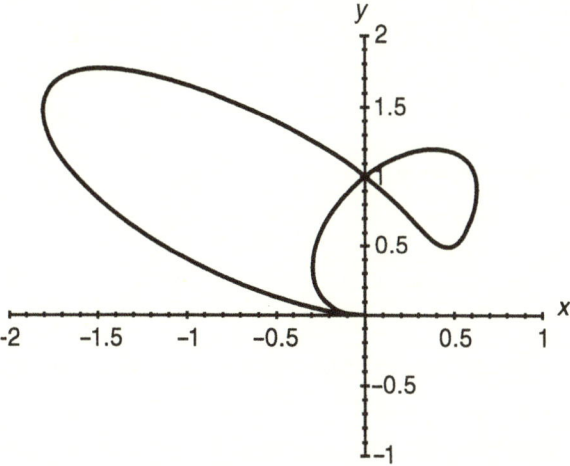

Figure 6.20. Parametrized quartic with cusp and crunode.

The gap in the curve near the point for which $s = 0$ (the finite parameter t/s tends to infinity) can be closed by reparametrization, but only at the cost (since this is a closed curve) of introducing a gap somewhere else.

6.5 Hermite Interpolation and Multiple Points

Interpolation has been a central theme to our discussions. Constraints on curves are just another way of describing interpolation conditions. Theorem 1.9 relates pointwise interpolation in $P_2(x, y)$ to conditions on conics. Theorem 1.10 is a generalization of this result for pointwise interpolation in $P_n(x, y)$. This result is a strong connection between interpolation and geometry.

The Hermite interpolant of a function of one variable interpolates the function value and derivatives up to some predetermined order. For functions of two variables, we will interpolate the function value and the values of the partial derivatives up to some predetermined order. This is different from merely interpolating the direction of a tangent.

We want to find out if there is a connection between the homogeneous Hermite problem and algebraic curves like the one between the pointwise interpolation problem and algebraic curves. In Section 5.2 we saw that the condition for a multiple point is that the function (the polynomial under consideration) and all its partial derivatives, up to some order, are zero at the point in question. The notion of multiple points is what enables us to relate two-variable Hermite interpolation in $P_n(x, y)$ to multiple points of the corresponding algebraic curve, and to generalize Theorem 1.10 so that it covers both pointwise (Lagrange) and Hermite interpolation.

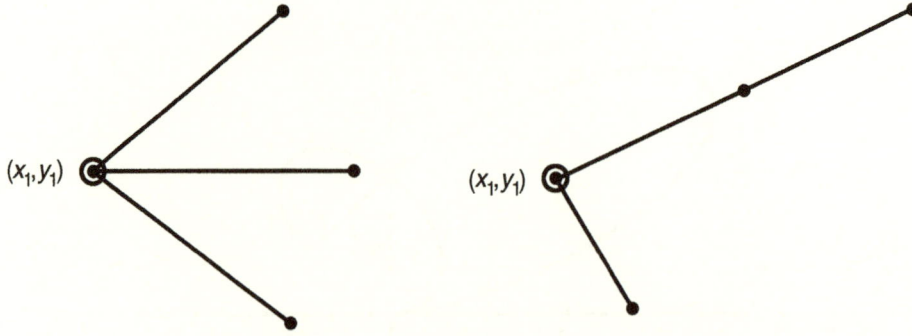

Figure 6.21. Quadratic interpolation using a double point.

In the following example, and in most of the work in this section, we use affine coordinates and avoid interpolation of properties on the line at infinity.

Example 6.10 *Consider the system*

$$f(x, y) \in P_2(x, y), \qquad f(x_i, y_i) = \alpha_i, \quad i = 1, 2, 3, 4,$$

and

$$f_x(x_1, y_1) = \beta_1, \qquad f_y(x_1, y_1) = \beta_2,$$

where (x_i, y_i) are points in general position (Figure 6.21).

(a) Show that, in general, the system possesses the finite linear interpolation property.

(b) Show that if (x_1, y_1) is collinear with two of the other given points, the system does not possess the finite linear interpolation property.

(c) Derive a normalized basis for the general case.

(d) Repeat (a) and (b) for the special case where $(x_1, y_1) = (1, 1)$, $(x_2, y_2) = (3, 2)$, $(x_3, y_3) = (0, -2)$, and (x_4, y_4) is as yet undefined. Derive the normalized basis when $(x_4, y_4) = (2, 0)$.

(a): We use Theorem 1.4 and investigate the homogeneous problem, that is we assume that $\alpha_i = \beta_i = 0$. This means that the conic corresponding to $f(x, y) = 0$ has a point of multiplicity two at (x_1, y_1) and simple points at the other three interpolation points. Since the line through the double point and one of the other three points meets the conic in three points (one double point and one simple point), we deduce from the Maclaurin–Bézout theorem that the conic must contain the line. The conic must, therefore, be a line pair. The other line must also pass through the double point. It can therefore only pass through one of the two remaining points in general position. Hence the line pair cannot pass through all the given points, and the only way the polynomial can satisfy all the homogeneous conditions is that it must be identically zero. Therefore, the only solution to the homogeneous problem is the zero solution, and hence, for points in general position, the system does posses the finite linear interpolation property.

(b): This follows directly from the discussion of (a). Suppose that the line through the double point and (x_2, y_2) also contains the point (x_3, y_3). The degenerate conic consisting of this line and the line through the double point and (x_4, y_4) satisfies all the homogeneous conditions but is not identically zero, and the system does not posses the finite linear interpolation property.

(c): We have to determine six quadratic polynomials $v_j(x, y)$ such that $v_j(x_i, y_i) = \delta_{ij}$, $i \in \{1, 2, 3, 4\}$, $j \in \{1, 2, 3, 4, 5, 6\}$ and

$$\left.\frac{\partial v_j}{\partial x}\right|_{(x_1; y_1)} = \delta_{5j}, \qquad \left.\frac{\partial v_j}{\partial y}\right|_{(x_1; y_1)} = \delta_{6j}.$$

For each $j \in \{2, 3, 4\}$ the basis function $v_j(x, y)$ has a double point at (x_1, y_1) and is zero at two of the other three points. Following the discussion of (a), we see that this polynomial must be the product of two linear forms, each associated with the line through the double point and one of the other two points. This product is a second-degree polynomial, and is normalized to ensure that it has the value one at the associated point (x_j, y_j).

For $j \in \{5, 6\}$ we consider the quadratic polynomials associated with the pencil of conics through the four points (Section 2.8). These polynomials will be of the form $\lambda C_1 + \mu C_2$, where C_1 and C_2 represent any two of the line pairs through the four points. To find λ and μ we use the appropriate interpolation conditions and solve the resulting system. For example, for $v_5(x, y)$ we solve the system

$$\begin{pmatrix} \dfrac{\partial}{\partial x}(C_1) & \dfrac{\partial}{\partial x}(C_2) \\[2mm] \dfrac{\partial}{\partial y}(C_1) & \dfrac{\partial}{\partial y}(C_2) \end{pmatrix} \begin{pmatrix} \lambda \\ \mu \end{pmatrix} = \begin{pmatrix} 1 \\ 0 \end{pmatrix},$$

where the partial derivatives are evaluated at the point (x_1, y_1).

The basis function $v_1(x, y)$ must have zero partial derivatives at (x_1, y_1), must have the value one at this point, and must evaluate to zero at the other three points. The polynomial is therefore (Section 2.10.3) of the form $\alpha l_2 l_3 + \beta l_3 l_4 + \gamma l_4 l_2$, where l_i represents the distinct line defined by (x_1, y_1) and the point (x_i, y_i), $i = 2, 3, 4$. To find α, β, and γ, we solve the appropriate system of interpolation conditions, namely

$$\begin{pmatrix} l_2 l_3 & l_3 l_4 & l_4 l_2 \\ \dfrac{\partial}{\partial x}(l_2 l_3) & \dfrac{\partial}{\partial x}(l_3 l_4) & \dfrac{\partial}{\partial x}(l_4 l_2) \\ \dfrac{\partial}{\partial y}(l_2 l_3) & \dfrac{\partial}{\partial y}(l_3 l_4) & \dfrac{\partial}{\partial y}(l_4 l_2) \end{pmatrix} \begin{pmatrix} \alpha \\ \beta \\ \gamma \end{pmatrix} = \begin{pmatrix} 1 \\ 0 \\ 0 \end{pmatrix},$$

where all the evaluations are at the point (x_1, y_1).

(d): Using the same arguments as for (a), we note that the quadratic polynomial that satisfies the homogeneous conditions must contain the line through $(1, 1)$ and $(3, 2)$, with associated linear form (Section 4.2)

$$\det \begin{pmatrix} x & y & 1 \\ 1 & 1 & 1 \\ 3 & 2 & 1 \end{pmatrix} = -x + 2y - 1,$$

as well as the line through $(1, 1)$ and $(0, -2)$, with linear form

$$\det \begin{pmatrix} x & y & 1 \\ 1 & 1 & 1 \\ 0 & -2 & 1 \end{pmatrix} = 3x - y - 2.$$

This means that the interpolant must have the form

$$f(x, y) = k(-x + 2y - 1)(3x - y - 2),$$

where k is some constant. If this function has to be zero at a point (x_4, y_4) that does not lie on one of the two lines, then the only solution is $k = 0$ and the system has the finite interpolation property. However, if (x_4, y_4) does lie on one of these lines, the homogeneous problem has a nonzero solution and the system does not possess the finite interpolation property.

Let us also approach the problem by using Theorems 1.2 and 1.3 directly, and try to construct a polynomial

$$f(x, y) = ax^2 + bxy + cy^2 + dx + ey + f$$

that satisfies the interpolation conditions. Consider the determinant of the interpolation matrix, namely

$$\det \begin{pmatrix} 1 & 1 & 1 & 1 & 1 & 1 \\ 2 & 1 & 0 & 1 & 0 & 0 \\ 0 & 1 & 2 & 0 & 1 & 0 \\ 9 & 6 & 4 & 3 & 2 & 1 \\ 0 & 0 & 4 & 0 & -2 & 1 \\ x_4^2 & x_4 y_4 & y_4^2 & x_4 & y_4 & 1 \end{pmatrix} = 5(2 - 3x_4 + y_4)(1 + x_4 - 2y_4).$$

This determinant is zero, and the system does not possess the finite interpolation property, if and only if the point (x_4, y_4) lies on one of the lines $2 - 3x + y = 0$ or $1 + x - 2y = 0$.

These are the lines through $(1, 1)$ and, respectively, $(0, -2)$ and $(3, 2)$ that we found with the first solution method. While the second approach is more straightforward, the first makes use of our geometrical insight and results in simpler algebra and direct interpretation of the result.

Using the results on (c), we conclude that

$$v_2(x, y) = \tfrac{1}{15}(3x - 2y - 2)(x + y - 2),$$

$$v_3(x, y) = -\tfrac{1}{20}(x + y - 2)(x - 2y + 1),$$

$$v_4(x, y) = \tfrac{1}{12}(x - 2y + 1)(3x - y - 2).$$

To determine $v_5(x, y)$ and $v_6(x, y)$ we choose $C_1 = (x - 2y + 1)(x - y - 2)$ and $C_2 = (3x - y - 2)(2x - y - 4)$, and solve the systems

$$\begin{pmatrix} -2 & -9 \\ 4 & 3 \end{pmatrix} \begin{pmatrix} \lambda \\ \mu \end{pmatrix} = \begin{pmatrix} 1 \\ 0 \end{pmatrix} \qquad \text{(for } v_5\text{)}$$

and

$$\begin{pmatrix} -2 & -9 \\ 4 & 3 \end{pmatrix} \begin{pmatrix} \lambda \\ \mu \end{pmatrix} = \begin{pmatrix} 0 \\ 1 \end{pmatrix} \qquad \text{(for } v_6\text{)}.$$

The solutions are, respectively,

$$\lambda = \tfrac{1}{10}, \quad \mu = -\tfrac{2}{15} \quad \text{and} \quad \lambda = \tfrac{3}{10}, \quad \mu = -\tfrac{1}{15},$$

so that

$$v_5(x, y) = \tfrac{1}{10}(x - 2y + 1)(x - y - 2) - \tfrac{2}{15}(3x - y - 2)(2x - y - 4)$$
$$= \tfrac{1}{30}(-21x^2 + 11xy + 2y^2 + 61x - 15y - 38)$$

and

$$v_6(x, y) = \tfrac{3}{10}(x - 2y + 1)(x - y - 2) - \tfrac{1}{15}(3x - y - 2)(2x - y - 4)$$
$$= \tfrac{1}{30}(-3x^2 - 17xy + 16y^2 + 23x + 15y - 34).$$

To find $v_1(x, y)$, we have to determine α, β, and γ so that the conic

$$v_1(x, y) = \alpha(x - y - 2)(2x - y - 4) + \beta(2x - y - 4)(4x - 3y - 6)$$
$$+ \gamma(4x - 3y - 6)(x - y - 2)$$

evaluates to one at $(1, 1)$ and has zero partial derivatives at this point. This means that we have to solve

$$\begin{pmatrix} 6 & 15 & 10 \\ -7 & -22 & -13 \\ 5 & 14 & 11 \end{pmatrix} \begin{pmatrix} \alpha \\ \beta \\ \gamma \end{pmatrix} = \begin{pmatrix} 1 \\ 0 \\ 0 \end{pmatrix},$$

which implies that

$$\alpha = 1, \qquad \beta = \gamma = -\tfrac{1}{5}.$$

Substitution of these values results in

$$v_1(x, y) = \tfrac{1}{5}(-2x^2 + 2xy - y^2 + 2x + 4).$$

This example shows us how to generalize Theorem 1.10. For a point to be a point of multiplicity r the polynomial and all partial derivatives up to and including those of order $r - 1$ must be zero at the point in question. This imposes $\tfrac{1}{2}r(r+1)$ linear conditions on the coefficients of the polynomial. Since a polynomial of degree n is determined by $\tfrac{1}{2}(n + 1)(n + 2)$ linear conditions, we have the following definition and theorem.

Definition 6.1 *Given a set of m distinct points $(x_i, y_i), i = 1, \ldots, m$, and a set of linear functionals*

$$L_{i,j,k}f(x, y) = \frac{\partial^{j+k} f(x_i, y_i)}{\partial x^j \partial y^k},$$

the interpolation problem defined by $L_{i,j,k}f(x, y) = \alpha_{i,j,k}$ is called the generalized pointwise interpolation problem.

Theorem 6.1 *Generalized pointwise interpolation in $P_n(x, y)$ at m distinct points (x_i, y_i) possesses the finite interpolation property if and only if there is no curve of the form $P_n(x, y) = 0$ which has the points (x_i, y_i) as points of multiplicity r_i, where $\tfrac{1}{2} \sum_{i=1}^{m} r_i(r_i + 1) = \tfrac{1}{2}(n + 1)(n + 2)$.*

Let us consider another example.

Example 6.11 *Determine under which conditions the generalized pointwise interpolation problem in the space $P_4(x, y)$ at the four distinct points $(x_1, y_1) = (2, 1), (x_2, y_2) = (1, 0), (x_3, y_3) = (3, 0)$ and a general point (x_4, y_4), with linear operators*

$$L_{i,j,k}f(x, y) = \frac{\partial^{j+k} f(x_i, y_i)}{\partial x^j \partial y^k}, \qquad 0 \le j + k \le 1, \quad 1 \le i \le 4,$$

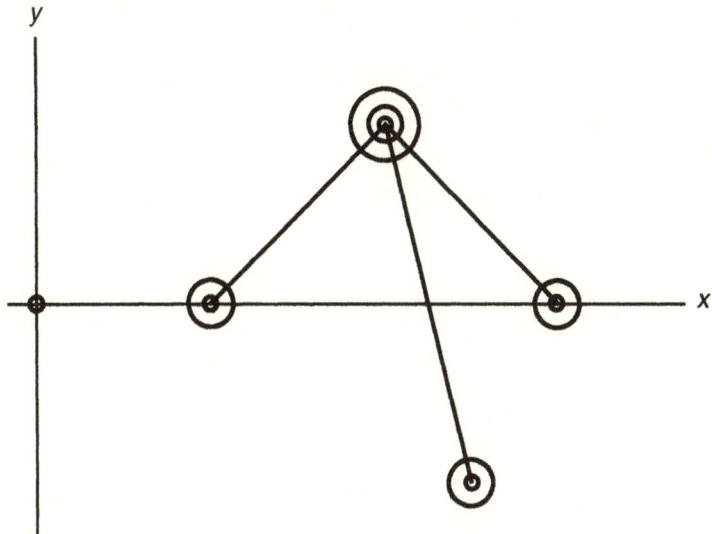

Figure 6.22. Quartic interpolation example.

and

$$L_{i,j,k} f(x, y) = \frac{\partial^{j+k} f(x_1, y_1)}{\partial x^j \partial y^k}, \qquad j + k = 2,$$

possesses the finite interpolation property (Figure 6.22).

We use Theorem 6.1, and investigate the existence of a quartic curve $f(x, y) = 0$ with a triple point at $(2,1)$ and double points at the other three points.

A line determined by the triple point and one of the double points meets the quartic at five points. Hence, by the Maclaurin–Bézout theorem, such a line must be a factor of the quartic. These lines are given by

$$x - y - 1 = 0, \qquad x + y - 3 = 0,$$

$$x(y_4 - 1) - y(x_4 - 2) + x_4 - 2y_4 = 0.$$

The quartic must be of the form

$$f(x, y) = (ax + by + c)(x - y - 1)(x + y - 3)$$

$$\times [x(y_4 - 1) - y(x_4 - 2) + x_4 - 2y_4].$$

So far, the curve has simple points at $(1, 0)$, $(3, 0)$ and (x_4, y_4), but these must be double points. Therefore, the remaining linear term $ax + by + c$ must be zero at these points. This will only be possible if (x_4, y_4) is collinear with $(1, 0)$

and $(3, 0)$. In this case the quartic is given by

$$f(x, y) = y(x - y - 1)(x + y - 3)[x + y(x_4 - 2) - x_4] = 0.$$

This quartic satisfies all the homogeneous interpolation conditions and, hence, the system does not possess the finite interpolation property.

If the point (x_4, y_4) lies on the line $x - y - 1 = 0$ (the other case is similar), any quartic of the form

$$f(x, y) = [a(x - 3) + by](x - y - 1)^2(x + y - 3) = 0,$$

will satisfy all the required homogeneous interpolation conditions for any line $a(x - 3) + by = 0$ through the point $(3, 0)$.

The given system therefore possesses the finite interpolation property if and only if (x_4, y_4) does not lie on any one of the lines $x - y - 1 = 0$, $x + y - 3 = 0$, or $y = 0$.

Two things stand out from these examples. In the first place, the geometrical approach allows us to tell, *a priori*, whether or not the system possesses the finite interpolation property. It also enables us to do this without any great amount of computation. Secondly, the geometrical approach gives us a more efficient way of solving the problem in those cases where the system does possess the finite interpolation property.

6.6 Parabolic Interpolation

In Chapter 1 we discussed parabolic interpolation with a parabola of the form $y = ax^2 + bx + c$. In Chapter 2 we discussed conics, which include parabolas, and in Chapter 4 we showed the invariance of parabolas under affine transformations. In Section 2.16 we constructed the parametrization of a particular parabola defined by a specific interpolation problem. In this section we examine a few different parabolic interpolation problems.

6.6.1 A Parabola through Four Points

We saw, in Section 2.8, that there is a pencil of conics through four affine points. In Section 2.6 we showed that the center of the general conic given by the equation

$$ax^2 + 2hxy + by^2 + 2gx + 2fy + c = 0, \tag{6.1}$$

is ill defined if

$$h^2 - ab = 0. \tag{6.2}$$

In this case the conic is a parabola. We can use this property as the condition that defines the parabola. In Chapter 2 we used affine coordinates. In the projective terminology developed in Chapters 3 and 4 we can say that the center of the parabola is on the ideal line. This gives us an alternative way of thinking about the parabola, for it shows that there is one direction, the direction given by the center, in which all lines will meet the parabola in precisely one affine point. This property was used in Section 2.16 to obtain a parametrization of a parabola. This means that by a rotation of axes in this direction the parabola can be written as a function. This is a special case of the more general statement that by using a rotation of axes any planar algebraic curve of order n with a point of multiplicity $n-1$ on the ideal line can be written as a function.

The pencil of conics through four points can be written in the form

$$\lambda l_1 l_3 + \mu l_2 l_4 = 0, \tag{6.3}$$

where the l_i, $i = 1, 2, 3, 4$, are the linear forms associated with the four lines that form the boundary of the quadrilateral defined by the four points. This equation is of the form of (6.1). It is linear in the ratio λ/μ. The condition (6.2) for this conic to be a parabola is quadratic in terms of the coefficients. In general, this condition will have two roots for the ratio λ/μ, each one corresponding to one member of the pencil. Hence, in general, there will be two parabolas through four points. Let us look at an example.

Example 6.12 *Determine the equations of all the parabolas through the four points* $(-1, 0)$, $(1, 0)$, $(2, 1)$, *and* $(-1, 3)$.

Using appropriate pairs of lines as indicated in (6.3), we find that the pencil of conics through the four points can be expressed in the form

$$\lambda y(2x + 3y - 7) + \mu(x - y - 1)(x + 1) = 0,$$

with quadratic form

$$\mu x^2 + (2\lambda - \mu)xy + 3\lambda y^2 - (7\lambda + \mu)y - \mu = 0.$$

The condition (6.2) for the conic to be a parabola now becomes

$$(2\lambda - \mu)^2 - 12\lambda\mu = 0, \quad \Longrightarrow \quad \mu = (8 \pm 2\sqrt{15})\lambda.$$

The two parabolas are, therefore,

$$[(\sqrt{5} + \sqrt{3})x - \sqrt{3}y]^2 - \sqrt{15}(\sqrt{15} + 2)y - 8 - 2\sqrt{15} = 0,$$

and

$$[(\sqrt{5} - \sqrt{3})x + \sqrt{3}y]^2 - \sqrt{15}(\sqrt{15} - 2)y - 8 + 2\sqrt{15} = 0.$$

These parabolas are shown in Figure 6.23.

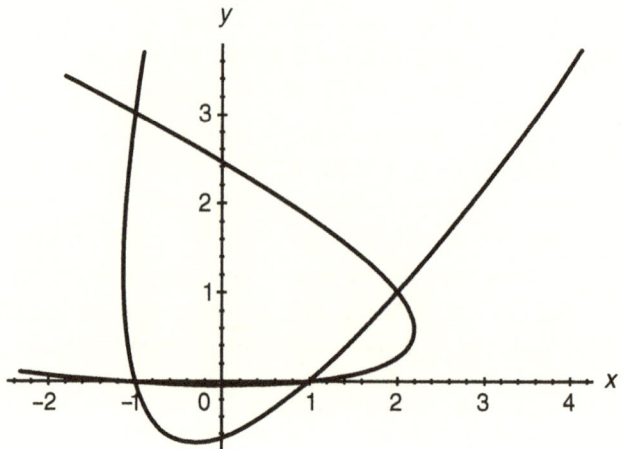

Figure 6.23. Two parabolas through four points.

Let us consider the possibility that there might be some orientation of the four points such that there is a unique parabola through the four points.

Let the linear forms in (6.3) be given by

$$l_i = a_i x + b_i y + c_i.$$

The condition (6.2) for the conic to be a parabola becomes, after simplification,

$$2\lambda\mu[(a_1b_3 + a_3b_1)(a_2b_4 + a_4b_2) - 2a_1a_3b_2b_4 - 2a_2a_4b_1b_3]$$
$$+ \lambda^2(a_1b_3 - a_3b_1)^2 + \mu^2(a_2b_4 - a_4b_2)^2 = 0.$$

This is the quadratic that will, in general, give us the two solutions for the ratio λ/μ. The discriminant of this quadratic is

$$4(a_1b_2 - a_2b_1)(a_2b_3 - a_3b_2)(a_3b_4 - a_4b_3)(a_4b_1 - a_1b_4).$$

This expression can be zero, with a unique parabola through the four points, only if one of the factors is zero. That cannot happen, because it would imply that the corresponding lines (neighboring sides of the boundary of the quadrilateral defined by the four points) were parallel. This, clearly, must lead to a contradiction, since if the two lines are parallel, their intersection must lie on the line at infinity, which is therefore not an affine point. However, if the lines that form opposite sides of the quadrilateral are parallel, then either the coefficient of λ^2 or the coefficient of μ^2 is zero. In this case the quadratic in μ and λ factors, with one of the factors being, either, μ or λ. The corresponding solution [see (6.3)] is a line pair. The other solution for the ratio λ/μ (or μ/λ, whichever is appropriate), leads to a parabola. We summarize this result by stating the following theorem.

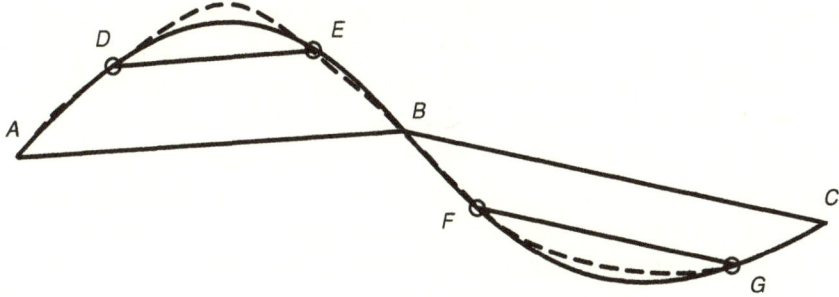

Figure 6.24. A parabolic curve approximation.

Theorem 6.2 *There is a unique parabola through four distinct points if and only if the four points form the vertices of a trapezoid.*

This result provides the following method for piecewise parabolic interpolation to a given curve.

Consider a given curve divided into segments by placing nodes at selected points. Since we are using parabolic segments for the approximation, node placements should include all inflection points of the given curve. Connect the selected nodes by straight line segments (AB and BC in Figure 6.24, where the broken curve is given). For each segment demarcated by two consecutive nodes, draw another line segment parallel to the one connecting the nodes and with its endpoints on the curve between the two selected nodes (DE and FG in Figure 6.24). Since the four points relevant to each segment form the vertices of a trapezoid, they define a unique interpolating parabola that can be used to form a continuous approximation of the given curve segment. If the constructed parallel line segment is translated until it is tangent to the given curve, the two additional points coalesce and the interpolation problem becomes that of constructing the (unique) parabola through three points, with tangent at the middle point parallel to the chord connecting the endpoints. This situation is discussed in more detail in the following subsection.

6.6.2 A Parabola through Three Points

In Section 2.16 we discussed a parametrization of the parabola based on projections along lines parallel to the axis of the parabola. This provides a geometrical interpretation of the transformation from a straight line to the parabola and, since a rational curve is a rational mapping of a line, a geometrical interpretation of the parametrization, with the parameter specifying a point on the line. The result was the determination of a parabola through three points. Although the parabola does have four degrees of freedom and can be determined by four linear conditions, these conditions need not be the interpolation at four points.

An alternative way to produce a parabolic interpolation through three points is to create a parametrization that goes through the endpoints of three consecutive points (two linear conditions) and then not only to force the third point to lie on the parabola, but also to prescribe the parameter value corresponding to this point. We will show that this is equivalent to imposing two more linear conditions.

The general equation of a conic in projective coordinates is

$$ax^2 + 2hxy + by^2 + 2gxz + 2fyz + cz^2 = 0. \tag{6.4}$$

This conic meets the line at infinity where $z = 0$, that is, where $ax^2 + 2hxy + by^2 = 0$. If the conic must be a parabola, then $h^2 = ab$ and the intersections at infinity can be expressed in the form $(\alpha x + \beta y)^2 = 0$, where $\alpha^2 = a$ and $\beta^2 = b$. Since a conic does not have a double point, a line can only meet a conic at two coincident points if it is a tangent. Hence the line at infinity is a tangent to the parabola. Furthermore, since a line meets a conic in precisely two points, the line at infinity cannot meet the parabola at any other point than the one where it is a tangent. As we discussed in Section 5.4, an asymptote of a curve is a line, other than the line at infinity, that is a tangent to the curve at infinity. Therefore, a parabola does not have an asymptote (real or complex). We have seen that any conic always allows a rational parametrization. Any zeros of the denominator would correspond to points on the line at infinity, and hence to potential asymptotes. The parabola, therefore, has a polynomial parametrization, which can be constructed as follows.

The general parabola $(\alpha x + \beta y)^2 + 2gxz + 2fyz + cz^2 = 0$ meets the line at infinity at the point $(\beta, -\alpha, 0)$. The general line through this point has the equation $\alpha x + \beta y + \gamma z = 0$, and meets the parabola where

$$\frac{x}{z} = \frac{\beta\gamma^2 - 2f\gamma + \beta c}{2(\alpha f - \beta g)} \quad \text{and} \quad \frac{y}{z} = -\frac{\alpha\gamma^2 - 2g\gamma + \alpha c}{2(\alpha f - \beta g)}.$$

This gives a general parametrization of the parabola that is polynomial in the parameter γ.

Example 6.13 *Use the method described above to find a polynomial parametrization of the parabola*

$$(3x - 2y)^2 - 9xz + 2yz - 18z^2 = 0.$$

The given parabola meets the line at infinity in the point $(2, 3, 0)$, and the pencil of lines through this point is given by

$$3x - 2y + \gamma z = 0.$$

Simultaneous solution of this line equation and the equation of the parabola yields the coordinates of their only affine intersection point, namely

$$x = \frac{z}{6}(\gamma^2 + \gamma - 18) \quad \text{and} \quad y = \frac{z}{4}(\gamma^2 + 3\gamma - 18).$$

Although the lines $3x - 2y + \gamma z = 0$ are all diameters of the parabola (Section 2.6), only one value of γ corresponds to the axis itself, namely $\gamma = -\frac{31}{26}$.

Let us consider the parabola through two given points. Let the points have coordinates (x_1, y_1) and (x_2, y_2). Given one polynomial parametrization of a curve in terms of a parameter γ, we can derive another one by a change of parameter of the form $t = p\gamma + q$. The coefficients p and q can be used to make two points on a curve with polynomial parametrization to correspond to two preselected values of the parameter. [For rational curves we may use the reparametrization $t = (p\gamma + q)/u\gamma + v, u^2 + v^2 \neq 0$, and select three points on a curve to correspond to three preselected values of the parameter.] Hence we can write the parabola through two points in the affine parametric form

$$x = \alpha t^2 + (x_2 - x_1 - \alpha)t + x_1,$$
$$y = \beta t^2 + (y_2 - y_1 - \beta)t + y_1.$$

The coefficients α and β in these equations are simply arbitrary constants needed to define the parabola. The slope of the axis of the parabola is given by the ratio β/α. The point (x_1, y_1) corresponds to the parameter value $t = 0$, while the point (x_2, y_2) corresponds to the parameter value $t = 1$.

This parabola has two degrees of freedom (α and β), which is just what we would expect for a parabola through two points. It would be awkward to use this form of the parabola to determine a parabola through four points, for we would have to deduce the values of the parameter that corresponded to the two other points. We cannot specify these values. However, we can determine a unique parabola through three points by specifying not only the condition for the third point to lie on the parabola but also at what parameter value this must happen.

Let the third point be (x_m, y_m), and let this point correspond to $t = \frac{1}{2}$. The parabola becomes

$$x = 2(x_1 + x_2 - 2x_m)t^2 - (x_2 + 3x_1 - 4x_m)t + x_1,$$
$$y = 2(y_1 + y_2 - 2y_m)t^2 - (y_2 + 3y_1 - 4y_m)t + y_1.$$

The slope of this parabola at the point (x_m, y_m) is $\dot{y}/\dot{x}$ evaluated at $t = \frac{1}{2}$. This is given by $(y_2 - y_1)/(x_2 - x_1)$, which is the slope of the line joining the first two

points. We see, therefore, that this interpolation (three points at three equally
spaced parameter values) is equivalent to the interpolation of the three points
and a slope, parallel to the line defined by the endpoints, at the middle point.
We know (Theorem 6.2) that there is a unique parabola through four points if
and only if the four points lie on a trapezoid. If two of these points (on the same
one of the parallel lines) coalesce, the interpolating conditions change, in the
limit, from interpolation at two distinct points to interpolation at the coalesced
point, with prescribed slope at that point given by the direction along which the
points coalesce.

The axis of the parabola has slope $(y_1 + y_2 - 2y_m)/(x_1 + x_2 - 2x_m)$, which
is the slope of the line defined by the midpoint of the chord connecting (x_1, y_1)
and (x_2, y_2), and the point (x_m, y_m). This is in agreement with the interpolation
discussed in Section 2.16.

6.6.3 A Parabola with Given Slopes at Two Points

Let us look at a method that will enable us to produce a smooth, slope-
continuous, parabolic interpolation to a given curve.

Let the broken line through the points A and B be the given curve, and let
CA and CB be the tangents at A and B, respectively (Figure 6.25). Let the
linear forms associated with the line AB and the tangents CA and CB be given
by $l = 0, t_1 = 0$, and $t_2 = 0$. Then the pencil of conics through the two
points and tangent to the two lines is given (as described in Section 2.11) by
$l^2 + \alpha t_1 t_2 = 0$. The condition for this conic to be a parabola will be quadratic
in the parameter α. In this particular case there is a very interesting solution.
Since we are working in affine space, we can apply any affine transformation
we wish. We therefore make the following simplifying transformation of coordinates. Without loss of generality we can consider the tangents to be the axes
and the line AB to have equation $1 - x - y = 0$. This is equivalent to mapping
the three points A, B, and C to the points with coordinates $(1, 0)$, $(0, 1)$, and
$(0, 0)$. The tangents cannot be parallel, or no such affine transformation will
exist. The equation of the pencil of conics now becomes

$$(1 - x - y)^2 + \alpha xy = 0.$$

The condition that this conic is a parabola is $\alpha(\alpha + 4) = 0$. The solution
$\alpha = 0$ corresponds to a line pair (a repeated line), which is degenerate and not
a parabola. The other solution $\alpha = -4$ gives the parabola

$$(x - y)^2 - 2x - 2y + 1 = 0.$$

Once again we have managed to find a unique parabola.

Since this technique of approximation can match prescribed slopes at the
endpoints of segments, it provides a piecewise parabolic interpolation technique

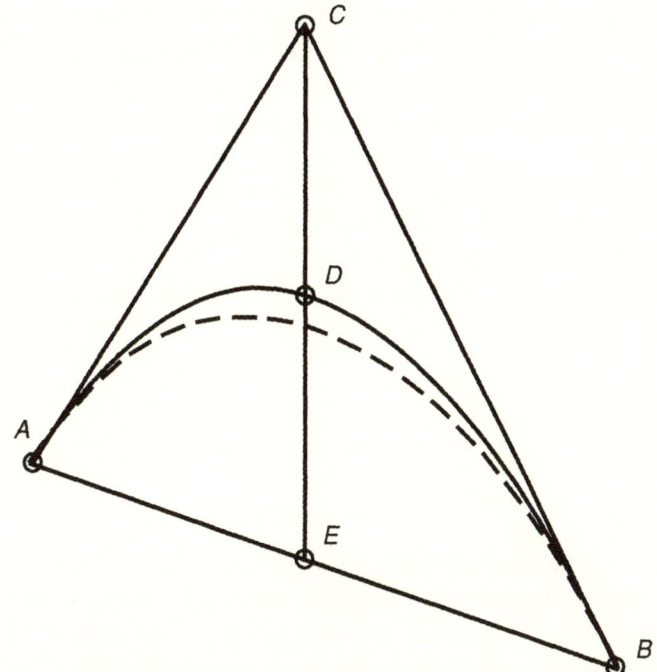

Figure 6.25. A parabola through two points with given slopes.

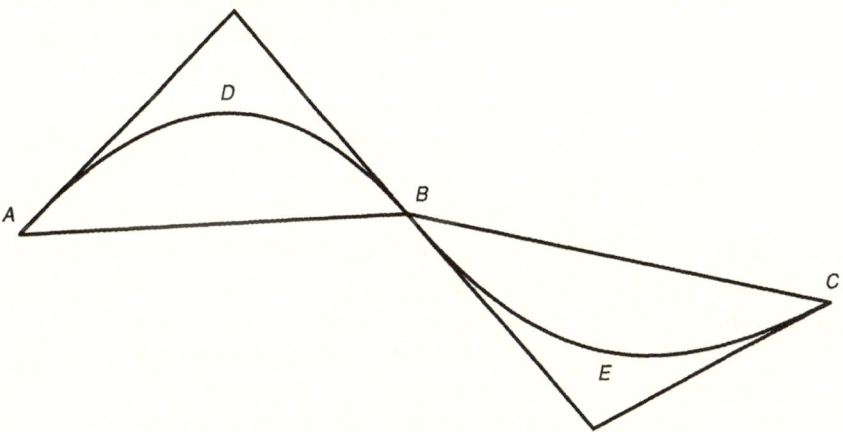

Figure 6.26. Piecewise parabolic C^1 interpolation.

that maintains continuity of slope between parabolic segments. Figure 6.26 shows two parabolic arcs (ADB and BEC) meeting with continuous slope at B.

We relate this method to the method of Section 6.6.2 by pointing out that the parabola so constructed is the same one that would have been constructed by selecting the point D (Figure 6.25) to be the midpoint of the line CE, where

E is the midpoint of AB, and then using the previous method to construct a parabola through the three points A, D, and B. This observation leads to the following equivalent algorithm.

Given two points and the slopes at these points:

1. Construct the point where the tangents meet (point C in Figure 6.25).
2. Construct point E as the midpoint of AB.
3. Construct point D as the midpoint of CE.
4. Use the points A, B, and D as the three points in the method given in Section 6.6.

6.7 Geometric Approximation

6.7.1 Function Metrics

In this section we take a brief excursion into the area of approximation. In many cases one uses interpolation as an approximation, but it is important to understand the distinction between the two. We have been using the word "interpolation" to refer to a procedure where some curve is forced to satisfy a set of conditions. That is, there are certain equations that must be satisfied by the constructed curve. The implication of the word "approximation" is that we are given some curve and are trying to construct another curve that is, in some sense, close to the given one. In many cases, a curve constructed via some interpolation procedure will be close to the given curve if the interpolation conditions are consistent with properties of the given curve. For example, in the previous section we discussed a method of producing a parabola that shared two points and two slopes with some given curve. In many situations such a parabola will be close to the given curve over a reasonably large segment. In this way the interpolation has resulted in an approximation. However, the focus in approximation is often a little different. When we talk about one curve being close to another curve we must have some meaning of "close." It is this concept of closeness that we now discuss.

Figure 6.27 illustrates an approximation $g(x)$ to a given function $f(x)$ over the closed interval $[a, b]$. Common measures of closeness are

$$L_1 \equiv \int_a^b |f(x) - g(x)|\, dx,$$

$$L_2 \equiv \left(\int_a^b [f(x) - g(x)]^2\, dx \right)^{1/2},$$

$$L_\infty \equiv \max_{a \le x \le b} |f(x) - g(x)|.$$

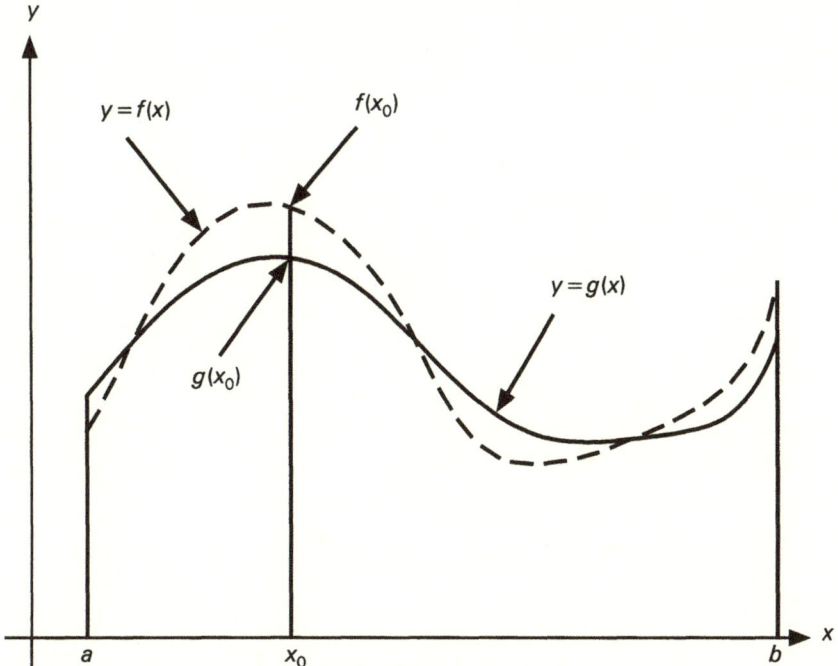

Figure 6.27. Function approximation.

We often try to produce the "best approximation" of a given function $f(x)$ in the sense of one of these measures. We try to find a $g(x)$ from some preselected class of functions, such that the measure of closeness between $f(x)$ and $g(x)$ is as small as possible. For example, we may decide to use polynomials of degree four as our class of approximating functions and then try to find that polynomial of degree four, $p_4(x)$, such that

$$\int_a^b |f(x) - p_4(x)|\, dx \le \int_a^b |f(x) - g(x)|\, dx \quad \text{for all} \quad g(x) \in P_4(x).$$

Each of the three examples of measures we have given possesses the following properties.

1. $L(f, g) \ge 0, L(f, g) = 0 \iff f = g,$
2. $L(f, g) = L(g, f),$
3. $L(f, g) + L(g, h) \ge L(f, h).$

Functions that satisfy these properties are called *distance functions* or *metrics*.

We have discussed the difficulties inherent in using curves that are functions in interpolation and approximation. The property of being a function is dependent on the particular orientation of the axes. For example, the curve

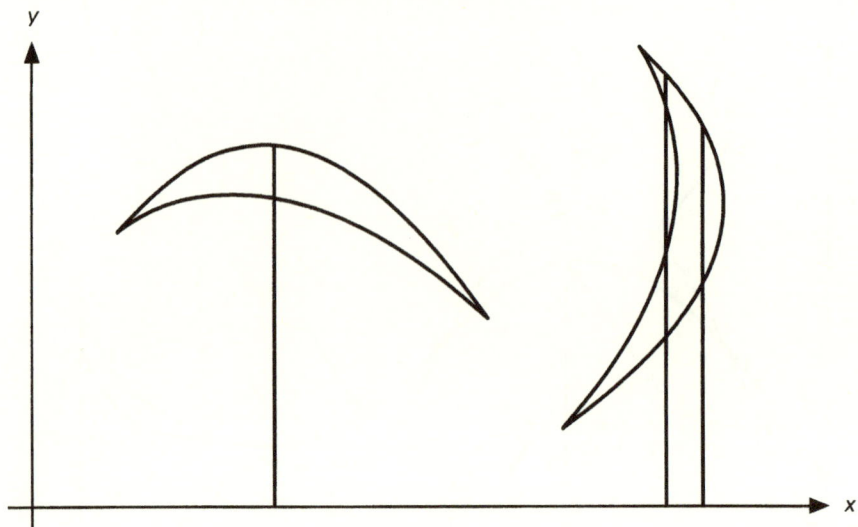

Figure 6.28. Distance between function and nonfunction curves.

$y = x^2$ is a curve that represents a function from x to y. The curve $y^2 = x$ is the same curve rotated clockwise through $90°$, but its equation does not represent a function of x; rather it represents a function from y to x. The curve $(x - y)^2 - 2(x + y) + 1 = 0$ does not represent either a function from x to y or from y to x. It does, however, represent a function from $x - y$ to $x + y$. However, there is no combination of variables that could be used to make the even simpler curve $x^2 + y^2 = 1$ a function.

Do the metrics we have given make any sense in the context of our geometrical approach to curve design and approximation? Consider Figure 6.28. On the left we get the usual picture of a distance function acting on two curves. We measure some distance parallel to the y-axis, and things work as expected. The distance gives us a workable measure of closeness. However, the curve pair on the right is simply the same two curves rotated through $90°$. These curves are just as close to each other as they were before, yet the metric we used in the first case will be quite meaningless. For some lines parallel to the y-axis there are not unique intersections with the appropriate curves. These metrics are not geometrical properties of the two curves.

6.7.2 Closeness as a Geometrical Property

Let us now seek some functions of a given and an approximating curve that will not only be meaningful measures of closeness but also geometrical invariants, invariant under a particular transformation group. The group associated with

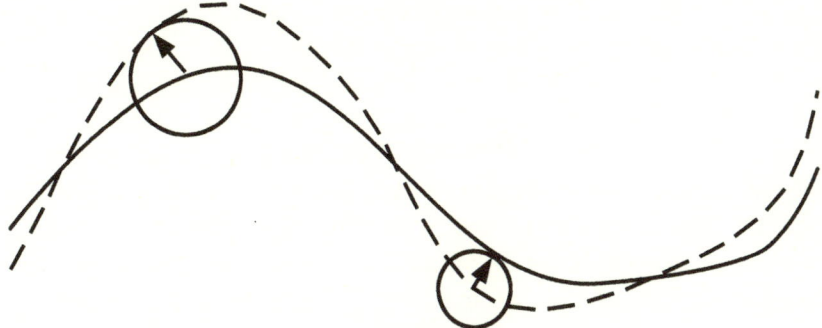

Figure 6.29. The Hausdorff distance.

Euclidean geometry is the group of rigid motions and reflections. Since this includes rotations, we see that measuring distance in a direction parallel to the y-axis cannot lead to even a Euclidean property. The curve approximation problem we address is the following: Given a curve C_g, a transformation group G, and a member C_a from a class of approximating curves, can we find some function $F(C_g, C_a)$ that will provide a meaningful measure of closeness between the two curves and is also invariant under the group G?

One example is the following. In Figure 6.29 we show the given curve as a broken line and the approximating curve as a solid line. For a point on one curve we find the minimum Euclidean distance to the other curve. This distance is the radius of the smallest circle, centered on the first point, that meets the second curve. The distance between the curves is then defined to be the maximum of these minimum distances as we move along the curve. This measure is a Euclidean invariant. It is also nonnegative. However, the measure is not symmetric, in the sense that the distance from the given curve to the approximating curve is not, in general, the same as the distance from the approximating curve to the given curve. One could calculate both distances and use the larger of the two values as the distance between the curves. This is a special case of what is called the Hausdorff distance function or metric.

There is great generality in this idea, in that it allows for a very broad class of approximating curves. However, its implementation requires much computation. We often find that simple curves can still be effective in approximation, and when we use simple curves, we may get the benefit of utilizing their special properties. Often we are interested in position, slope, and perhaps curvature, but it is rare that we are interested in approximating any properties that involve higher-order derivatives than the second. It is also rare that we would desire our approximation to have cusps, nodes, or higher-order singularities. However, the use of high-order curves may result in such undesirable properties occurring. This is one of the reasons that it is usually advisable to use simple curves. The following application illustrates this idea.

6.7.3 Parabolic Approximation

The parabola has a unique direction associated with it, namely the direction of its axis. This means that any line in Euclidean space that is parallel to the axis of the parabola will meet the parabola in a single point. Therefore, a line through some point of a given curve and parallel to the axis of the approximating parabola will meet the parabola in a unique point, and the distance between these points will serve as a measure. We illustrate this concept in Figure 6.30.

For convenience we will assume that the given curve is expressed in parametric form. This is not a requirement for the method to work, but simply allows a neater notation. Let the given curve segment be expressed by the equations

$$x = A(t), \quad y = B(t), \qquad t \in [0, 1],$$

and the parabola by

$$x = as^2 + (x_2 - x_1 - a)s + x_1,$$
$$y = bs^2 + (y_2 - y_1 - b)s + y_1, \qquad s \in [0, 1].$$

As we have seen, we can always select two points to correspond to two predetermined values of the parameter, so this representation is quite general. The slope of the axis of this parabola is b/a. The equation of the line through the

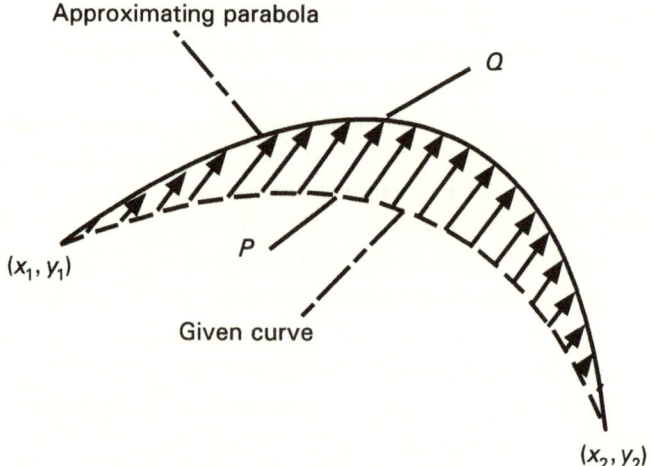

Figure 6.30. Parabolic approximation.

point $(A(t), B(t))$ and parallel to the axis of the parabola is

$$a[y - B(t)] = b[x - A(t)],$$

and this meets the parabola where

$$\tilde{s} = s(t) = \frac{a[B(t) - y_1] - b[A(t) - x_1]}{a(y_2 - y_1) - b(x_2 - x_1)}.$$

Hence, if P represents an arbitrary point $(A(t), B(t))$ on the given curve and Q is the corresponding point on the parabola, then Q is given by $(x(\tilde{s}), y(\tilde{s}))$, and the distance between these two points provides a measure that can be used to define the closeness of the two curves. There is no guarantee that the value of $\tilde{s}$ will be in the interval $[0,1]$ for all $t \in [0, 1]$, but that will not affect the method. One can constrain the slope of the axis of the parabola so that the intersections are in the required interval, reparametrize the parabola so that the extreme values of $\tilde{s}$ correspond to the parameter values zero and one, respectively, or simply use the corresponding segment of the parabola as the approximating segment, which is equivalent to ignoring any restriction on the parameter range of the parabola.

The distance PQ between the points is given by

$$d(t) = \{[A(t) - x(\tilde{s})]^2 + [B(t) - y(\tilde{s})]^2\}^{1/2}.$$

We could now use any suitable metric as our measure of closeness, for example,

1. $\max d(t), t \in [0, 1]$,
2. $\int_0^1 [d(t)]^2\, dt$.

This approach does not yield a measure that is symmetric, since if $m(C_1, C_2)$ represents the measure of closeness from curve C_1 to the parabola C_2, then $m(C_1, C_2) \neq m(C_2, C_1)$. In fact, $m(C_2, C_1)$ may not even be defined.

Example 6.14 *Use the second measure of closeness given above to determine the closeness of the cubic curve segment $x = 3t^2 - 1$, $y = -9t^3 + 9t$, $0 \leq t \leq 1$, to the parabola that has the same tangents as the given segment at its endpoints.*

Elimination of the parameter t in the parametric equations of the given curve shows that the given segment is the upper half of the closed loop of the cubic curve $y^2 - 3(x - 2)^2(x + 1) = 0$. The tangent at $t = 0$ [the point $(-1, 0)$] is vertical, whereas the tangent at the double point $(2, 0)$, which is reached when $t = 1$ (also when $t = -1$, but this is outside the segment under discussion) has slope -3. We follow the construction procedure for the unique parabola

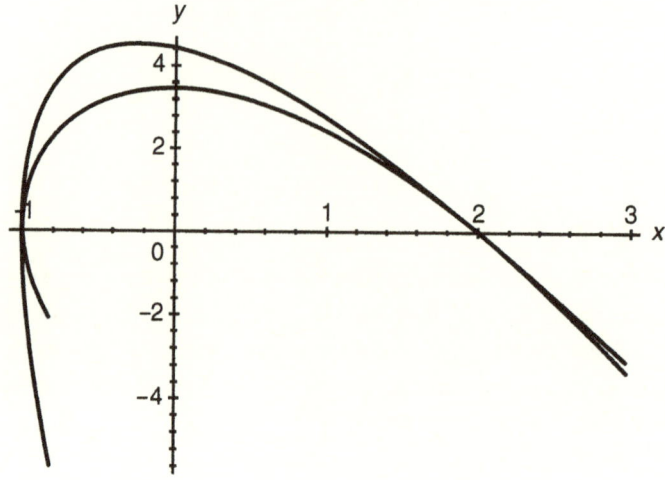

Figure 6.31. Parabolic approximation of cubic.

with given slopes at two interpolation points, and find that the parabola has the algebraic form

$$(6x + y)^2 - 36x + 12y - 72 = 0,$$

with associated polynomial parametrization

$$x = 3s^2 - 1, \quad y = -18s^2 + 18s, \qquad 0 \le s \le 1.$$

Segments of the two curves are shown in Figure 6.31.

For any selected value of $t \in [0, 1]$, the line parallel to the slope of the axis of the parabola through the point P with coordinates $x_P = 3t^2 - 1$, $y_P = -9t^3 + 9t$ is given by $y + 9t^3 - 9t = -6(x - 3t^2 + 1)$. Substitution of the parametric equations of the parabola for x and y in the line equation shows that the line intersects the parabola when $\tilde{s} = \frac{1}{2}t(t + 1)$. For $t \in [0, 1]$, the value of $\tilde{s}$ will also be in the designated interval. The coordinates of the point Q on the approximating parabola are, after substitution and simplification,

$$x_Q = \tfrac{3}{4}t^4 + \tfrac{3}{2}t^3 + \tfrac{3}{4}t^2 - 1, \qquad y_Q = -\tfrac{9}{2}t^4 - 9t^3 + \tfrac{9}{2}t^2 + 9t,$$

so that

$$x_P - x_Q = -\tfrac{3}{4}t^4 - \tfrac{3}{2}t^3 + \tfrac{9}{4}t^2, \qquad y_P - y_Q = \tfrac{9}{2}t^4 - \tfrac{9}{2}t^2.$$

The required measure of closeness can now be calculated, namely

$$\int_0^1 [(x_P - x_Q)^2 + (y_P - y_Q)^2]\,dt = \frac{131}{224} \approx 0.585.$$

6.7.4 Geometrical Approximation with Special Curves

The idea behind the parabolic approximation described in the previous subsection is: since a conic can only have simple points, we can use any point on it as the common point of the pencil of parametrizing lines. In our application we used the unique point of the parabola on the line at infinity as the common point of the lines. We showed (Section 5.3) that an irreducible curve of order n with point of maximum multiplicity $n - 1$ always has a rational parametrization, and described how this parametrization can be obtained by using lines through the multiple point and calculating the one remaining intersection with the curve in terms of the slope of the line (the parameter).

Consider that subset of rational curves of order n with the property that every curve in the set has a point of multiplicity $n - 1$ on the ideal line. Such curves can be used to approximate given parametrized curve segments in much the same way that we used parabolas.

In order to derive a general equation for a rational curve of order n with point of multiplicity $n - 1$ at infinity, let us first consider such a curve with the multiple point at the origin. From Section 5.4 we know that the curve has the general form

$$\sum_{i=0}^{n} a_i x^i y^{n-i} + z \sum_{i=0}^{n-1} b_i x^i y^{n-i-1} = 0.$$

Since the multiplicity of a point is invariant under projective transformations (Theorem 5.6), we can transform the point $(0, 0, 1)$ to the point $(a, b, 0)$ to find the required general equation. We are therefore applying the theory conversely to the applications in Section 5.4 – rather than shift a multiple point to the origin so that we can read off its multiplicity, we prescribe the multiplicity at the origin, write down the equation of the curve, and shift the origin to the point at infinity. One way of doing this is to first interchange the lines $y = 0$ and $z = 0$, so that $(0, 0, 1)$ maps to $(0, 1, 0)$, and then apply an affine transformation that maps $(0, 1, 0)$ to $(a, b, 0)$. The projective transformation matrix is

$$\begin{pmatrix} b & a & 0 \\ 0 & b & 0 \\ 0 & 0 & b \end{pmatrix} \begin{pmatrix} 1 & 0 & 0 \\ 0 & 0 & 1 \\ 0 & 1 & 0 \end{pmatrix} = \begin{pmatrix} b & 0 & a \\ 0 & 0 & b \\ 0 & b & 0 \end{pmatrix},$$

and the general form of the curve becomes

$$\sum_{i=0}^{n} \tilde{a}_i (bx - ay)^i z^{n-i} + y \sum_{i=0}^{n-1} \tilde{b}_i (bx - ay)^i z^{n-i-1} = 0, \qquad (6.5)$$

where the $\tilde{a}_i$ and $\tilde{b}_i$ are still arbitrary (we have incorporated powers of b into the original a_i and b_i). The parametric equation of the pencil of lines through the multiple point can be written in the form $x = a + \alpha t$, $y = b$, and $z = \beta t$. One of these lines meets the curve where

$$t^{n-1}\left(t \sum_{i=0}^{n} \tilde{a}_i b^i \alpha^i \beta^{n-i} + b \sum_{i=0}^{n-1} \tilde{b}_i b^i \alpha^i \beta^{n-i} \right) = 0.$$

Since $t = 0$ corresponds to the multiple point, the homogeneous coordinates of the variable point of the curve is given by

$$x = a \sum_{i=0}^{n} \tilde{a}_i b^i \gamma^i - a\gamma \sum_{i=0}^{n-1} \tilde{b}_i b^i \gamma^i,$$

$$y = -b \sum_{i=0}^{n} \tilde{a}_i b^i \gamma^i,$$

$$z = -b \sum_{i=0}^{n-1} \tilde{b}_i b^i \gamma^i,$$

where $\gamma = \alpha/\beta$. Since the coefficients $\tilde{a}_i$ and $\tilde{b}_i$ are arbitrary, this can be further simplified to

$$x = a \sum_{i=0}^{n} A_i \gamma^i + a\gamma \sum_{i=0}^{n-1} B_i \gamma^i,$$

$$y = -b \sum_{i=0}^{n} A_i \gamma^i,$$

$$z = b \sum_{i=0}^{n-1} B_i \gamma^i.$$

This parametrized form of the approximating curve can now be used as in Section 6.7. The following example illustrates the method.

Example 6.15 *Consider, as the approximating curve, the cubic curve with a double point at $(1, 2, 0)$ and passing through the points $(1, 0)$ and $(0, 1)$.*

This curve will have an equation (in affine coordinates) of the form

$$\sum_{i=0}^{3} \tilde{a}_i (2x - y)^i + y \sum_{i=0}^{2} \tilde{b}_i (2x - y)^i = 0,$$

where

$$\sum_{i=0}^{3} 2^i \tilde{a}_i = 0 \quad \text{and} \quad \sum_{i=0}^{3}(-1)^i \tilde{a}_i + \sum_{i=0}^{2}(-1)^i \tilde{b}_i = 0.$$

For the point $P(A, B)$ on the given curve, the corresponding point on the cubic is given by

$$x = \tfrac{1}{2}[y + (2B - A)],$$

where

$$y = -\frac{\sum_{i=0}^{3} \tilde{a}_i (2B - A)^i}{\sum_{i=0}^{2} \tilde{b}_i (2B - A)^i}.$$

We can now proceed to calculate whichever measure of closeness we wish as in Section 6.7.

In this subsection we have found a sufficient condition on the coefficients of the parametrization for the general rational curve to be free of singularities in the affine plane. The general nth-order rational curve with parametrization

$$x = X_n(t), \qquad y = Y_n(t), \qquad z = Z_n(t)$$

will have a point of multiplicity $n - 1$ at $(a, b, 0)$ and therefore cannot have any singular points in the affine plane if the parametrization is of the form

$$x = a \sum_{i=0}^{n} a_i t^i + at \sum_{i=0}^{n-1} b_i t^i,$$

$$y = -b \sum_{i=0}^{n} a_i t^i,$$

$$z = b \sum_{i=0}^{n-1} b_i t^i.$$

6.7.5 Geometrical Optimization

In most of the cases we have discussed above, the curves used for the geometric approximation were not uniquely defined by the interpolation at the endpoints of the given curve segment. For example, if we use a parabola that has two points in common with a given curve, the approximating parabola still has two free parameters, or degrees of freedom. In one case we required that the parabola have the same tangent as the given curve at each of the shared endpoints.

Alternatively, we could have selected an axial direction and another point on the parabola.

Instead of preselecting the remaining two parameters, we may also calculate the distance between the given and approximating curves as a function of these remaining parameters, and then try to find a combination of parameter values for which the distance will be minimal. We perform an optimization over the free parameters and find the values of the parameters that make the approximating curve the closest possible to the given curve in the sense of the particular distance measure we are using. When approximating with parabolas, we may decide to use interpolation at three points and then find the slope of the axis such that the distance between the curves (measured by one of the distances given above) is minimized. If we use higher-order curves, we shall have more parameters that we can use in the optimization.

6.8 Geometry and Differential Equations

6.8.1 Interpolation and Difference Equations

In this section we give a brief discussion of an application of the geometrical approach to an area that may appear to be little connected to geometry – the solution of differential equations. In particular, we will consider the initial-value problem.

The scalar initial-value problem is given by

$$y' \equiv \frac{dy}{dx} = f(x, y), \qquad y(x_0) = y_0,$$

and has a unique solution $y(x)$ if $f(x, y)$ is a continuous function of x and Lipschitz in y. This solution is given by

$$y(x) = y_0 + \int_{x_0}^{x} f(t, y(t)) \, dt,$$

and, since $y(x)$ is unknown, the problem cannot, in general, be solved by analytical integration. Since $y(x)$ must be differentiable, it must represent an integrable function through the point (x_0, y_0) (Figure 6.32).

Since it may be impossible to find the values of y for all values of x, we select a set of equidistant points $\{x_n = x_0 + nh\}$ and approximate the solution $y(x_n)$ at these points.

An approach that is frequently used is to approximate the derivative at a point $(x_n, y(x_n))$ by the slope of the straight line segment connecting this point and

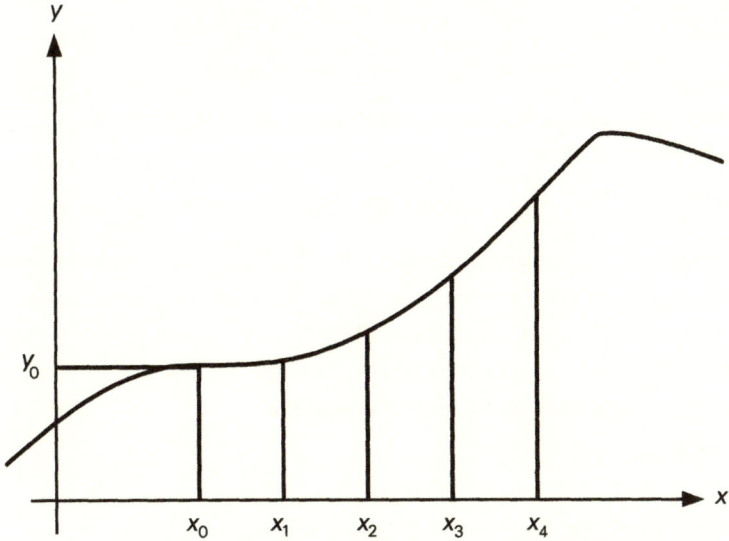

Figure 6.32. Discretization of the initial-value problem.

the point $(x_{n+1}, y(x_{n+1}))$, that is,

$$y'(x_n) \approx \frac{y(x_{n+1}) - y(x_n)}{h}.$$

We assume that we have an approximation $y_n \approx y(x_n)$ and use the derivative approximation to calculate an approximation to $y(x_{n+1})$, namely

$$y_{n+1} \approx y(x_{n+1}) \approx y(x_n) + hy'(x_n)$$
$$= y(x_n) + hf(x_n, y(x_n)) \approx y_n + hf(x_n, y_n).$$

We calculate successive approximations to the solution by using the iterative scheme

$$y_{n+1} = y_n + hf(x_n, y_n) \qquad \text{with } x_0, y_0 \text{ given.} \tag{6.6}$$

This scheme is identical to calculating $\int_{x_n}^{x_n+h} f(t, y(t))\, dt$ by means of the left-endpoint quadrature rule

$$y_{n+1} \approx y(x_{n+1}) = y(x_n) + \int_{x_n}^{x_n+h} f(t, y(t))\, dt \approx y_n + hf(x_n, y_n).$$

Let us relate this to interpolation. Consider the function

$$I(x) = ax + b \in P_1(x)$$

that satisfies the Hermite interpolation conditions

$$I(x_n) = y_n, \qquad I'(x_n) = f(x_n, y_n),$$

namely

$$I(x) = y_n + (x - x_n) f(x_n, y_n).$$

If we now demand that this function also interpolate the point (x_{n+1}, y_{n+1}), we introduce the condition

$$I(x_{n+1}) = y_n + (x_{n+1} - x_n) f(x_n, y_n) = y_n + hf(x_n, y_n) = y_{n+1},$$

which is exactly the same formula as (6.6). The formula connects information about approximate function values at x_n and x_{n+1}, as well as the value of the derivative at x_n. The same formula will result if we use any two of the three interpolation conditions and demand that the third be satisfied.

Let us now consider the interpolation problem

$$I(x) = ax + b \in P_1(x),$$

$$I(x_n) = y_n, \qquad I'(x_{n+1}) = f(x_{n+1}, y_{n+1}),$$

with solution

$$I(x) = y_n + (x - x_n) f(x_{n+1}, y_{n+1}).$$

If we impose the condition that the function must also interpolate the point (x_{n+2}, y_{n+2}), we introduce the condition

$$\begin{aligned} I(x_{n+2}) &= y_n + (x_{n+2} - x_n) f(x_{n+1}, y_{n+1}) \\ &= y_n + 2hf(x_{n+1}, y_{n+1}) = y_{n+2}, \end{aligned}$$

so that

$$y_{n+2} = y_n + 2hf(x_{n+1}, y_{n+1}).$$

This is the midpoint rule.

For our next example let us consider the interpolation problem

$$I(x) = ax^3 + bx^2 + cx + d \in P_3(x),$$

$$I(x_n) = y_n, \qquad I(x_{n+2}) = y_{n+2}, \qquad I'(x_n) = f_n,$$

$$I'(x_{n+2}) = f_{n+2},$$

where we use the abbreviated notation f_n to indicate $f(x_n, y_n)$. The algebraic manipulations required to solve the problem are simplified if we rewrite the interpolant in the form

$$I(x) = A(x - x_{n+1})^3 + B(x - x_{n+1})^2 + C(x - x_{n+1}) + D$$

and keep in mind that $x_{n+2} - x_{n+1} = h$ and $x_n - x_{n+1} = -h$. The interpolation conditions become

$$-Ah^3 + Bh^2 - Ch + D = y_n,$$

$$Ah^3 + Bh^2 + Ch + D = y_{n+2},$$

$$3Ah^2 - 2Bh + C = f_n,$$

$$3Ah^2 + 2Bh + C = f_{n+2},$$

and the solution can be calculated. In particular, we find that

$$4Ch = 3(y_{n+2} - y_n) - h(f_{n+2} + f_n).$$

Let us demand that the interpolant satisfy the additional (overspecified) condition $I'(x_{n+1}) = f_{n+1}$. We have to satisfy another equation, namely $C = f_{n+1}$. We substitute this in the last equation to obtain the rule

$$y_{n+2} = y_n + \frac{h}{3}(f_n + 4f_{n+1} + f_{n+2}), \tag{6.7}$$

which the reader will recognize as Simpson's rule.

All these examples are special cases of the general *linear multistep method*. This method is defined by the difference equation

$$\sum_{i=0}^{k} \alpha_i y_{n+i} = h \sum_{i=0}^{k} \beta_i f_{n+i}, \qquad \alpha_k = 1, \quad \alpha_0^2 + \beta_0^2 \neq 0. \tag{6.8}$$

This equation relates information over k intervals and is referred to as a linear k-step method. In general, any solution to a finite linear interpolation problem can be used to generate a linear multistep method if one additional condition is imposed. The method can be formulated as follows.

Given a set of $k + 1$ distinct points $x_{n+i} = x_n + ih, i = 0, \ldots, k \leq N$, and a set of $N + 1$ linear functionals $\{L_i, M_j\}$, determine a function $I(x) \in P_N(x)$ such that

$$L_i(I) \equiv I(x_{n+i}) = \sum_{m=0}^{k} \alpha_{i,m} y_{n+m} = A_i,$$

$$M_j(I) \equiv I'(x_{n+j}) = \sum_{m=0}^{k} \beta_{j,m} f_{n+m} = B_j,$$

assuming that the system possesses the finite interpolation property. Some of the coefficients $\alpha_{i,m}$ and $\beta_{j,m}$ may be zero. If we want to use the method to solve an ordinary differential equation, there must be at least one nonzero $M_j(I)$. Let

$\{\phi_i(x)\}$ be an orthonormal basis for $\{L_i\}$, and let $\{\psi_j\}$ be an orthonormal basis for $\{M_j\}$. Then the solution can be expressed in the form

$$I(x) = \sum A_i\phi_i(x) + \sum B_j\psi_j(x),$$

where the coefficients A_i and B_j are linear combinations of function values y_{n+m} and derivative values f_{n+m}, respectively. We now demand that one additional interpolation constraint must be satisfied. For example, if the chosen functionals did not include function evaluation at x_{n+k}, we may require that

$$I(x_{n+k}) = \sum_i A_i\phi_i(x_{n+m}) + \sum_j B_j\psi_j(x_{n+m}) = y_{n+k}.$$

This results in a relationship that can be used to form a rule for a linear multistep method, since

$$y_{n+k} = \sum_i A_i\phi_i(x_{n+m}) + \sum_j B_j\psi_j(x_{n+m})$$

$$= \sum_i \sum_{m=0}^{k-1} \alpha_{i,m} y_{n+m}\phi_i(x_{n+m}) + \sum_j \sum_{m=0}^{k} \beta_{j,m} f_{n+m}\psi_j(x_{n+m})$$

$$= \sum_{m=0}^{k-1} \left(\sum_i \alpha_{i,m}\phi_i(x_{n+m})\right) y_{n+m} + \sum_{m=0}^{k} \left(\sum_j \beta_{j,m}\psi_j(x_{n+m})\right) f_{n+m},$$

which is of the general form (6.8).

Example 6.16 *Relate this general description of the derivation of linear multistep methods to the derivation of Simpson's rule (6.7).*

The four relevant functionals are function and derivative evaluation at x_n and x_{n+2}, with $N = 3$ and $k = 2$. This means that $\alpha_{0,0} = \alpha_{2,2} = 1$ and $\beta_{0,0} = \beta_{2,2} = 1$, while the rest of the functional coefficients are zero. Therefore, $A_0 = y_n$, $A_2 = y_{n+2}$, $B_0 = f_n$, and $B_2 = f_{n+2}$. The basis functions are the piecewise cubic Hermite interpolation basis functions discussed in Section 1.5, namely

$$\phi_1 = \frac{1}{4h^3}(x-x_n)^3 - \frac{3}{4h^2}(x-x_n)^2 + 1,$$

$$\phi_2 = -\frac{1}{4h^3}(x-x_{n+2})^3 - \frac{3}{4h^2}(x-x_{n+2})^2 + 1,$$

$$\psi_1 = \frac{1}{4h^2}(x-x_n)^3 - \frac{1}{h}(x-x_n)^2 + (x-x_n),$$

$$\psi_2 = \frac{1}{4h^2}(x-x_{n+2})^3 + \frac{1}{h}(x-x_{n+2})^2 + (x-x_{n+2}).$$

Finally, when we demand that $I'(x_{n+1}) = f_{n+1}$, we have to satisfy the condition

$$f_{n+1} = y_n \phi_1'(x_{n+1}) + y_{n+2} \phi_2'(x_{n+1}) + f_n \psi_1'(x_{n+1}) + f_{n+2} \psi_2'(x_{n+1})$$

$$= -\frac{3}{4h} y_n + \frac{3}{4h} y_{n+2} - \tfrac{1}{4} f_n - \tfrac{1}{4} f_{n+2}.$$

As before, we have derived Simpson's rule

$$y_{n+2} = y_n + \frac{h}{3}(f_n + 4f_{n+1} + f_{n+2}).$$

One difficulty in this method of deriving linear multistep methods is that it yields only methods of maximal order, which are known to be zero-unstable when the step number exceeds one (for an explicit method) or two (for an implicit method). For example, if the interpolant problem is

$$I(x) = ax^3 + bx^2 + cx + d \in P_3(x),$$

$$I(x_n) = y_n, \qquad I(x_{n+1}) = y_{n+1}, \qquad I'(x_n) = f_n, \qquad I'(x_{n+1}) = f_{n+1},$$

and the enforced condition is

$$I(x_{n+2}) = y_{n+2},$$

we obtain the third order two-step explicit method

$$y_{n+2} = 5y_n - 4y_{n+1} + 2h(f_n + 2f_{n+1}),$$

which is zero-unstable. This difficulty can be overcome by replacing the cubic interpolant by one that is quadratic and changing the interpolation problem to

$$I(x) = ax^2 + bx + c \in P_2(x),$$

$$\alpha I(x_n) + (1 - \alpha)I(x_{n+1}) = \alpha y_n + (1 - \alpha)y_{n+1},$$

$$I'(x_n) = f_n, \qquad I'(x_{n+1}) = f_{n+1}.$$

The interpolant can now be calculated, and the demand

$$I(x_{n+2}) = y_{n+2},$$

extrapolates to the second-order method

$$y_{n+2} = \alpha y_n - (\alpha - 1)y_{n+1} + \frac{h}{2}[(\alpha - 1)f_n + (\alpha + 3)f_{n+1}],$$

in which the parameter α can be chosen to avoid zero instability. For $\alpha = 5$, we have the third-order method described above, which is zero-unstable. For $\alpha = 0$ (the Adams–Bashford method) $I(x)$ interpolates y_{n+1} but not y_n. For $\alpha = 1$ (the midpoint rule) $I(x)$ interpolates y_n but not y_{n+1}. For all other values of α the function $I(x)$ interpolates neither y_n nor y_{n+1}.

6.8.2 Trajectory Problems

A differential equation which does not explicitly involve the independent variable is called *autonomous*. The second-order scalar equation

$$\ddot{x} + f(x, \dot{x})\dot{x} + g(x) = 0, \qquad x = x(t), \qquad \dot{x} = \frac{dx}{dt},$$

is an example of an autonomous equation. This equation can be reduced to a system of two first-order equations by writing $\dot{x}(t) = y(t)$ to obtain

$$\dot{x} = y, \qquad \dot{y} = -f(x, y)y - g(x),$$

or, in more general form,

$$\frac{dx}{p(x, y)} = \frac{dy}{q(x, y)}. \tag{6.9}$$

The (x, y) plane is called the *phase plane* of the problem.

Example 6.17 *Consider the simple harmonic oscillator described by*

$$\ddot{x} + x = 0,$$

as a phase-plane problem.

This equation has solutions of the form $x(t) = a \sin(t + b)$. The corresponding system of first-order equations in the phase plane is

$$\dot{x} = y, \quad \dot{y} = -x, \qquad \text{or} \qquad \frac{dx}{y} = -\frac{dy}{x},$$

with solutions that consist of concentric circles around the origin of the phase plane.

We use the phrase *phase-plane problem* to refer to any problem of the form (6.9) where the aim is to find, not solutions in the (x, t) and (y, t) planes, but the trajectories, that is, the integral curves in the (x, y) plane. In many cases these trajectories cannot be found by analytical methods. As we will discuss in the next subsection, interpolation can be used to develop numerical methods that are naturally suited to this problem.

The scalar initial-value problem

$$y' \equiv \frac{dy}{dx} = f(x, y) = \frac{q(x, y)}{p(x, y)}, \qquad y(x_0) = y_0,$$

can be replaced by the phase-plane problem

$$\dot{x} = p(x, y), \quad \dot{y} = q(x, y), \qquad x(t_0) = x_0, \quad y(t_0) = y_0.$$

Example 6.18 *(Competing Species) Discuss the competing-species problem in terms of its phase plane.*

Consider two species with population densities given by x and y, respectively. The simplest model for the growth rate of a species is to assume that its rate of growth is proportional to the current density, that is,

$$\dot{x} = ax.$$

This leads to exponential growth, which is a reasonable model only in circumstances where there is an infinite supply of food and no competition or predators. The next model assumes that there is some steady-state population to which the given population approaches. Let us suppose that this steady-state population has density x_0. We expect that the rate of growth will diminish as the population gets close to the steady-state value but will behave almost like exponential growth when far away from this value. The ratio of growth rate to density cannot, therefore, be a constant, but must be a function that diminishes as x approaches x_0. A model that behaves like this is given by

$$\dot{x} = a(x_0 - x)x.$$

Suppose that there is a second species in competition with the first. This will also affect the growth rate, even if the density is far from the steady state. Once more, the proportionality must be diminished by some term that depends on the density of the competing species. A simple model that takes all these aspects into account is obtained by assuming that the proportionality is given by $a(x_0 - x - by)$, giving the differential growth-rate equations

$$\dot{x} = a(x_0 - x - by)x,$$

$$\dot{y} = c(y_0 - y - dx)y.$$

This is a reasonable model for a pair of competing species, and for stable coexistence of the species we must have

$$x_0 > by_0, \qquad dx_0 < y_0,$$

where all the coefficients are positive. This is a phase-plane problem as depicted in Figure 6.33.

The lines in the figure divide the plane into four regions depending on whether or not the individual species are increasing or decreasing in density. This gives an easy way to get an intuitive feel for the trajectory through any particular point. It does not, however, give us any way of actually computing that integral curve. For this we shall, in general, need some numerical method of solving the initial-value problem.

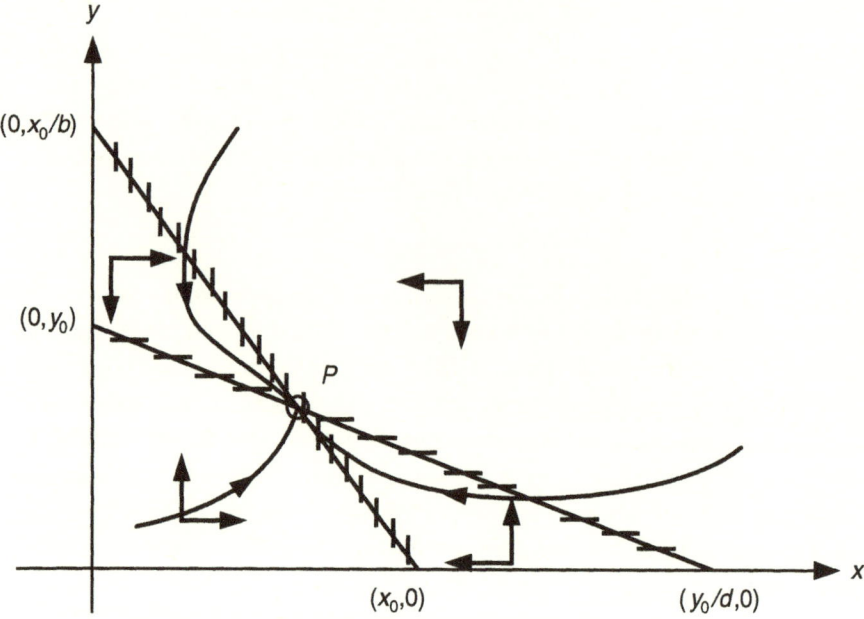

Figure 6.33. The competing-species phase-plane model.

6.8.3 A Geometrical Approach to Difference Equations

We consider the autonomous system of ordinary differential equations

$$\frac{d\mathbf{y}}{dt} = \mathbf{f}(\mathbf{y}),$$

where $\mathbf{y}$ and $\mathbf{f}$ are vectors in R^m. We will construct two methods, one a predictor and the other a corrector. Both methods will, apart from roundoff error, yield exact answers for any differential equation whose integral curve, passing through the given initial condition, is a circle. We call such a method *circularly exact*. We will do the construction in two-space, but the resulting method will apply in any dimension. The solution vector in two dimensions is given by $\mathbf{y} = (x, y)$, and the slope vector by $\mathbf{f}(\mathbf{y}) = (f_1(x, y), f_2(x, y))$. There is a unique circle C_n through the points (x_n, y_n) and (x_{n+1}, y_{n+1}) with slope $\mathbf{f}_{n+1}$ at (x_{n+1}, y_{n+1}). We select a third point on this circle by starting from (x_n, y_n) and drawing a line parallel to the tangent at (x_{n+1}, y_{n+1}) until this line intersects the circle (Figure 6.34). Since we already know one of the intersections, namely (x_n, y_n), we know, *a priori*, that we shall be able to obtain an explicit formula for the other intersection point. We will call this point (x_{n+2}, y_{n+2}). The algebra required in

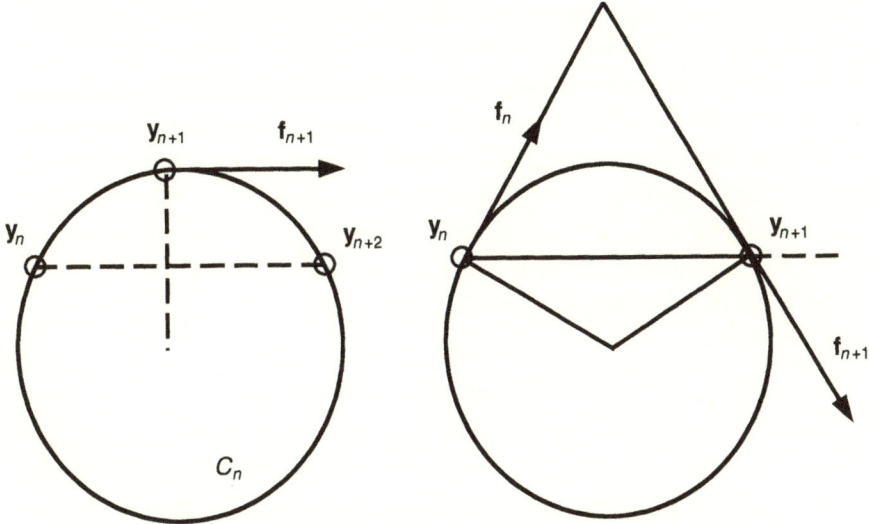

Figure 6.34. The circularly exact predictor–corrector method.

the construction of this formula is left to the reader. The formula is

$$\mathbf{y}_{n+2} = \mathbf{y}_n + 2\left[\mathbf{F}_{n+1}^T(\mathbf{y}_{n+1} - \mathbf{y}_n)\right]\mathbf{F}_{n+1},\qquad(6.10)$$

where

$$\mathbf{F}_{n+1} = \frac{\mathbf{f}_{n+1}}{\|\mathbf{f}_{n+1}\|}.$$

Since $\mathbf{y}_{n+2}$ is given explicitly in terms of previous values, this method is explicit, that is, it is a predictor.

Let us describe another circle by the same two points (x_n, y_n) and (x_{n+1}, y_{n+1}), but using the slope at (x_n, y_n). For our extra condition we now require that the slope at (x_{n+1}, y_{n+1}) be $\mathbf{f}_{n+1}$. This yields the method

$$\mathbf{y}_{n+1} - \mathbf{y}_n = \frac{k}{2}(\mathbf{F}_n + \mathbf{F}_{n+1}),\qquad(6.11)$$

where $\mathbf{F}_n$ and $\mathbf{F}_{n+1}$ are defined as above. The new value $\mathbf{y}_{n+1}$ depends on the slope at that new point and hence this method is implicit, that is, it is a corrector. The parameter k is simply the step size, and we notice that this method is nothing more than the trapezoidal rule applied to the autonomous system we obtain when we scale the direction vector to have unit length (that is, when we deal with $\mathbf{F}$ instead of $\mathbf{f}$). This may seem like a small advantage. However, the trapezoidal rule is only circularly exact if we correct to convergence, whereas the predictor–corrector pair we have constructed here is circularly exact even with a single application.

We can construct difference methods that will be exact for any class of curve we want. All we need do is to define such a curve using interpolation and a collection of points and slopes (including at least one slope) and then to pick one additional point or slope to produce the difference equation. Methods derived this way, while exact may have poor stability properties. The particular pair given above happens to have good stability properties, but no general theory of the stability of such methods is known.

6.8.4 Plotting Algebraic Curves

Let us consider the use of simple curves to produce plots of more complicated curves. The plotting of functions is simple because we can evaluate points on the curve representing the function. For the same reason the plotting of parametric curves is also easy. However, the plotting of general curves is not so straight-forward. One crude way to approach this problem on a computer screen is to evaluate the expression representing the curve at every pixel and display the pixels where the expression evaluates to zero, or nearly to zero. However, we can also use the geometrically derived differential-equation solvers to plot curves.

By differentiation of an implicit curve equation $g(x, y) = 0$, we obtain the associated differential equation

$$g_x(x, y) + g_y(x, y)\frac{dy}{dx} = 0, \qquad \text{or} \quad \frac{dy}{dx} = f(x, y),$$

with phase-plane formulation

$$\dot{x} = g_y(x, y), \qquad \dot{y} = -g_x(x, y).$$

We can now use one of the methods described above to obtain the trajectory. In this particular case, however, there is one additional advantage – we know the solution to the problem. We can now solve the differential equation as an initial-value problem, with $y(x_0) = y_0$, where (x_0, y_0) is a point on the curve $g(x, y) = 0$. If the solver is exact for the particular integral curve, which is the curve we want to plot, the solution points will, apart from roundoff, lie on the curve. If not, we have an exceedingly good initial guess for a new solution of the equation $g(x, y) = 0$, and any nonlinear-equation solver can be used to correct the position of the point so that it lies on the required curve. One step would suffice in most cases. Obviously, some additional care will have to be taken so that we identify singularities as we approach them and take appropriate action.

This approach was used to plot Figures 6.35 to 6.38. The circularly exact predictor (6.10) was used to compute a point near, or on, a given curve, and then

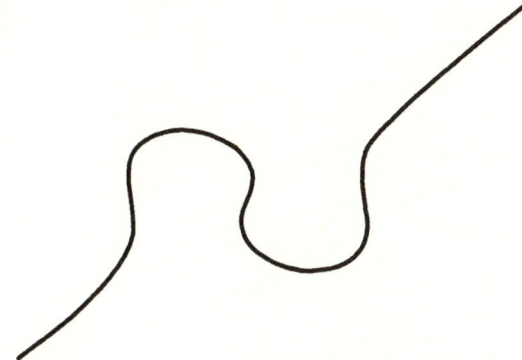

Figure 6.35. A cubic with three inflections and an asymptote.

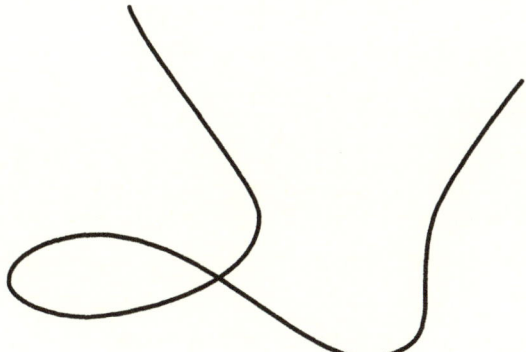

Figure 6.36. A quartic curve with a node.

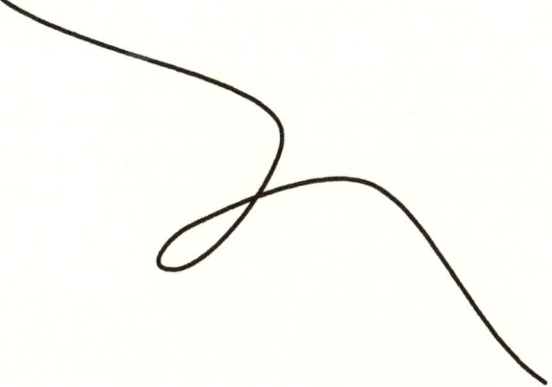

Figure 6.37. A quintic curve with one node and an asymptote.

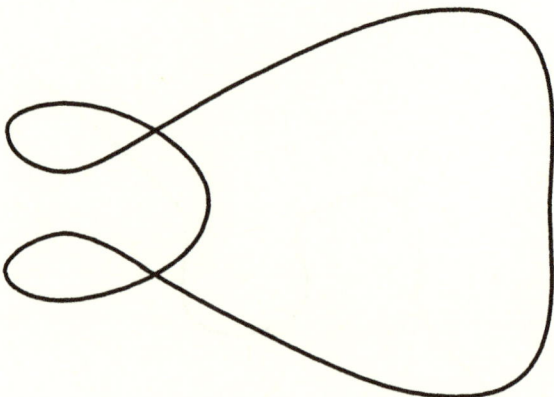

Figure 6.38. A quintic curve with two nodes.

the Newton–Raphson method was used to correct to the curve. The algebraic equations of the four depicted curves are as follows:

Figure 6.35:　　　　　　　　$x^3 - y^3 - 4x + y = 0;$

Figure 6.36:　　$x^4 + x^2y^2 - 2xy^2 - y^3 - x^2 + y^2 = 0;$

Figure 6.37:　　　　$x^5 + y^5 - 2x^2 + 5xy - 2y^2 = 0;$

Figure 6.38:　　$x^5 + y^4 - 2x^3 - 2y^3 - x^2 + y^2 = 0.$

6.9 Bibliographical Notes

It is the rational curves, rather than algebraic curves, that have been used most often in applications. Virtually all parametric splines are rational curves, though the numerical-analysis books rarely mention the geometry of such curves. For application in CAD the book *Rational Curves and Surfaces: Applications to CAD* by J. C. Fiorot may be of interest. Very little on construction methods for algebraic curves, and even less on the construction of rational curves, is contained in available literature. We have extended the approach for the construction of conics given by Liming to more general curves. *Mathematics for Computer Graphics* by R. Liming, published by Aero Publishers (1979), is of interest in that it discusses one of the earliest applications of pencils of conics to curve generation, a method known as Liming's method. *Computational Geometry for Design and Manufacture*, by I. D. Faux and M. J. Pratt, first published by Ellis Horwood (1979), is highly recommended. *Introduction to Computing with Geometry* by A. Bowyer and J. Woodwark, published by A. Rowe (1993), is also a nice book. For those seeking further information on the stability of multistep methods, almost any text on numerical methods for initial-value problems will be of use. *Computational Methods in Ordinary Differential Equations* by J. D. Lambert, published by Wiley (1974), is particularly recommended, as it has a very readable explanation without going into great mathematical detail. *Mathematical Ideas in Biology* by J. M. Smith, published by Cambridge University Press (1968), gives a simple but quite adequate introduction to phase planes.

Exercises

6.1. Investigate the conditions on the coefficients in the parametrization for the rational cubic that will ensure that there are no affine singular points.

6.2. Complete the classification of the rational cubics with a cusp.

6.3. Construct an example of a cubic with an affine cusp, an affine point of inflection, and three asymptotes. Determine a rational parametrization, and give the positions of the cusp and inflection point, as well as the equations of the asymptotes.

6.4. Show that a cubic with a crunode can never have exactly two asymptotes.

6.5. Construct an example of a cubic with an affine acnode, three affine points of inflection, and three asymptotes. Determine a rational parametrization, and give the positions of the acnode and the inflection points. Check that the inflection points are collinear.

6.6. Write down the general form of an algebraic cubic with a crunode at the origin of the affine plane and the coordinate axes as tangents. Calculate the Hessian, and hence show that the three points of inflection will be collinear. They will therefore remain collinear under any projective transformation.

6.7. Construct a quartic with a prescribed double point and passing through eleven other simple points in general position by using products of cubics and lines.

6.8. Using curves of lower order as described in Section 6.3, construct quintic curves with

   (a) four double points and eight simple points,

   (b) six double points and two simple points,

   (c) a triple point, three double points, and five simple points,

   where all points are prescribed but assumed to be in general position.

6.9. Discuss the construction of a curve of order six, with two triple and four double points and passing through three other prescribed points, all points being in general position.

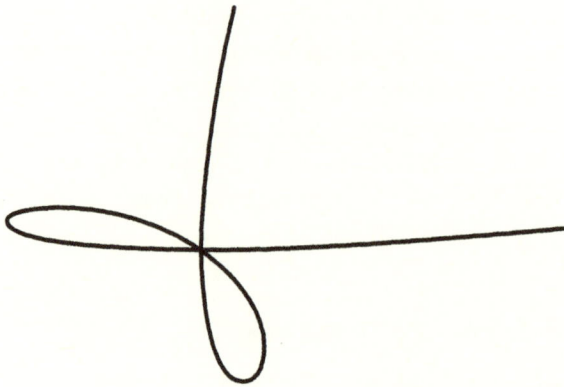

Figure 6.39. A quartic with a triple point.

6.10.   Construct a polynomial parametrization of a quartic with an ordinary triple point at $(0, 0, 1)$, with tangents there given by $x = 0$, $y = 0$, and $x + y = 0$. Graph the curve. Adjust the remaining parameter(s) until your curve looks like a bow tie (Figure 6.39).

6.11.   Construct a rational quartic with a crunode at $(0, 0, 1)$ and two cusps on the ideal line.

6.12.   Find the solution to the following Hermite interpolation problem: determine a function $f(x, y) \in P_3(x, y)$ such that

$$
\begin{aligned}
f(1, 1) &= 2, & f_x(1, 1) &= 0, & f_y(1, 1) &= 4, \\
f(0, -1) &= 0, & f_x(0, -1) &= 0, & f_y(0, -1) &= 1, \\
f(1, 0) &= 1, & f_x(1, 0) &= 2, & f_y(1, 0) &= -1, \\
f(0, 0) &= 1. & & & &
\end{aligned}
$$

6.13.   Complete the details of Example 6.11.

6.14.   Given a triple point, two double points, and two simple points in general position, derive the condition(s) on the position of a third simple point so that the finite interpolation problem cannot be satisfied by any function $f(x, y) \in P_4(x, y)$.

6.15.   Find the vertex of each of the parabolas in Example 6.12.

6.16.   Calculate:
   (a) the two parabolas through the four points $(-1, 0)$, $(1, 0)$, $(2, -1)$, and $(-2, 2)$,
   (b) the unique parabola through the four points $(-1, 0)$, $(1, 0)$, $(2, -1)$, and $(2, -3)$.

6.17.   Find a parametrization of the parabola

$$[(\sqrt{5} + \sqrt{3})x - \sqrt{3}y]^2 - \sqrt{15}(\sqrt{15} + 2)y - 8 - 2\sqrt{15} = 0.$$

6.18.   Show that the axis of the parabola $(\alpha x + \beta y)^2 + 2gx + 2fy + c = 0$
is given by $\alpha x + \beta y + \gamma = 0$, where

$$\gamma = \frac{\alpha g + \beta f}{\alpha^2 + \beta^2}.$$

6.19.   Show that the slope of the axis of the parabola

$$x = at^2 + (x_2 - x_1 - a)t + x_1,$$
$$y = bt^2 + (y_2 - y_1 - b)t + y_1$$

is given by the ratio b/a.

6.20.   Given a pencil of lines with a given (fixed) slope, use the theory developed in Section 2.10 to investigate the condition that a member of the pencil is tangent to the general conic. Under which circumstances will there be exactly one such tangent line?

6.21.   Show that $CD = DE$ in Figure 6.25.

6.22.   Use the three function metrics L_1, L_2, and L_∞ to calculate the difference between the functions

$$f(x) = x(1-x)^2, \qquad 0 \le x \le 1,$$

and

$$g(x) = ax(1-x), \qquad 0 \le x \le 1,$$

where a is a positive constant. For which value of a will $g(x)$ be a "best" approximation of $f(x)$ in each of the metrics?

6.23.   Show, both theoretically and by using an example, that the measure of the distance from one curve to another used in the definition of the Hausdorff distance is not symmetric.

6.24.   Explain the comment that, if the distance between points on a given curve and a parabolic approximating curve is measured from a point on the parabola to the given curve in the direction of the axis of the parabola, then the distance may not even be defined.

6.25.   Instead of the approximation parabola used in Example 6.14, use the parabola with

$$x(0) = -1, \qquad y(0) = 0, \qquad x(1) = 2, \qquad y(1) = 0,$$

and

$$x(1/\sqrt{3}) = 0, \qquad y(1/\sqrt{3}) = 2\sqrt{3}$$

to approximate the cubic segment

$$x = 3t^2 - 1, \quad y = -9t^3 + 9t, \qquad 0 \le t \le 1,$$

322 Examples and Applications

and calculate (numerically, if necessary) both measures of closeness described in that subsection.

6.26. Given the cubic segment

$$x = 8t^3 - 16t^2 + 10t - 1, \quad y = 8t(t-1)^2, \qquad 0 \le t \le 1,$$

and the approximating parabola

$$x = as(s-1) + 2s - 1, \qquad y = bs(s-1)$$

that intersects the cubic segment at its endpoints $(-1, 0)$ and $(1, 0)$, show that there exist parabolic slopes b/a for which the lines parallel to the slope and through the points $(x(t), y(t))$, with $0 \le t \le 1$, on the given curve segment will not all intersect the parabola for $0 \le s \le 1$. Also show that for any parabola that is a function, as well as for those with axial slope equal to one, the intersections will fall in the interval $0 \le s \le 1$.

6.27. Use the interpolation problem

$$I(x) = ax^2 + bx + c \in P_2(x),$$

$$I(x_n) = y_n, \qquad I(x_{n+1}) = y_{n+1}, \qquad I'(x_n) = f_n,$$

to derive the trapezoidal rule for solving the initial-value problem $y' = f(x, y)$, $y(x_0) = y_0$.

6.28. Complete the algebra for the predictor

$$\mathbf{y}_{n+2} = \mathbf{y}_n + 2\left[\mathbf{F}_{n+1}^T(\mathbf{y}_{n+1} - \mathbf{y}_n)\right]\mathbf{F}_{n+1}, \qquad \mathbf{F}_{n+1} = \frac{\mathbf{f}_{n+1}}{\|\mathbf{f}_{n+1}\|}.$$

6.29. Complete the algebra for the corrector

$$\mathbf{y}_{n+1} - \mathbf{y}_n = \frac{k}{2}(\mathbf{F}_n + \mathbf{F}_{n+1}).$$

6.30. Construct a conically exact difference method, using ideas similar to those used for constructing the circularly exact methods.

7

Surfaces

7.1 Introduction

In this, the final chapter, we discuss local interpolation in two dimensions. To introduce this we will recall several one-dimensional interpolation methods that we discussed in Chapter 1. The two-dimensional problem is much more difficult than the one-dimensional one because, instead of having only a single finite domain, it has infinitely many finite domains, which are not equivalent under linear transformations. Nonetheless, we shall manage to construct an interpolation basis construction technique that will enable us to perform local interpolation on a very wide class of such domains. We shall have to discuss interpolation on a curve and analyze the dimension of the interpolation problem. This and the construction of the individual basis functions form the main part of the theory of the chapter. After this has been accomplished, the application of the theory to a wide class of examples becomes a straightforward matter.

We will discuss a selection of the most important examples and apply the same procedures to three-dimensional interpolation. The situation is very simple when the local domains are bounded by lines and planar surfaces, but when they are not we must make use of the algebraic geometry we have studied in previous chapters. The application of these techniques enables us to construct interpolants for extremely complex curved domains and is a fitting climax to our main theme of applications of algebraic geometry to interpolation.

However, as a closing segment, we show how the interpolants we have developed can be used to generate rational surfaces. These surfaces are fascinating, but we only touch upon a few properties of the simplest of them. We look at the Steiner surface, a rational quartic with three double lines that meet at a triple point; the cubic surface and its twenty-seven lines; and the quadric.

324 Surfaces

7.2 Local Interpolation

By now we can change origin and scale with ease. Similarly, we can easily
map results that refer to one plane to those of another plane that is an affine
map of the first. An interpolation problem on the one-dimensional intervals
$x \in [2, 6]$ or $x \in [-2, 6]$ or $x \in [a, b]$ (where a and b are finite) can easily
be mapped onto the interval $x \in [0, 1]$ and then mapped back again. In one
dimension there are therefore only three regions we must consider: $x \in [0, 1]$,
$x \in (-\infty, 0]$, and $x \in (-\infty, \infty)$. This is a result of the fact that a region is
bounded by sets of lower dimension, and hence a region in one-dimensional
space is bounded by points, and there are only two kinds of points in affine
space: finite and infinite.

In two or more dimensions the boundaries of regions are no longer discrete
points. A two-dimensional region is bounded by curves. There are an infinite
number of curves that are not affinely equivalent; hence the regions they bound
will not be affinely equivalent. All triangles are affinely equivalent, and so are
all ellipses and all parabolas. However, an ellipse is not affinely equivalent to
a parabola, and neither is affinely equivalent to a cubic, and so on. Thus, while
there is only one finite region in one dimension, there are infinitely many in two
or more dimensions. To appreciate what this will mean to our study, we consider
the difference between local and global interpolation. Since a solution of an
interpolation problem can always be rewritten in terms of some basis (usually
a normalized basis), the properties of the solution of the interpolation problem
can often be deduced from the properties of the individual basis functions.
There are, therefore, three interacting aspects of the problem which we need to
understand. They are:

the interpolation to data, which so far have been point values but will now
 be curves and surfaces,
the aspect of global or local interpolation, and
the required properties of the corresponding individual basis functons.

These are not difficult concepts, but, because of the important ramifications
in what follows, we will first discuss them briefly in the context of the work in
Chapter 1.

Consider pointwise interpolation in one variable at the points $x = -3$,
$-2, -1, 0, 1, 2$, and 3. The solution to the interpolation problem can be written
in the form

$$I(x) = \sum_{i=-3}^{3} y_i \phi_i(x),$$

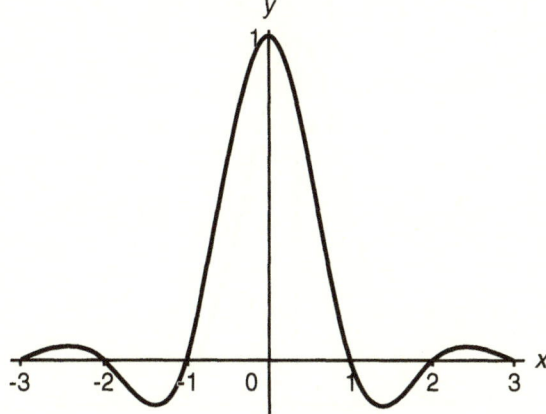

Figure 7.1. A natural cubic spline.

where the $\phi_i(x)$, for $i = -3, \ldots, 3$, are the normalized basis functions. In Chapter 1 we discussed a wide variety of possible choices for these basis functions. For example, the basis function $\phi_0(x)$ for the natural cubic splines is shown in Figure 7.1. This basis function satisfies the biorthonormalized conditions $\phi_0(x_j) = \delta_{0j}$ and is zero only at a finite number of points throughout the entire domain $x \in [-3, 3]$. Therefore, any change in the coefficient of this function in the interpolant, that is, the data value y_0, will cause a change throughout the entire region, apart from at the six other data points where it remains zero.

This property results in this basis being of limited value. The natural splines are called *global* because each basis function is defined in terms of the entire set of x-values of the data points.

If we merely wish to have a technique for drawing smooth free-form curves, then a method based on natural splines might meet the requirements. However, if the interpolation method has to provide a good approximation to some known shape, then this method is inappropriate because a change in any one data value will affect the entire curve. Suppose that we try to approximate a given curve and our approximating curve matches very well for most of the data interval with only one small segment requiring change. A change in the data in that one segment will affect the whole curve and may change a good approximation to a poor approximation. It was difficulties like these that prompted the investigation of local methods.

Piecewise interpolation on a domain involves performing interpolation on subdomains and then piecing together the different interpolants. The corresponding basis must have local, not global, support, that is, must be defined in terms of information associated only with a subdomain. This requirement led to

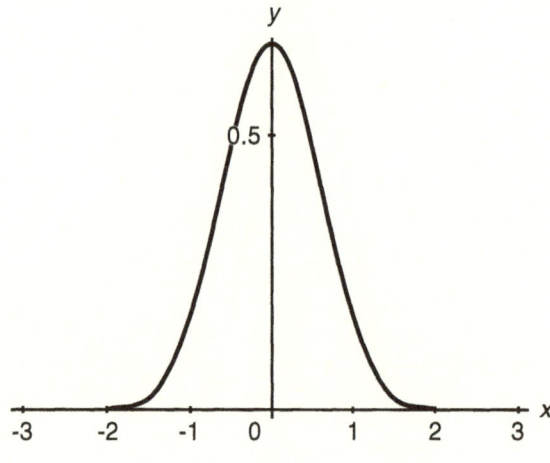

Figure 7.2. A cubic B-spline.

the construction of B-splines (Figure 7.2), which were discussed in Section 5.3 of Chapter 1. This basis function is nonzero only in the interval $x \in [-2, 2]$. It is usual for this property to be described by saying that the B-spline basis has *local* support. This is true, but only in a rather limited sense. We might think, from Figure 7.2, that such a basis would be suitable for piecewise interpolation on subintervals involving five data points, in our example the integer points in the interval $x \in [-2, 2]$. This is not the case. It is the symmetry of the figure and the fact that we have drawn only one of the basis functions that are misleading. Let us call the illustrated spline $B_0(x)$. What does $B_{-2}(x)$ or $B_1(x)$ look like?

These functions are shown in Figure 7.3 together with $B_0(x)$. The definition and the domain of support of the two new functions extend beyond the interval $x \in [-2, 2]$. Hence the behavior of the interpolant within the interval $x \in [-2, 2]$ is not fully determined by information local to that interval. Therefore, although the B-splines are "more local" than the natural splines, they still fail to provide a local basis for piecewise interpolation. Although each B-spline has a support of only four intervals, the B-splines associated with any four-interval segment do not share the same four-interval support, but extend over eight intervals. If we then think of considering interpolation over these eight intervals, we find that the interpolant will depend on information over twelve intervals, and so on.

The Lagrange cubic basis, shown in Figure 7.4, is defined using four data points. However, unlike the cubic B-splines, each basis function uses precisely the same four data points. This means that we can consider a subregion of four consecutive data points as a single entity and perform a completely local

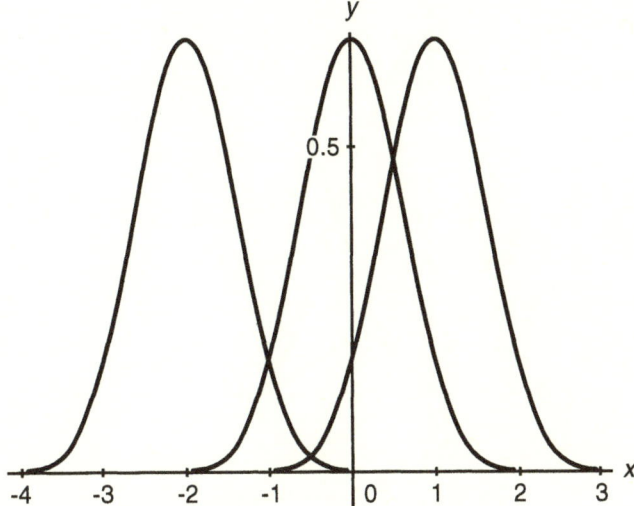

Figure 7.3. The B-splines $B_{-2}(x)$, $B_0(x)$, and $B_1(x)$.

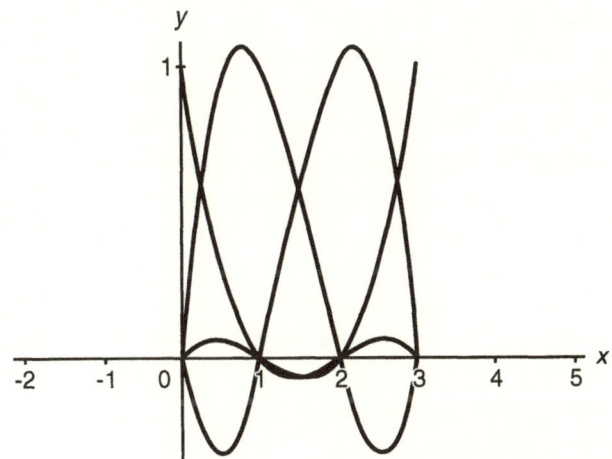

Figure 7.4. The Lagrange cubic basis.

interpolation over this subdomain. In this way the interpolant within this subregion will depend only on the data within and at the boundaries (the endpoints) of this region and will not be affected by, nor affect, the interpolant in any other part of the domain.

The cubic Hermite interpolants also form a basis for a local interpolation method as can be deduced from the sketch of the complete basis shown in Figure 7.5. In this case the subdomain over which local interpolation can be done comprises the closed interval defined by two consecutive data points. The

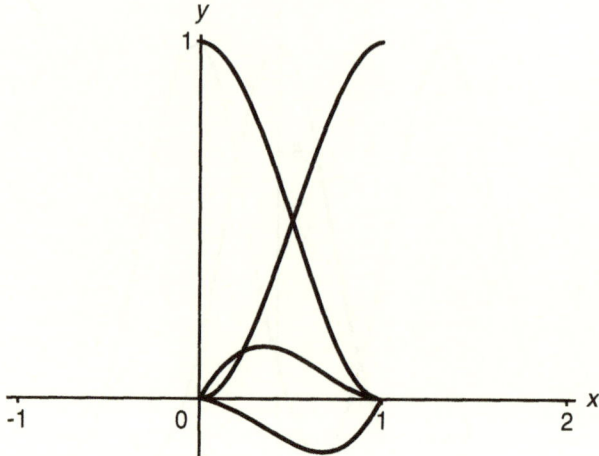

Figure 7.5. The cubic Hermite basis.

interpolant within one subregion will not affect, nor be affected by, the data in any other subregion.

Each individual basis function can be associated with a point. This is a result of the biorthonormalization process. In the case of the Lagrange interpolation, two of these points are interior to the subregion and the remaining two points form the boundaries of the subregion. In the Hermite case there are only two points and each of the four basis functions is associated with one of these points. Each basis function is zero at all the data points except the one associated with that basis function. In particular, the basis function associated with a particular point is zero on the subdomain boundaries that do not contain that point. For example, the Lagrange basis functions associated with the two interior points are both zero at both of the boundary points. This statement may sound rather trivial, but it has important implications when we consider two- and three-dimensional local interpolation.

Consider pointwise interpolation in the plane, and associate each basis function with its corresponding point or node. Local interpolation in the subregion R_1 of the region displayed in Figure 7.6 implies that the interpolant can be expressed in the form

$$I_1(x, y) = \sum_{i=1}^{4} \alpha_i A_i(x, y) + \sum_{i=1}^{4} \beta_i B_i(x, y) + \sum_{i=1}^{2} \gamma_i C_i(x, y),$$

where the functions $\{A_i\}$ are the basis functions associated with the boundary nodes $\{a_i\}$ common to the two subregions, the $\{B_i\}$ are the basis functions

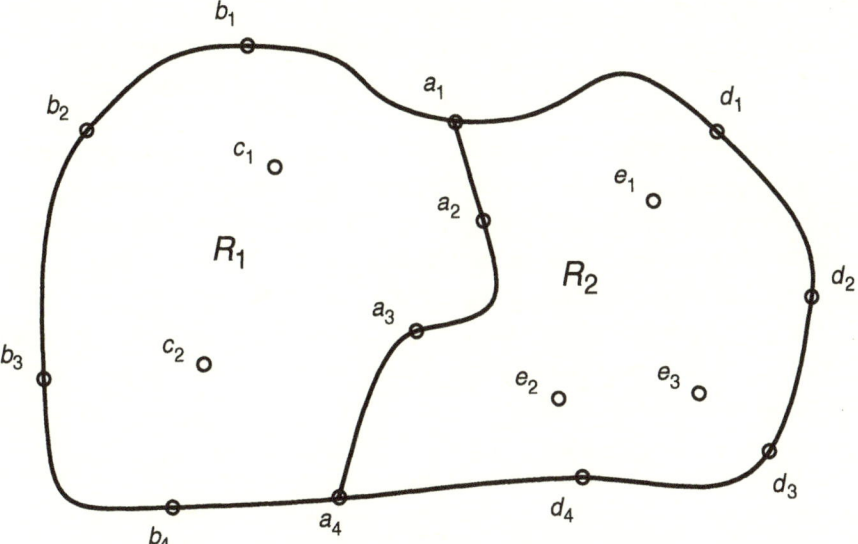

Figure 7.6. Local interpolation.

associated with the boundary nodes $\{b_i\}$, and the $\{C_i\}$ are the basis functions associated with the interior nodes $\{c_i\}$. Similarly, the local interpolant in the subregion R_2 can be expressed in the form

$$I_2(x, y) = \sum_{i=1}^{4} \alpha_i A_i(x, y) + \sum_{i=1}^{4} \delta_i D_i(x, y) + \sum_{i=1}^{3} \varepsilon_i E_i(x, y),$$

where the $\{D_i\}$ are the basis functions associated with the boundary nodes $\{d_i\}$, and the $\{E_i\}$ are the basis functions associated with the interior nodes $\{e_i\}$.

In virtually all interpolation problems it is desirable that the interpolant be continuous, and we will assume that this is the case in our discussion. If the individual basis functions are continuous, then the interpolant in each subregion will be continuous, but this does not imply that the interpolant will be continuous as we move from one subregion to another. We will illustrate this for the regions in Figure 7.6.

Assume that the two local interpolants I_1 and I_2 are such that we have continuity across the boundary between the two subregions. Let the data in region R_2 change in such a way that one of the coefficients, for example ε_3, changes to ε_3^*. This means that the interpolant in this region has now changed. The new interpolant is given by

$$I_2^*(x, y) = I_2(x, y) + (\varepsilon_3^* - \varepsilon_3)E_3(x, y).$$

Nothing has changed in region R_1; hence the values of $I_1(x, y)$ on the interface have not changed. If continuity must be maintained, then it is essential that

$$(\varepsilon_3^* - \varepsilon_3)E_3(x, y) = 0$$

on the interface between regions R_1 and R_2. Since it is our assumption that $\varepsilon_3^* \neq \varepsilon_3$, we deduce that $E_3(x, y) = 0$ on the interface between regions R_1 and R_2. Since, for continuity, this must be true for any basis function associated with a data point that is not on the interface between the two subregions we have the following essential requirement for local interpolation.

For global continuity in a local piecewise interpolation problem, the basis functions must be zero on all boundaries that do not contain the node corresponding to that basis function.

Definition 7.1 *By the **piecewise local interpolation problem** we mean an interpolation problem defined on a collection of nonoverlapping subregions where the interpolant on each subregion is defined solely in terms of the geometry of that subregion together with the data given within that subregion and on its boundary.*

We have, in almost all of our discussion so far, considered the interpolation conditions to be linear functionals. We will continue to do so. In most cases these linear functionals will correspond to the interpolation of the function value or the value of a derivative of the function at some point within, or on the boundary of, the region. We know that the solution to such interpolation problems can always be rewritten in terms of a biorthonormalized basis. Therefore, we will assume that this is the type of basis that has to be calculated. The point corresponding to such a normalized basis function will be referred to as the node of the basis function.

Definition 7.2 *The **node** of a biorthonormalized basis function $\phi_j(\underline{x})$ for a pointwise interpolation problem is the point $\underline{x}_j$ such that $L(\phi_j(\underline{x}))|_{\underline{x}_j} = 1$, where L denotes the linear functional under consideration (often L denotes function evaluation and is given by $L(\phi_j(\underline{x})) = \phi_j(\underline{x}_j)$).*

Definition 7.3 *The subregion on which one set of local basis functions are defined is the **local domain** for that basis.*

Definition 7.4 *A **boundary segment** of a local domain is either a part of the boundary that has a different defining equation than any other part, or a part of the boundary that is common to another local domain.*

Definition 7.5 *The* **continuity condition** *for local interpolation is the requirement that each basis function must be zero on all local domain boundary segments that do not contain its node.*

Definition 7.6 *A basis satisfies the* **local interpolation property** *if it is a suitable basis for the local interpolation problem and satisfies the continuity condition.*

Figure 7.7 shows some examples of nodes, local domains, and boundary segments. The rectangular domain $ABCD$ is subdivided into two local domains $ABED$ (local domain 1) and $BCDE$ (local domain 2) by the curve BED. Two nodes A and E are indicated. The boundary segments of the first local domain $ABED$ are the distinct straight line segments AB and AD and the curve segment BED, whereas the boundary segments of the second local domain $BCDE$ are the distinct straight line segments CB and CD and the curve segment BED, which is common to both local domains. In order to satisfy the continuity condition, the basis function associated with node A and defined on the first local domain must be zero at all points on the curved boundary segment BED and must, therefore, contain the expression defining this curve as a factor. The basis function associated with node E and defined on the *first* local domain must be zero at all points on the straight line segments AB and AD and must contain both linear expressions defining these two lines as factors. However, node E is also a node of the *second* local domain, and in this context it has

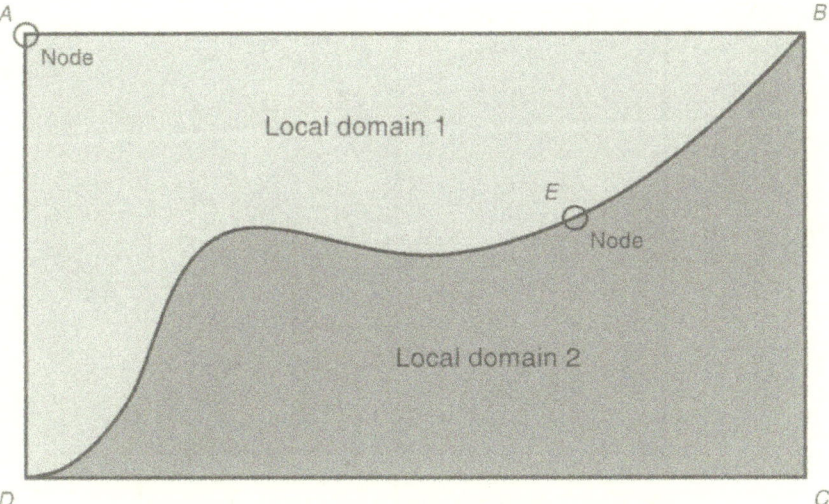

Figure 7.7. Nodes, local domains, and boundary segments.

another basis function, defined on the second local domain, associated with it. This basis function must be zero at all points on the straight line segments CB and CD and must contain both linear expressions defining these two lines as factors.

The definitions given above are not restricted to a particular space dimension. For interpolation in one dimension the nodes are points on a single straight line segment (the local domain), boundary segments are the endpoints of such a line segment, and continuity is achieved by imposing conditions at these endpoints. The basis functions are functions of one variable and can be interpreted as curves. In two dimensions the nodes are points in, or on the boundary of, a single planar region (the local domain), boundary segments are continuous curves, and continuity of the interpolant is achieved by imposing conditions along these curves. The basis functions are functions of two variables and can be interpreted as surfaces. In three dimensions the nodes are points in, or on the boundary of, a single three-dimensional region or volume (the local domain), boundary segments are continuous surface patches, and continuity of the interpolant is achieved by imposing conditions on these surface patches. The basis functions are functions of three variables and cannot be interpreted graphically in our restricted universe.

Let us now consider the visual aspects of basis functions associated with particular nodes and local domains. Figure 7.8 shows a quarter-circle local domain with one curved and two straight boundaries. Any normalized basis

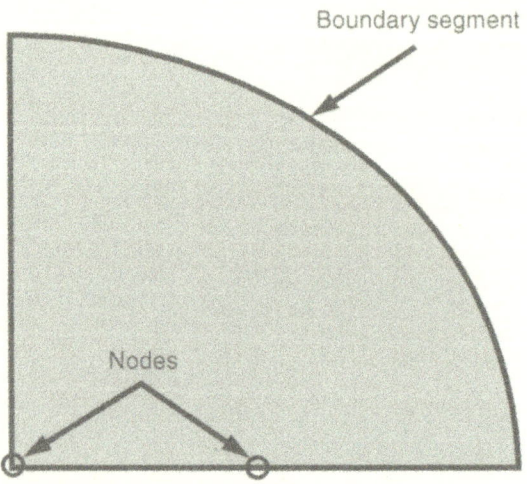

Figure 7.8. A local domain with two nodes.

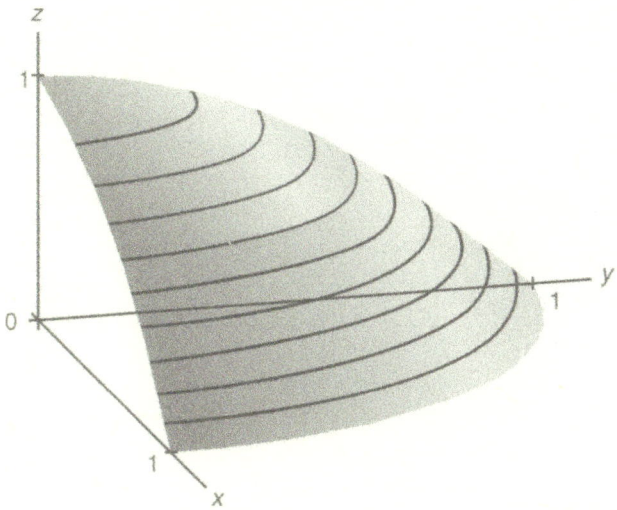

Figure 7.9. A basis function that is zero on a boundary segment.

function associated with the node at the intersection of the straight boundary segments (the origin) must be zero on the curved side in order to satisfy the continuity condition. The curved boundary is the first quadrant segment of the unit circle in the plane $z = 0$.

One such basis function, namely

$$\phi_0(x, y) = 1 - x^2 - y^2,$$

is shown in Figure 7.9. The surface represented by this function is not a segment of a sphere, although the unit sphere with center at the origin, $z^2 = 1 - x^2 - y^2$, will indeed have the same intersection curve, namely the unit circle, with the plane $z = 0$.

Consider the construction of a basis function associated with the other node shown in Figure 7.8, namely the point $P_1 = (\frac{1}{2}, 0)$ in the plane $z = 0$. We require a function that will assume the value one at the node, and that will evaluate to zero at every point on the curved edge, as well as on the other straight edge. All such functions that satisfy the continuity condition must be of the general form

$$\phi(x, y) = g(x, y)x(1 - x^2 - y^2),$$

where $g(x, y)$ is an arbitrary function. Furthermore, we require that

$$\phi(\tfrac{1}{2}, 0) = g(\tfrac{1}{2}, 0)(\tfrac{1}{2})(\tfrac{3}{4}) = 1,$$

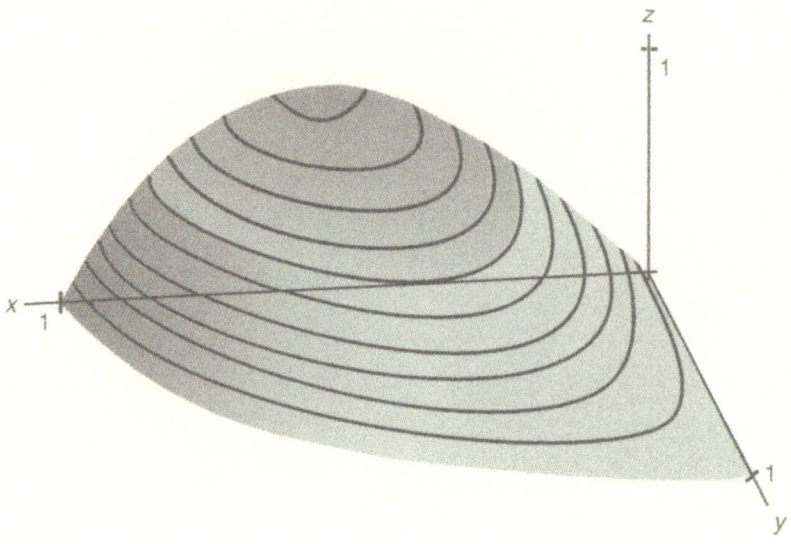

Figure 7.10. A basis function that is zero on two boundary segments.

so that

$$g\left(\tfrac{1}{2}, 0\right) = \tfrac{8}{3}.$$

The simplest way to achieve this is to choose $g(x, y) \equiv \tfrac{8}{3}$, so that

$$\phi_1(x, y) = \tfrac{8}{3}x(1 - x^2 - y^2).$$

This surface function is illustrated in Figure 7.10.

Let us next consider a more interesting three-sided local domain (Figure 7.11), with boundaries

$$\begin{aligned}
1 - x^2 - y^2 &= 0, & 0 \le x, y \le 1, \\
x^2 + y^2 + x - y &= 0, & 0 \le x, y \le 1, \\
y &= 0, & 0 \le x \le 1.
\end{aligned}$$

Two nodes are shown, namely $P_1 = (\tfrac{1}{2}, 0)$ and $P_2 = (\tfrac{1}{2}, \tfrac{1}{2})$. For the interior node P_2 the continuity condition requires that the associated basis function must have the general form

$$\phi(x, y) = g(x, y)y(1 - x^2 - y^2)(x^2 + y^2 + x - y),$$

where $g(x, y)$ is an arbitrary function. Furthermore, we require that

$$\phi\left(\tfrac{1}{2}, \tfrac{1}{2}\right) = g\left(\tfrac{1}{2}, \tfrac{1}{2}\right)\left(\tfrac{1}{2}\right)\left(\tfrac{1}{2}\right)\left(\tfrac{1}{2}\right) = 1,$$

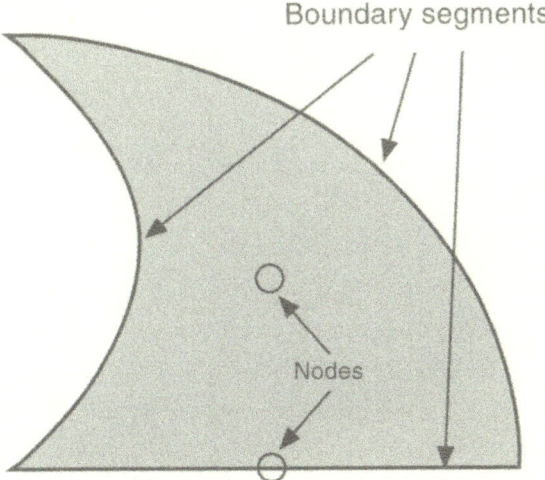

Figure 7.11. A local domain with two curved boundary segments.

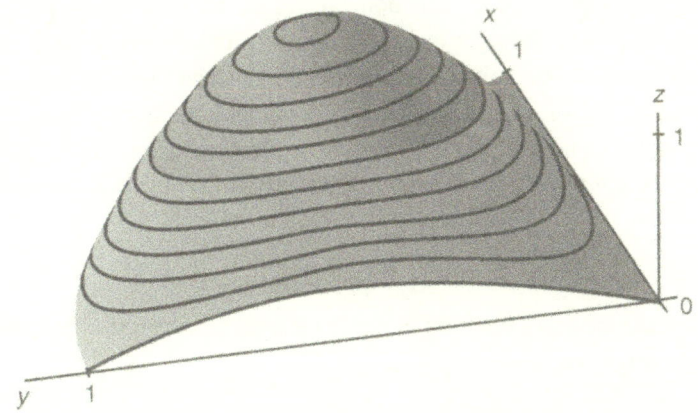

Figure 7.12. A basis function that is zero on three boundary segments.

so that

$$g\left(\tfrac{1}{2}, \tfrac{1}{2}\right) = 8.$$

We may, for example, choose $g(x, y) \equiv 8$, so that

$$\phi_2(x, y) = 8y(1 - x^2 - y^2)(x^2 + y^2 + x - y).$$

This surface function is illustrated in Figure 7.12.

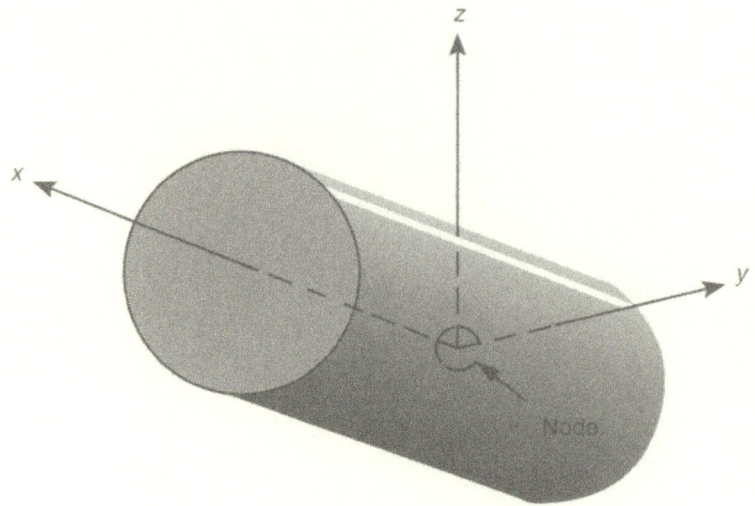

Figure 7.13. Local domain in three dimensions.

As a final example of a local domain, let us consider a domain in three dimensions, namely the right circular cylinder with axis along the x-axis and with cross section the unit circle (Figure 7.13). The cylinder extends over the interval $x \in [-1, 1]$. To satisfy the continuity condition, a basis function associated with a node at the origin must be identically zero on the boundary of the domain, namely the surface of the cylinder. Therefore, it must be of the general form

$$\phi(x, y, z) = g(x, y, z)(1 - x^2)(1 - y^2 - z^2).$$

As before, the arbitrary function $g(x, y, z)$ is used to normalize the basis function at the origin, that is,

$$\phi(0, 0, 0) = g(0, 0, 0)(1)(1) = 1,$$

so that

$$g(0, 0, 0) = 1,$$

and a suitable basis function is

$$\phi_0(x, y, z) = (1 - x^2)(1 - y^2 - z^2).$$

7.3 Interpolation to Curves

Our discussion of the local interpolation property shows that the basis functions must be zero on boundary segments that do not contain the function's associated node. It does not, however, tell us anything about the behavior of the function on boundary segments that do contain the node. In this section we discuss this aspect of basis functions from two different but equivalent viewpoints.

The usual approach to interpolation is to select, *a priori*, a basis which behaves in a certain way. We often require that the basis span polynomials of some degree. Basis construction in two dimensions will not be quite as straightforward as it was in one dimension. The following example illustrates this point.

Consider the local domain of Figure 7.8, and let us construct a basis possessing the local interpolation property and spanning polynomials of degree one. Immediately, we are in trouble. Consider the basis function corresponding to the node at the origin. The continuity condition implies that the basis function must be of the general form

$$z(x, y) = \phi(x, y)(1 - x^2 - y^2).$$

If we expected a polynomial basis function of degree one, then our hopes are dashed as a result of the continuity condition. We can no longer demand that the function itself be a linear polynomial over the entire local domain, but we would still like it to be linear on adjacent boundary segments . Is this possible? In many cases it will be, and we will discuss this more fully in Section 7.4. For now, let us approach a solution in the following way.

Since we are seeking a normalized basis function, we will require $z(0, 0) = 1$. We also know that we must have $z(0, 1) = 0$ and $z(1, 0) = 0$. The requirement that the function be linear on the adjacent boundary segments then implies $z = 1 - y$ when $x = 0$, and $z = 1 - x$ when $y = 0$. Therefore,

$$\phi(0, y)(1 - y^2) = 1 - y \quad \text{and} \quad \phi(x, 0)(1 - x^2) = 1 - x.$$

On removal of the common factors, we see that this implies

$$\phi(0, y) = \frac{1}{1 + y} \quad \text{and} \quad \phi(x, 0) = \frac{1}{1 + x}.$$

One function $\phi(x, y)$ that satisfies these conditions is $\phi(x, y) = (1 + x + y)^{-1}$ which implies that the normalized basis function is given by

$$z(x, y) = \frac{1 - x^2 - y^2}{1 + x + y}.$$

This is certainly not the only function which will meet the requirements of having the local interpolation property and having linear behavior on adjacent boundary segments but it is the simplest. This function is shown in Figure 7.14 and should be compared with the function shown in Figure 7.9.

As a second example let us consider the local domain shown in Figure 7.11 and construct a basis function for the node at the point $(\frac{1}{2}, 0)$. We want this basis

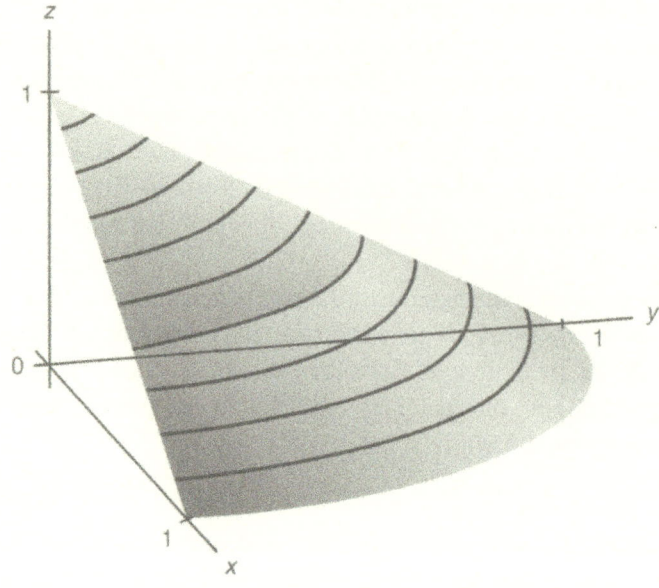

Figure 7.14. A basis function linear on adjacent boundaries.

function to behave like a second-degree polynomial. As we might expect from
the example we have just considered, this will be impossible, for the continuity
condition implies that

$$z(x, y) = \phi(x, y)(1 - x^2 - y^2)(x^2 + y^2 + x - y),$$

and this is certainly not quadratic. We impose the following conditions on the
basis function $z(x, y)$:

$$z\left(\tfrac{1}{2}, 0\right) = 1, \qquad z(0, 0) = 0, \quad \text{and} \quad z(1, 0) = 0.$$

We require that the function must be quadratic on the boundary segment $y = 0$.
This implies that $z(x, 0) = 4x(1 - x)$, so that

$$\phi(x, 0)(1 - x^2)(x^2 + x) = 4x(1 - x).$$

On removal of the common factors we see that

$$\phi(x, 0) = \frac{4}{(x + 1)^2}.$$

One function $\phi(x, y)$ that satisfies all these conditions is given by

$$\phi(x, y) = \frac{4}{x^2 + y^2 + 2x + y + 1},$$

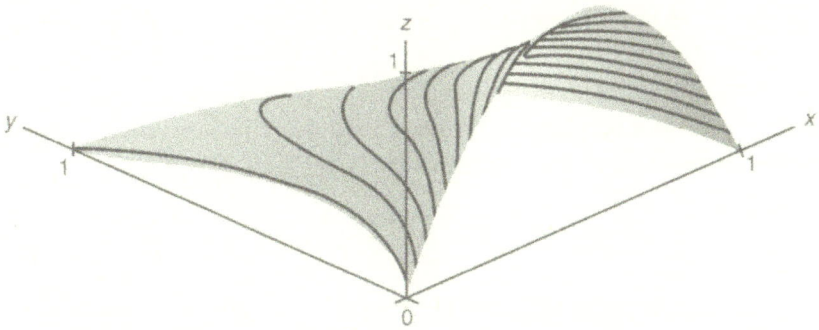

Figure 7.15. A rational basis function, quadratic on a line segment – first view.

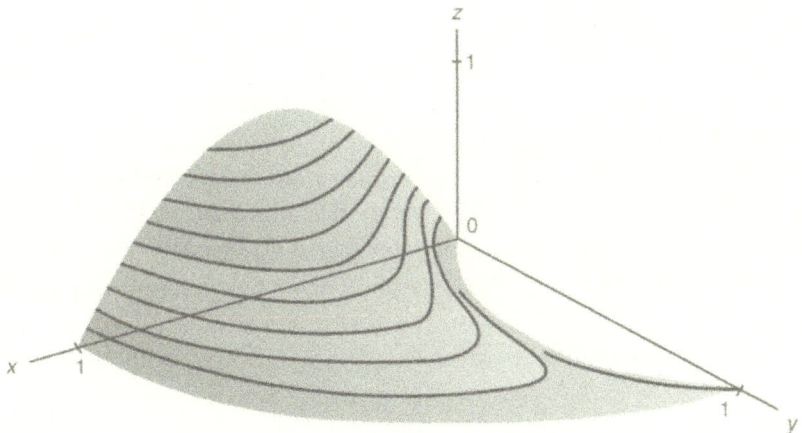

Figure 7.16. A rational basis function, quadratic on a line segment – second view.

and the normalized basis function is given by

$$z(x, y) = \frac{4(1 - x^2 - y^2)(x^2 + y^2 + x - y)}{x^2 + y^2 + 2x + y + 1}.$$

Two views of this surface function are shown in Figures 7.15 and 7.16.

In each of these examples the function was required to have a particular polymonial behavior along a straight line. However, our boundary segments are not always straight lines. We must therefore define the phrase "polynomial behavior on a curve." We will discuss this more fully in the next section, where we will determine the dimension of the implied space, but for the present we will discuss the geometrical and visual interpretation.

Functions of two variables are represented by surfaces in three-dimensional space. Therefore, when we say a function exhibits *polynomial behavior*, we mean that it can be represented as a surface of the form

$$z = f(x, y),$$

where $f(x, y)$ is a polynomial. When we say that z *behaves like a rational function* we mean that z can be represented in the form

$$z = \frac{f(x, y)}{g(x, y)},$$

where f and g are polynomials. When we say z is *algebraic*, we mean that it satisfies an equation of the form

$$f(x, y, z) = 0$$

where f is a polynomial. By z *on a surface* $h(x, y, z) = 0$ we mean the restriction of the surface that determines z to the surface $h(x, y, z) = 0$. Geometrically, this restriction is the intersection of the two surfaces.

When considered as a surface in three-dimensional space, a curve $h(x, y) = 0$ in the plane is a right cylinder. By saying that z *has polynomial behavior on h* we mean that the intersection of the surface that represents z with the cylinder $h(x, y) = 0$ is the same as the intersection curve of some polynomial surface [a surface of the form $z = f(x, y)$ with f a polynomial] with the cylinder $h(x, y) = 0$.

Let a polynomial surface be given by

$$p = -2x^2 + xy + y^2 + 3x - 1,$$

and let a local boundary segment be given by the segment, in the first quadrant, of the curve

$$x^2 + y^2 + x + 2y - 5 = 0.$$

The polynomial surface is shown in Figure 7.17, and the curved boundary segment is the cylinder shown in Figure 7.18.

A function behaves like $p = -2x^2 + xy + y^2 + 3x - 1$ on the curve $x^2 + y^2 + x + 2y - 5 = 0$ if it intersects the cylinder $x^2 + y^2 + x + 2y - 5 = 0$ in the same curve as p does. The polynomial surface and the cylinder are shown, from different viewpoints, in Figures 7.19 and 7.20.

It is a particularly important concept. We have, on the one hand, expressed the desire to have a basis which spans polynomials. On the other hand, we have seen that even in simple situations we have not been able to find polynomial functions that satisfy the required conditions. We will have to use a function that is *not* a polynomial but which has polynomial behavior on the boundary segments.

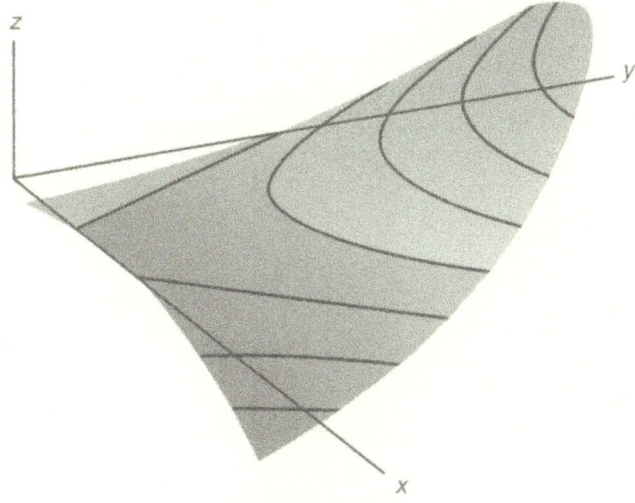

Figure 7.17. The polynomial surface.

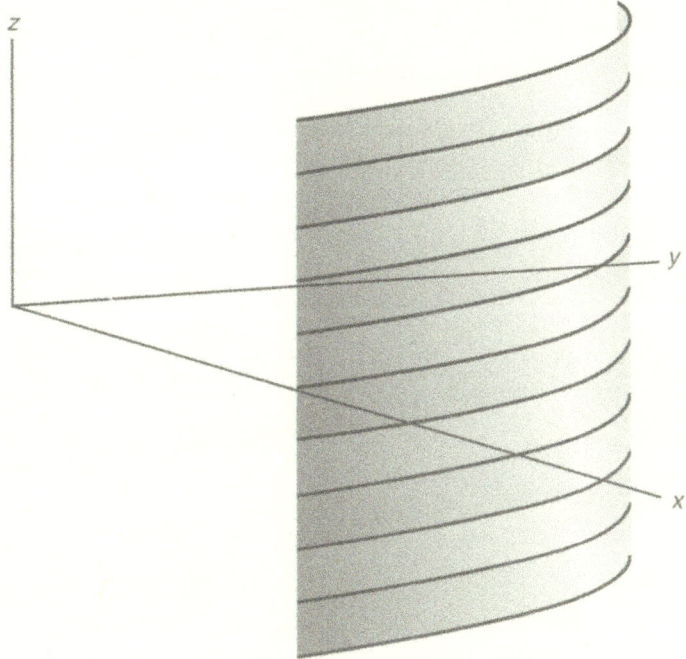

Figure 7.18. The curved cylindrical boundary.

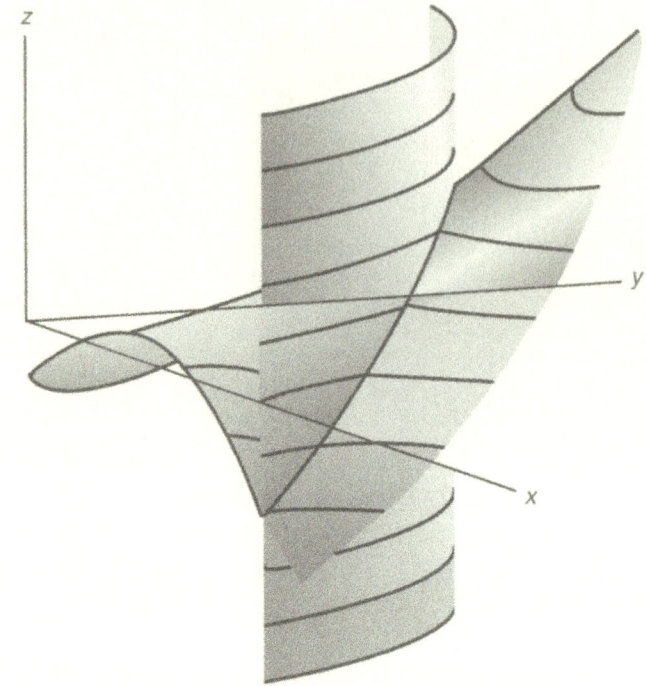

Figure 7.19. The intersection curve – first view.

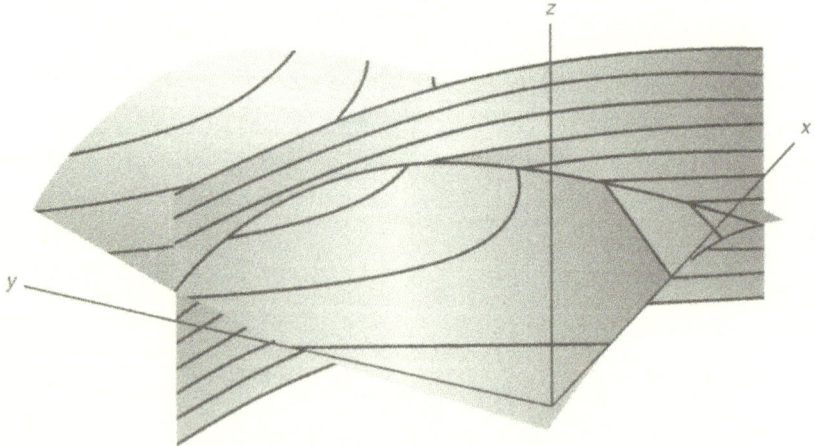

Figure 7.20. The intersection curve – second view.

Example 7.1 *Let the local domain in the first quadrant be bounded by*

$$x = 0, \qquad y = 0, \quad and \quad x^2 + 2y^2 - y - 1 = 0.$$

Find a function that behaves like $p = x + 2y - 1$ *on the curved boundary segment.*

The vertices of the local domain are $(0, 0)$, $(1, 0)$, and $(0, 1)$. The polynomial takes the values zero at $(1, 0)$ and one at $(0, 1)$. Let us look for a function that has unit value at the origin. Since it makes no sense to consider discontinuous functions, we will look for a function z that satisfies the conditions

$$
\begin{aligned}
z &= 1 && \text{when} \quad x = 0, \\
z &= 1 - x && \text{when} \quad y = 0, \\
z &= x + 2y - 1 && \text{when} \quad x^2 + 2y^2 - y - 1 = 0.
\end{aligned}
$$

One function that satisfies all these conditions is given by

$$z = \frac{-x^2 + 4xy + 2y + 1}{x + 2y + 1}.$$

This example is illustrated in Figures 7.21, 7.22, and 7.23 by three different views of the surfaces involved in the solution, namely, the right cylinder defined by the curved boundary, the plane $z = x + 2y - 1$ that determines the linear behavior of the function on the curved boundary, and the rational function. These surfaces intersect in a common space curve.

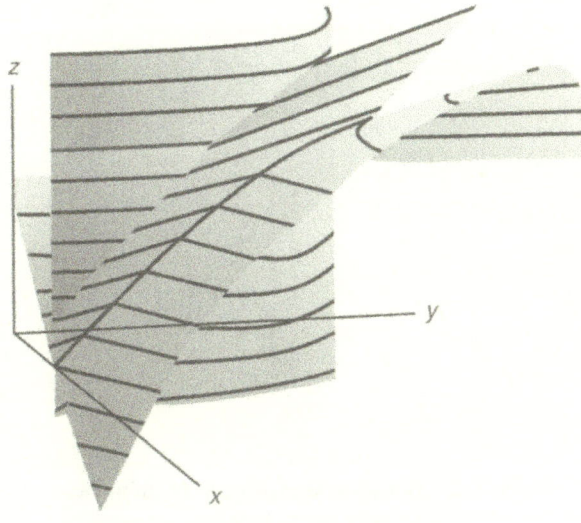

Figure 7.21. Surfaces through a curve – first view.

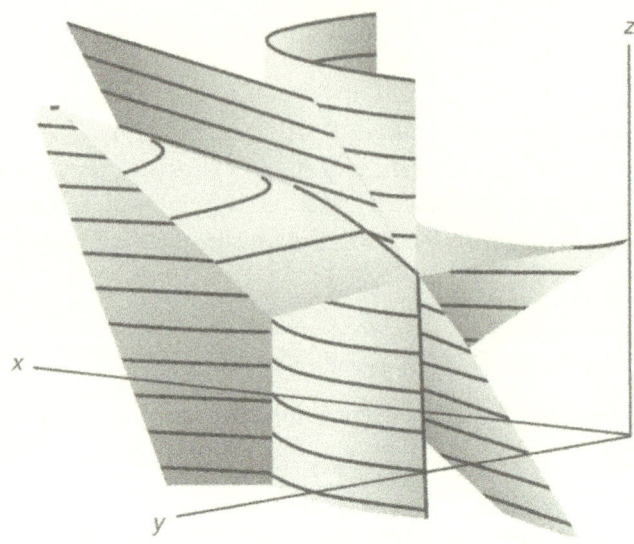

Figure 7.22. Surfaces through a curve – second view.

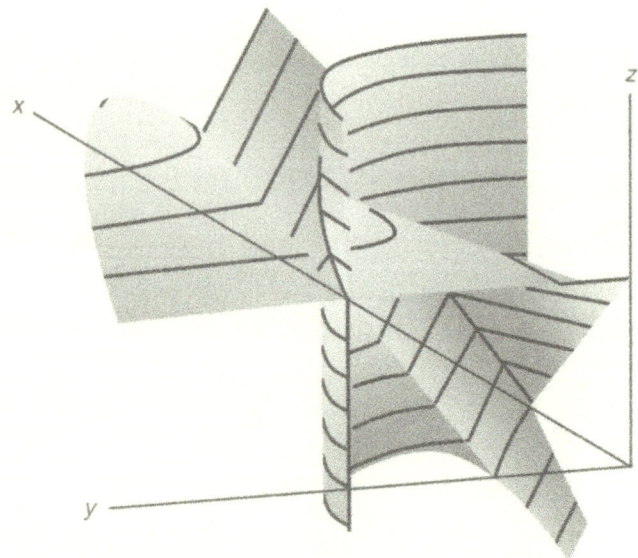

Figure 7.23. Surfaces through a curve – third view.

In three dimensions the concept is the same, though it is not so easy to construct a visualization. A polynomial surface is a hypersurface of the form $w = f(x, y, z)$, where f is a polynomial. A boundary segment is a surface of the form $g(x, y, z) = 0$, where g is a polynomial. A function $h(x, y, z)$, not necessarily a polynomial, has polynomial behavior on the surface $g(x, y, z) = 0$ if it has the same intersection with the hypercylinder $g(x, y, z) = 0$ that the hypersurface $w = f(x, y, z)$ has with that hypercylinder.

We finish this section by emphasizing the connection between a function having a particular behavior on a boundary segment and the corresponding surface passing through a particular curve. This gives us two different ways of stating the same requirement, and it is often quite useful to be able to think of a situation in more than one way. For example, instead of saying that we seek a function $z(x, y)$ such that $z = 1 - y$ when $x = 0$, we can say that we wish to find a surface that passes through the curve given by the intersections of the surfaces $x = 0$ and $y + z - 1 = 0$. This provides a more consistent view of the interpolation problem. We have already seen that the continuity condition implies that the function must be zero on all boundary segments that do not contain its node. This is the same thing as saying that the function (or surface) interpolates to the boundary segments that do not contain the node. We extend this idea by giving the other curves to which the surface must interpolate. From continuity requirements, this sequence of space curves must be continuous. We have found that the local interpolation problem in two dimensions is identical to the problem of finding a three-dimensional surface that passes through a piecewise defined, continuous space curve.

Example 7.2 *Find a surface function $z = F(x, y)$ that passes through the curves given by the intersections of the surfaces*

$$z = 0 \quad and \quad f(x, y) \equiv ax^2 + bxy + cy^2 - (1 + a)x - (1 + c)y + 1,$$

$$x = 0 \quad and \quad y + z - 1 = 0,$$

$$y = 0 \quad and \quad x + z - 1 = 0.$$

The first curve is the equation of the general conic through the points $(1, 0)$ and $(0, 1)$ in the (x, y) plane such that the conic does not pass through the origin. This problem could equally well have been stated in the form: find the continuous basis function that interpolates to the general conic through $(1, 0)$ and $(0, 1)$, is linear on $y = 0$ and on $x = 0$, and evaluates to one at the origin. The surface must have the general form

$$z(x, y) = \phi(x, y)[ax^2 + bxy + cy^2 - (1 + a)x - (1 + c)y + 1],$$

subject to

$$\phi(0, y)[cy^2 - (1+c)y + 1] = 1 - y,$$
$$\phi(x, 0)[ax^2 - (1+a)x + 1] = 1 - x.$$

This simplifies to the requirements

$$\phi(0, y) = \frac{1}{1 - cy},$$

$$\phi(x, 0) = \frac{1}{1 - ax}.$$

One surface satisfying all the curve interpolation requirements is

$$z(x, y) = \frac{ax^2 + bxy + cy^2 - (1+a)x - (1+c)y + 1}{1 - ax - cy}.$$

Example 7.3 *Construct a surface function $z = F(x, y)$ that interpolates to the curves defined by the following pairs of surfaces.*

$$z = 0 \quad and \quad Cyl_1 = 0,$$
$$z = Quad \quad and \quad Cyl_2 = 0,$$
$$z = Quad \quad and \quad x = 0,$$
$$y = 0 \quad and \quad z = 0,$$

where $Quad(x, y) = \frac{1}{4}y^2 - \frac{1}{2}x^2 + 14x$, *and where* $Cyl_i = 0, i = 1, 2,$ *denote cylinders perpendicular to the (x, y) plane with elliptic generating curves and are therefore quadric surfaces. The generating ellipses pass through the common points $(3, 2)$, $(5, 1)$, $(3, -3)$, and $(-1, -3)$. For the first cylinder the generating ellipse C_1 also passes through $(1, 0)$, while the second, C_2, passes through $(0, 2)$.*

The required surface has to pass through a straight line and an ellipse in the (x, y) plane. In the plane $x = 0$ [the (y, z) plane] it must pass through the parabola $Quad(0, y) = \frac{1}{4}y^2$, and it must pass through the intersection curve of the surface $z = Quad(x, y)$ with the second cylinder. This is, therefore, an example of an interpolating surface that is required to have quadratic behavior on both a straight line segment and a conic. The boundary curve segments are continuous, and the interpolating surface must evaluate to unity at the point $(0, 2)$. The conics C_1 and C_2 each has to pass through five given points

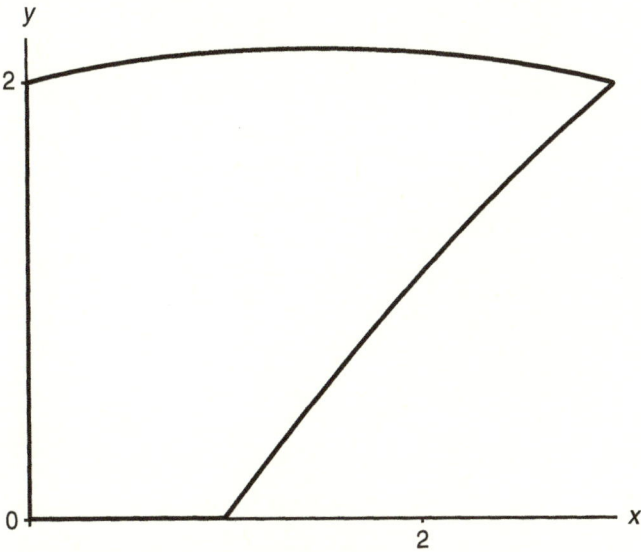

Figure 7.24. Local domain with two conic boundaries.

and can be generated by the methods discussed in Chapter 6. Their equations
are

$$C_1: \qquad 9x^2 - 11xy + 5y^2 - 51x + 38y + 42 = 0,$$
$$C_2: \qquad 5x^2 - xy + 13y^2 - 13x + 16y - 84 = 0.$$

The local domain of the interpolating function is shown in Figure 7.24.

The details of the construction of an interpolating surface will be discussed
when we have developed the required mathematical tools. One such surface, possibly the simplest to pass through all the required curves, is shown
in Figure 7.25, together with the interpolated curves and the second cylinder.

7.4 The Dimension of the Interpolation Problem

We have defined the local interpolation problem, discussed some of its implications for the interpolating basis, and suggested thinking of the basis functions, or
the complete interpolant, as a surface passing through particular space curves.
However, we have not yet considered the number of basis functions needed to
span polynomials of some given degree on a local domain.

The dimension of the polynomial space $P_n(x)$ is $n + 1$, and the dimension
of $P_n(x, y)$ is $\frac{1}{2}(n + 1)(n + 2)$. In general, the dimension $D_{n,d}$ of the space of
nth-degree polynomials $P_n(x_1, x_2, \ldots, x_d)$ in d variables is given by

$$D_{n,d} = \frac{(n + d)!}{n!d!}.$$

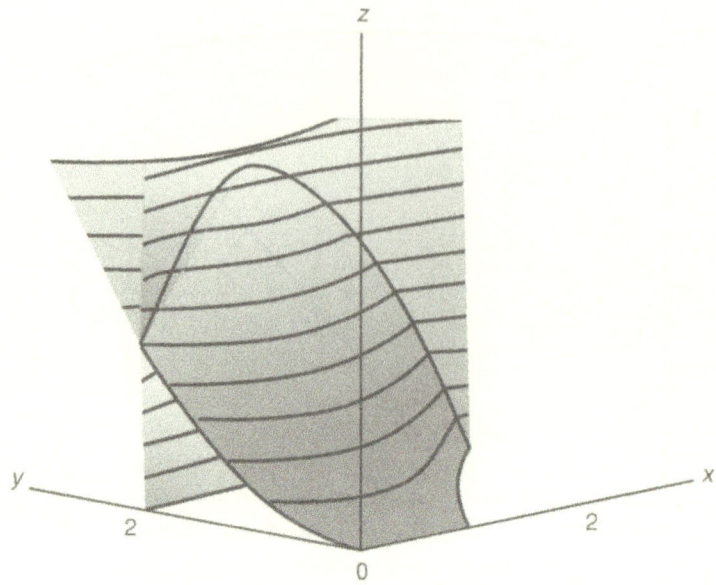

Figure 7.25. Surface interpolating to four curves.

This is the required dimension of an interpolating basis for a global interpolation problem, as shown in the following example.

Example 7.4 *Construct a global basis for the polynomial space $P_2(x, y)$.*

The simplest solution would be to say that all the polynomials in the polynomial space $P_2(x, y)$ consist of combinations of the $D_{2,2} = (2 + 2)!/2!2! = 6$ terms $1, x, y, x^2, xy$, and y^2. This is certainly a basis for the space. However, let us construct the orthonormal basis associated with the six nodes $P_1(0, 0)$, $P_2(1, 0)$, $P_3(2, 0)$, $P_4(0, 1)$, $P_5(0, -1)$, and $P_6(1, 1)$. These six nodes do not lie on a conic – the first five determine the degenerate conic $xy = 0$, which certainly does not contain the sixth node – and hence, from the equivalence theorem (Theorem 1.4), the system possesses the finite interpolation property.

The basis function $\phi_1(x, y)$ associated with the node P_1 must be zero at the other five nodes. We therefore construct the quadratic function associated with the conic through these points, namely

$$C_1(x, y) = x^2 + 2xy - 2y^2 - 3x + 2,$$

with $C_1(0, 0) = 2$. Therefore,

$$\phi_1(x, y) = \tfrac{1}{2}x^2 + xy - y^2 - \tfrac{3}{2}x + 1.$$

Similarly,

$$\phi_2(x, y) = x(2 - x - y),$$

$$\phi_3(x, y) = \tfrac{1}{2}x(x - 1),$$

$$\phi_4(x, y) = \tfrac{1}{2}y(1 - 2x + y),$$

$$\phi_5(x, y) = \tfrac{1}{2}y(y - 1),$$

$$\phi_6(x, y) = xy.$$

The use of the MacLaurin–Bézout theorem, discussed in Section 5.9, makes the construction of these five basis functions almost trivial: if a conic is zero at three points on a line, it must contain the line as a factor.

The dimension of a basis for local interpolation cannot be less than the number $D_{n,d}$. Before we can determine what the minimum dimension is for the appropriate interpolation over a specified local domain, we must determine the minimum dimension required to span the required-degree polynomials over the boundary segments. If the basis spans some polynomial space over the entire local domain, it must also span it over the boundary segments. Let us consider, therefore, the required dimension to span polynomials over certain curves. In generalized form, we have to determine the dimension of a d-space polynomial restricted to a subspace of dimension $d - 1$.

Consider polynomials $P_n(x)$ of degree n in one dimension ($d = 1$). This polynomial space has dimension $n + 1$. A basis for this space is $\{1, x, x^2, x^3, \ldots, x^n\}$. Consider a subspace of this one-dimensional domain space, for example, the point $x = 2$. Restricted to the subspace given by $x = 2$, this basis is $\{1, 2, 4, 8, \ldots, 2^n\}$. Each of the restricted basis functions is a linear multiple of the first one, and hence the dimension of the restricted space is 1.

Let us next consider polynomials $P_n(x, y)$ of degree n in two dimensions ($d = 2$). This polynomial space has dimension $\tfrac{1}{2}(n + 1)(n + 2)$. A basis for this space is

$$1, x, y, x^2, xy, y^2, \ldots, x^n, x^{n-1}y, \ldots, xy^{n-1}, y^n. \tag{7.1}$$

Consider the one-dimensional subspace of the domain space consisting of those points for which $x = 0$ (the y-axis). Restricted to this subspace, the basis is

$$1, 0, y, 0, 0, y^2, 0, 0, 0, y^3, 0, \ldots \ldots, 0, y^n,$$

and hence the dimension of the basis restricted to the line $x = 0$ is $n + 1$. This result is not dependent on the specific line to which we restrict the local domain. To show this, let us consider the general linear one-dimensional subspace for which $ax + by + c = 0$, with a, b not both zero. Assume that $a \neq 0$. Then we can express x in terms of y:

$$x = -\frac{by + c}{a} \equiv \alpha y + \beta,$$

and the basis restricted to this line becomes

$$1, \alpha y + \beta, y, (\alpha y + \beta)^2, (\alpha y + \beta)y, y^2, \ldots,$$
$$(\alpha y + \beta)^n, (\alpha y + \beta)^{n-1}y, \ldots, (\alpha y + \beta)y^{n-1}, y^n.$$

The second term is a linear combination of the first and third terms, the fourth and fifth terms are linear combinations of the first, third, and sixth terms (i.e., of 1, y, y^2), and so on, with 1, y, y^2, $\ldots$, y^n the only remaining independent basis functions on the restricted subspace.

Let us display this elimination scheme in a more systematic way. The basis (7.1) can be written schematically in the triangular form

$$
\begin{array}{ccccccccc}
 & & & & 1 & & & & \\
 & & & x & & y & & & \\
 & & x^2 & & xy & & y^2 & & \\
 & x^3 & & x^2y & & xy^2 & & y^3 & \\
 \ddots & & & & & & & & \ddots \\
 x^n & x^{n-1}y & & \cdots & & \cdots & \cdots & xy^{n-1} & y^n
\end{array}
$$

$$\tag{7.2}$$

For the case $x = 0$ all the terms except those on the right of each row disappear. For the general one-dimensional linear subspace $ax + by + c = 0$, $a \neq 0$, the triangle becomes

$$
\begin{array}{ccccc}
 & & \boxed{1} & & \\
 & \alpha y + \beta & & \boxed{y} & \\
 (\alpha y + \beta)^2 & & (\alpha y + \beta)y & & \boxed{y^2} \\
 \ddots \quad \vdots & & \vdots & & \vdots \quad \ddots
\end{array}
$$

The boxed terms represent the linearly independent basis functions that remain when we restrict the plane to a straight line. All the other elements of the triangle can be expressed in terms of these remaining basis functions. Another way of thinking about it is to say that we can use the boxed terms to eliminate all the others.

The one-dimensional subspace need not be linear. Let us use a conic as a boundary segment of a local domain. This is still a one-dimensional subspace of the plane (a planar curve has only one independent variable), but it is no longer linear. Since at least one of the second-order terms of the quadratic equation representing the conic must be nonzero, we can eliminate this term wherever it occurs in the basis functions.

Example 7.5 *Determine the dimension of the polynomial basis of $P_2(x, y)$ when restricted to the circle $x^2 + y^2 - 1 = 0$, and also when restricted to the coordinate axes.*

We have already discussed the restriction of the general polynomial bases to straight lines. Therefore, restricted to the axes – we often say *modulo* the axes – the basis is of dimension 3. On the x-axis we use $1, x, x^2$, and on the y-axis we use $1, y, y^2$.

Consider the restriction to the circle. In the first three rows of the triangular array that contains the global basis for $P_2(x, y)$, we can only replace one of the terms. For example, we can replace y^2 by $1 - x^2$, to obtain

$$\boxed{1}$$
$$\boxed{x} \qquad \boxed{y}$$
$$\boxed{x^2} \qquad \boxed{xy} \qquad 1 - x^2$$

There are five remaining basis functions, and the dimension of the space modulo the circle is 5.

Example 7.6 *Determine the dimension of the polynomial basis of $P_2(x, y)$ restricted to the cubic $y - x^3 = 0$.*

We can write $y = x^3$, so that, modulo the cubic, the basis becomes

$$1$$
$$x \qquad x^3$$
$$x^2 \qquad x^4 \qquad x^6$$

Since none of the terms are linear combinations of the others, the dimension of the basis remains 6. Another way of looking at the problem is to write $x^3 = y$; since the second-order basis does not contain a third-degree term, no substitution or elimination is possible.

Example 7.7 *Determine the dimension of the polynomial basis of $P_3(x, y)$, restricted to a general conic and restricted to the coordinate axes.*

Modulo the coordinate axes, the dimension decreases from 10 (global dimension $D_{3,2}$) to 4. Assume that the conic has a term in x^2 (if not, use y^2

or xy). Then

$$x^2 = f(x, y) = a_0 + a_1 x + a_2 y + a_3 xy + a_4 y^2,$$
$$x^3 = xf(x, y) = a_0 x + a_1 x^2 + a_2 xy + a_3 x^2 y + a_4 xy^2$$
$$= a_0 x + a_1 f + a_2 xy + a_3 fy + a_4 xy^2$$
$$= b_0 + b_1 x + b_2 y + b_3 xy + b_4 y^2 + b_5 xy^2 + b_6 y^3,$$
$$x^2 y = yf(x, y) = a_0 y + a_1 xy + a_2 y^2 + a_3 xy^2 + a_4 y^3,$$

and the triangular representation of the basis becomes

$$\boxed{1}$$
$$\boxed{x} \qquad \boxed{y}$$
$$f \qquad \boxed{xy} \qquad \boxed{y^2}$$
$$xf \qquad yf \qquad \boxed{xy^2} \qquad \boxed{y^3}$$

Since f, xf, and yf can be expressed as linear combinations of the boxed terms (alternatively, they can be eliminated by using the boxed basis functions), we see that the dimension of $P_3(x, y)$ restricted to a general conic is 7.

We summarize these results in the following theorem.

Theorem 7.1 *The dimension $D_{n,2}$ mod f_m of the polynomial space $P_n(x, y)$ restricted to a curve $f_m(x, y) = 0$ of order m is given by*

$$D_{n,2} \bmod f_m = D_{n,2} - H(n - m)\frac{(n - m + 2)!}{(n - m)!2!},$$

where $H(x)$ is the Heaviside unit step function

$$H(x) = \begin{cases} 1 & for \quad x \geq 0, \\ 0 & for \quad x < 0. \end{cases}$$

Proof. If the order of the curve is greater than the degree of the interpolating polynomials, no decrease in the number of basis functions is possible, since the elimination process does not result in any simplification, as demonstrated in Example 7.6. Therefore, $D_{n,2} \bmod f_m = D_{n,2}$ for $m > n$.

Consider the cases for which $m \leq n$. We have to show that

$$D_{n,2} \bmod f_m = \frac{(n + 2)!}{n!2!} - \frac{(n - m + 2)!}{(n - m)!2!}$$

$$= \tfrac{1}{2}m(2n + 3 - m) \qquad \text{for} \quad n \geq m.$$

Consider the right half of the triangular representation of the global basis, namely

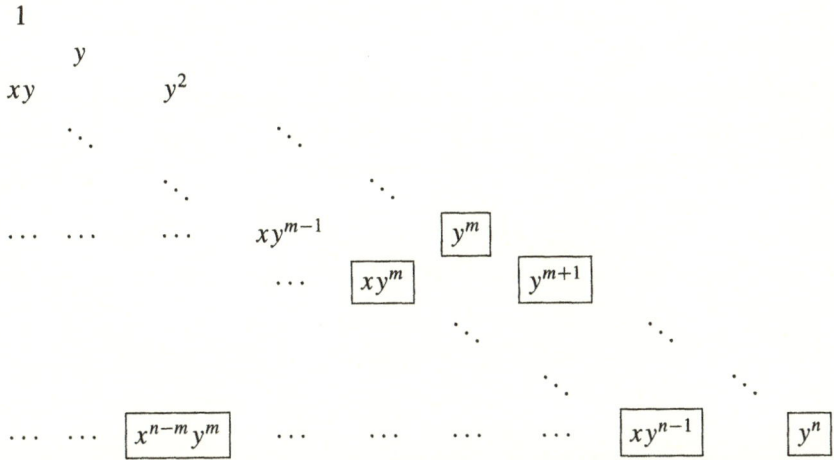

where the boxed terms are those that can be expressed as linear combinations of the other basis functions if we assume that the equation of the curve $f_m(x, y) = 0$ contains a nonzero y^m term. (If not, we can always use one of the other mth order terms, since at least one of them must be nonzero.) The number of boxed basis functions is

$$1 + 2 + \cdots + (n - m + 1) = \tfrac{1}{2}(n - m + 1)(n - m + 2)$$
$$= \frac{(n - m + 2)!}{(n - m)!2!}.$$

These are the basis functions that can be eliminated modulo the curve f_m. Therefore,

$$D_{n,2} \bmod f_m = D_{n,2} - \frac{(n - m + 2)!}{(n - m)!2!} \quad \text{for} \quad n \geq m. \qquad \blacksquare$$

A bounded local domain R in two dimensions consists of a region of the (x, y) plane enclosed by a continuous boundary, consisting of one or more boundary segments. Let the boundary segments form part of the set of curves $F = \{f_{i,m_i}(x, y) = 0, i = 1, 2, \ldots, k\}$, where m_i denotes the order of the ith boundary segment. We denote by $D_{n,2} \mid R$ (read as "$D\,n2$ on R") the dimension of the linear interpolation problem over the local domain for polynomial interpolation of order n. We select those intersections of the boundary segments that belong to R as nodes associated with basis functions. This results in the use of the smallest total number of basis functions, since two adjoining boundary segments "share" that node, which counts as one of the nodes required to comply

354   Surfaces

with the dimension requirements modulo each of the two adjoining boundary segments. This choice means that

$$D_{n,2} \mid R = \max\left(D_{n,2}, \sum_{i=1}^{k} D_{n,2} \bmod f_{i,m_i} - k\right).$$

Example 7.8 *Determine the dimension of the linear interpolation problem over the local domain in the first quadrant bounded by the coordinate axes and the unit circle, for polynomial interpolation of order 2.*

We have three boundary segments, two of which are of order one, and one of order two, while the polynomial space is of global dimension $D_{2,2} = 6$. Therefore,

$$F = \{f_{1,1} : x = 0, \ f_{2,1} : y = 0, \ f_{3,2} : x^2 + y^2 - 1 = 0\}$$

and

$$D_{2,2} \mid R = \max\left(6, \sum_{i=1}^{3} D_{2,2} \bmod f_{i,m_i} - 3\right)$$

$$= \max[6, (3 + 3 + 5) - 3] \qquad \text{(see Example 7.5)}$$

$$= 8.$$

We shall need eight basis functions to span the space of polynomials of degree two over the given local domain. Three of these must be associated with nodes on each of the straight line segments, and five with nodes on the circle. This is possible only if three of the nodes are placed at the intersection points of the boundary segments, that is, at $(0, 0)$, $(1, 0)$, and $(0, 1)$. Figure 7.26 demonstrates possible node placement for linear functionals that are function evaluation at the nodes.

Example 7.9 *Determine the dimension of the linear interpolation problem over the local domain in the first quadrant bounded by the coordinate axes and a general conic segment, for polynomial interpolation of order 3.*

The global dimension of the polynomial space is 10. We have

$$F = \{f_{1,1} : x = 0, \ f_{2,1} : y = 0, \ f_{3,2} : p_2(x, y) = 0\},$$

$$D_{3,2} \mid R = \max\left(10, \sum_{i=1}^{3} D_{3,2} \bmod f_{i,m_i} - 3\right)$$

$$= \max[10, (4 + 4 + 7) - 3] \qquad \text{(see Example 7.7)}$$

$$= 12.$$

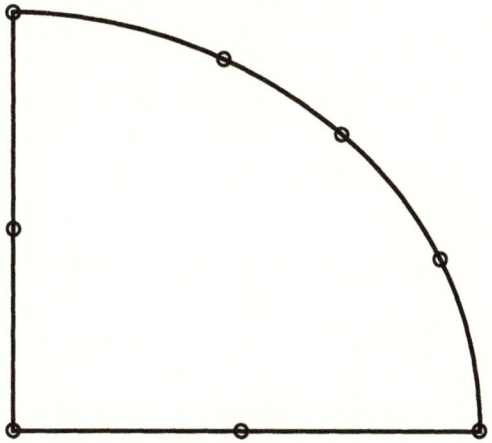

Figure 7.26. Second-order interpolation.

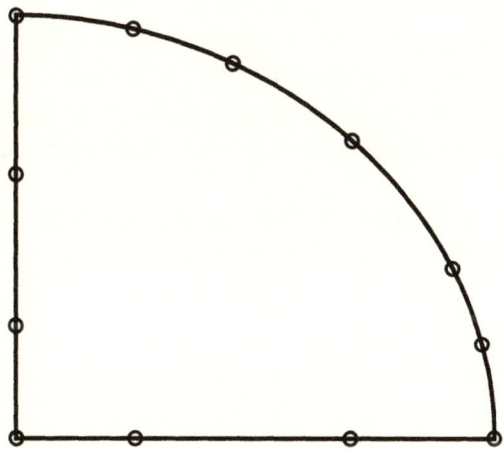

Figure 7.27. Cubic interpolation, one conic side.

We require twelve basis functions: one associated with each vertex, two more associated with each straight boundary segment, and five more associated with the curved boundary. Node placement for function evaluation at nodes is shown in Figure 7.27.

Example 7.10 *Determine the dimension of the linear interpolation problem over the local domain in the first quadrant bounded by the coordinate axes and the line* $1 - x - y = 0$, *for polynomial interpolation of order* 3.

We now have

$$F = \{f_{1,1} : x = 0, \ f_{2,1} : y = 0, \ f_{3,1} : 1 - x - y = 0\},$$

$$D_{3,2} \mid R = \max\left(10, \sum_{i=1}^{3} D_{3,2} \bmod f_{i,m_i} - 3\right)$$

$$= \max[10, (4 + 4 + 4) - 3] \qquad \text{(see Example 7.7)}$$

$$= 10.$$

We require ten basis functions, of which one is associated with each vertex and two more are associated with each of the three straight boundary segments, which gives the nine basis functions required to comply with the dimensions of the space modulo the boundary segments. However, these nine basis functions cannot span the cubic polynomial space. We need one more node, which cannot be placed on the boundary of the domain (otherwise we violate the dimension requirements modulo that boundary segment), and the tenth basis function must therefore be associated with an interior point of the local domain. Node placement for function evaluation at nodes is shown in Figure 7.28.

When is $D_{n,2} \mid R = D_{n,2}$? We have one example, with a triangle as local domain (Example 7.10), for which this is the case when $n = 3$. In fact, with a

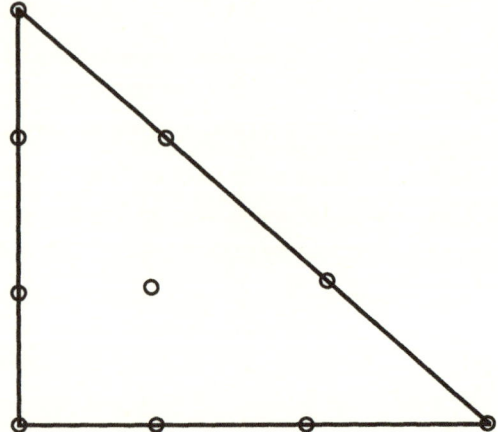

Figure 7.28. Cubic interpolation over a triangle.

triangle as the local domain R we have

$$D_{n,2} \mid R = \max \left[D_{n,2}, \sum_{i=1}^{3} D_{n,2} \bmod f_{i,1} - 3 \right]$$

$$= \max \left[D_{n,2}, \sum_{i=1}^{3} \left[D_{n,2} - \tfrac{1}{2} n(n+1) \right] - 3 \right]$$

$$= \max \left[\tfrac{1}{2}(n+1)(n+2), 3 \left[\tfrac{1}{2}(n+1)(n+2) - \tfrac{1}{2} n(n+1) \right] - 3 \right]$$

$$= \max \left[\tfrac{1}{2}(n+1)(n+2), 3n \right],$$

and, since $\tfrac{1}{2}(n+1)(n+2) - 3n = \tfrac{1}{2}(n-1)(n-2) \geq 0$ for all integer values of n, $D_{n,2} \mid R = D_{n,2}$.

The triangle is not the only local domain for which $D_{n,2} \mid R = D_{n,2}$. When one of the sides of the triangle changes to a conic, we require an additional n nodes on that side. The original triangle required $3n$ nodes on the boundary segments, and hence there were $\tfrac{1}{2}(n-1)(n-2)$ remaining nodes which we position inside the domain. Now, some of these nodes can move to the curved segment. We shall have enough to satisfy the required dimension, provided that

$$\tfrac{1}{2}(n-1)(n-2) > n.$$

This is the case when $n \geq 5$. Hence, in this case $D_{n,2} \mid R = D_{n,2}$. If two of the sides of the triangle now become conics, we require

$$\tfrac{1}{2}(n-1)(n-2) > 2n,$$

and this will be the case if $n \geq 7$. If all three segments are conics, we require $n \geq 9$ before $D_{n,2} \mid R = D_{n,2}$. When this is not the case, the additional nodes required on the curved segments are sufficient to force $D_{n,2} \mid R > D_{n,2}$.

Our discussion of dimension is not connected to the kind of interpolation being done. We have been using Lagrange interpolation, even though we have not yet fully discussed the actual construction of the basis, but the arguments also apply to Hermite interpolation.

Example 7.11 *Discuss polynomial interpolation of order 3 over a triangle, and consider the effect on the interpolation problem when one side of the triangle becomes a conic segment.*

This is the case discussed in Example 7.10, with $D_{3,2} \mid R = D_{3,2} = 10$ and $D_{3,2} \bmod l = 4$, where l represents any straight line boundary segment. This means that the dimension of the interpolation problem restricted to each of

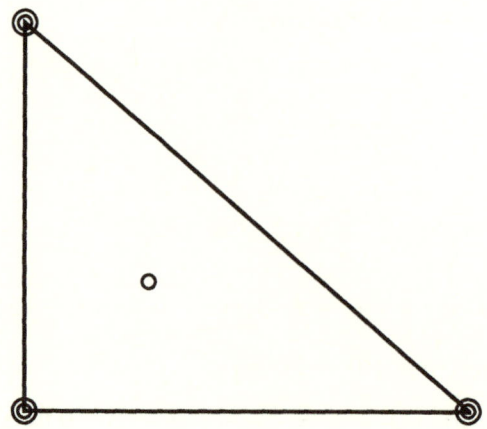

Figure 7.29. Cubic Hermite interpolation over a triangle.

the three triangle sides is 4, whereas the dimension of the local interpolation problem is 10. Therefore, node placement as illustrated in Figure 7.28, with Lagrange interpolation at each of the nodes, will satisfy all the dimension requirements. However, this is not the only interpolation system that will satisfy these requirements.

Consider the node configuration illustrated in Figure 7.29, where two concentric circles at a vertex indicate, as before, that we are interpolating both the first partial derivatives, as well as function value, at this point. We showed in Section 5.2 that geometrically this is equivalent to a double point at this node. Hence each of the lines does have four points on it, but, since each double point is equivalent to three conditions, the total number of conditions is, as in the Lagrange case, ten. At the interior node we interpolate only the function value. If one of the boundary segments is a conic, the dimension is 12. The node placement for cubic Hermite interpolation is shown in Figure 7.30. There are double points at the vertices (where we interpolate the function value and the first partial derivatives), and there are three additional nodes, at which we interpolate only the function value, on the conic segment. This gives a total of twelve interpolation conditions, and we see that we have, as in the Lagrange case, the equivalent of seven nodes on the conic segment.

7.5 Basis Construction for Simple Local Domains

We have discussed the local interpolation problem, the idea of thinking of a two-dimensional interpolant as a surface, and, in the preceeding section, the dimension of the interpolation problem. We know the requirements on the basis functions and how many we need. We are therefore ready to discuss

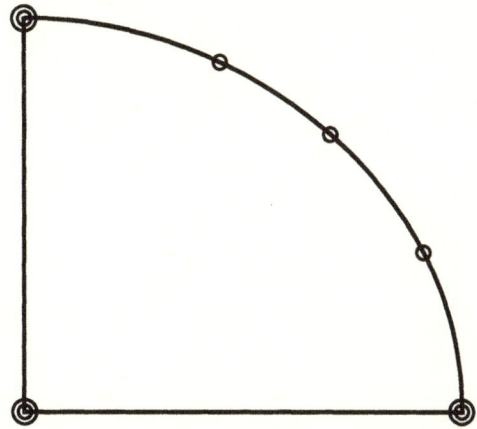

Figure 7.30. Cubic Hermite interpolation, one conic side.

the construction of these basis functions. When the local domain is a triangle, a parallelogram, a tetrahedron, or a parallelopiped, the constructions are
straightforward, as illustrated in the following examples.

When we refer to linear interpolation, quadratic interpolation, and so on, we
are referring to the *order* of the polynomial space spanned by the basis and
not to the form of the individual basis functions. Only in the simplest cases
shall we manage to solve the local interpolation problem using polynomial
basis functions.

7.5.1 Lagrange Interpolation on a Triangle

Let us first consider linear interpolation on a triangle. We shall require a minimum of three basis functions. The node placement is illustrated in Figure 7.31.
If the nodes are at the points $(0, 0)$, $(1, 0)$, and $(0, 1)$, the basis functions must
be of the form

$$\phi_1(x, y) = f(x, y)(1 - x - y),$$

$$\phi_2(x, y) = g(x, y)y,$$

$$\phi_3(x, y) = h(x, y)x.$$

Furthermore, we know that these basis functions must represent surfaces that
pass through the curves given by

$$x = 0, \quad y + z - 1 = 0 \text{ and } y = 0, \quad x + z - 1 = 0,$$

$$x = 0, \quad y - z = 0 \quad \text{and } x + y - 1 = 0, \quad y - z = 0,$$

$$y = 0, \quad x - z = 0 \quad \text{and } x + y - 1 = 0, \quad x - z = 0,$$

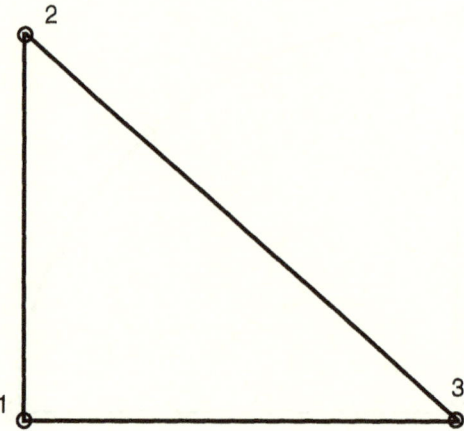

Figure 7.31. Linear interpolation on a triangle.

respectively. The simplest basis functions which satisfy these conditions are:

$$\phi_1(x, y) = 1 - x - y,$$

$$\phi_2(x, y) = y,$$

$$\phi_3(x, y) = x.$$

In this case, the basis itself comprises linear polynomials.

Let us look at this example in another way. Let us seek, from the outset, basis functions that are linear polynomials. Consider $\phi_1(x, y)$. This function must be zero at nodes 2 and 3. That is, it must be zero at two points of a line. Therefore, the locus of all zeros of this function must contain the nodes 2 and 3. The function is linear which means that the locus of all its zeros is a line. Therefore, this line must have two zeros in common with the line $1 - x - y = 0$. The Maclaurin–Bézout theorem tells us that algebraic curves of orders m and n meet either in mn points or in infinitely many points, that is, they must have a factor in common. Therefore, $\phi_1(x, y) = \alpha(1 - x - y)$. Normalization yields $\alpha = 1$. Let us use this alternative approach in another example.

Consider quartic interpolation on the standard triangle, with vertices at $(1, 0)$, $(0, 1)$, and $(0, 0)$. The situation is as depicted in Figure 7.32. $D_{4,2} \mid R = 15$, and we select fifteen nodes as shown. Assume that the nodes are regularly spaced on the sides and in the interior. Let us construct a basis, each member of which is a quartic polynomial. The basis function $\phi_1(x, y)$ associated with node 1 must be zero at nodes 5, 6, 7, 8, and 9. These five points are on a line. Therefore, since $\phi_1(x, y)$ is quartic,

$$\phi_1(x, y) = P_3(x, y)(1 - x - y).$$

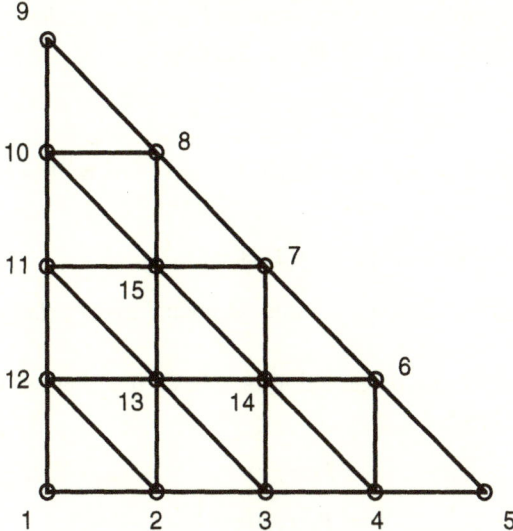

Figure 7.32. Quartic interpolation on a triangle.

Since $\phi_1(x, y)$ is also zero at the four collinear points 4, 14, 15, and 10, and since $1 - x - y$ is not zero at these points, $P_3(x, y)$ must be zero at these points. Therefore, $P_3(x, y)$ is a polynomial of degree three that is zero at four points of a line, and by the Maclaurin–Bézout theorem,

$$P_3(x, y) = P_2(x, y)(4x + 4y - 3),$$

so that

$$\phi_1(x, y) = P_2(x, y)(4x + 4y - 3)(1 - x - y).$$

Repeating this argument for points 3, 13, and 11 shows us that

$$\phi_1(x, y) = P_1(x, y)(2x + 2y - 1)(4x + 4y - 3)(1 - x - y),$$

and, using the same argument once again, we see that

$$\phi_1(x, y) = \alpha(4x + 4y - 1)(2x + 2y - 1)(4x + 4y - 3)(1 - x - y).$$

Finally, we select α such that $\phi_1(0, 0) = 1$, which results in

$$\phi_1(x, y) = -\tfrac{1}{3}(4x + 4y - 1)(2x + 2y - 1)(4x + 4y - 3)(1 - x - y).$$

We strongly recommend the reader review this last argument until the technique becomes second nature.

Let us give the argument sequence for the construction of two more of the basis functions.

The basis function $\phi_3(x, y)$ is quartic and is zero at nodes 5, 6, 7, 8, and 9, as well as at nodes 9, 10, 11, 12, and 1. Therefore,

$$\phi_3(x, y) = P_2(x, y)x(1 - x - y).$$

Since $P_2(x, y)$ must be zero at nodes 4, 14, and 15 as well as at nodes 15, 13, and 2,

$$\phi_3(x, y) = \alpha(4x - 1)(4x + 4y - 3)x(1 - x - y).$$

Normalizing the basis function to unity at node 3 gives $\phi_3(x, y) = -4(4x - 1)(4x + 4y - 3)x(1 - x - y)$.

As an example of the construction of a basis function associated with an interior node we have

$$\phi_{15}(x, y) = P_4(x, y)$$
$$= P_1(x, y)xy(1 - x - y)$$
$$= \alpha(4y - 1)xy(1 - x - y)$$
$$= 32(4y - 1)xy(1 - x - y).$$

7.5.2 Hermite Interpolation on a Triangle

We have mentioned the use of Hermite interpolation to reduce the number of nodes involved in the calculation of bases for local domains. This does not imply that the number of basis functions is decreased, since this number is determined by the dimension of the interpolation problem. Let us consider quintic interpolation on the standard triangle. For this case $D_{5,2}|R = 21$, and we require the equivalent of six points on each edge of the triangle. If we place a triple point at each of the vertices (Figure 7.33), we have six points on each side. A triple point is equivalent to interpolating the function value and first- and second-order partial derivatives (six conditions). Choosing three points in the interior brings the number of interpolating conditions to 21.

Let U_i, $V_{i,x}$, $V_{i,y}$, $W_{i,xx}$, $W_{i,xy}$, and $W_{i,yy}$ denote the normalized basis functions that interpolate the function value and the first and second partial derivatives. We seek a solution that is a quintic polynomial. Then the basis function associated with node 1 must be of the form

$$U_1(x, y) = P_4(x, y)(1 - x - y).$$

The quartic $P_4(x, y)$ must be unity at node 1, and both first and second partial derivatives must be zero at this point. Furthermore, since the basis function has triple points at nodes 2 and 3, and since the factor $1 - x - y$ accounts

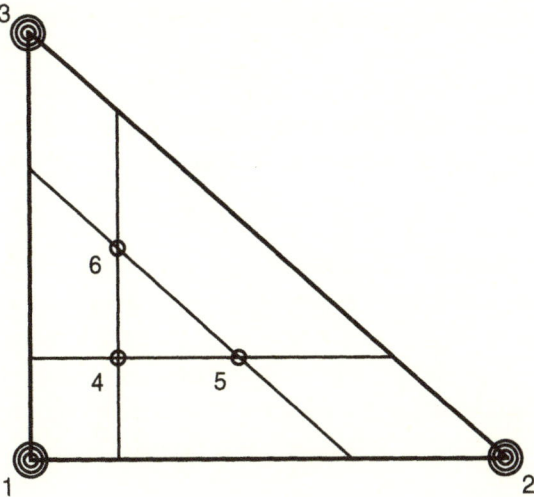

Figure 7.33. Hermite quintic interpolation on a triangle.

for only simple points, $P_4(x, y)$ must have double points at these nodes. We use the symmetry of the situation and find that $P_4(x, y)$ must be of the general form

$$P_4(x, y) = 1 + a(x + y) + b(x^2 + y^2) + cxy + d(x^3 + y^3)$$
$$+ exy(x + y) + f(x^4 + y^4) + gxy(x^2 + y^2) + hx^2y^2.$$

The derivative conditions for double points at nodes 2 and 3 imply that

$$a + b + d + f + 1 = 0,$$
$$3a + 2b + d + 4 = 0,$$
$$a + c + e + g = 0.$$

The fact that we have to satisfy only three equations to ensure double points and not six is the result of the built-in symmetry. This is, unfortunately, as far as we can go without involving the remaining linear term of $U_1(x, y)$. We replace d, f, and g and use the complete expression for $U_1(x, y)$ in all further calculations. The requirement that the first and second partial derivatives at the origin must be zero results in $a = b = 1$ and $c = 2$, so that only two coefficients, namely e and h, remain undetermined. If nodes 4, 5, and 6 have coordinates $(\frac{1}{4}, \frac{1}{4})$, $(\frac{1}{2}, \frac{1}{4})$, and $(\frac{1}{4}, \frac{1}{2})$, respectively, then requiring the function to be zero at these points finally gives $e = -\frac{594}{5}$ and $h = \frac{1654}{5}$. The basis function is

$$U_1(x, y) = \tfrac{1}{5}(1 - x - y)[5(1 + x + y + (x + y)^2) - 45(x^3 + y^3)$$
$$- 594xy(x + y) + 30(x^4 + y^4) + 579xy(x^2 + y^2) + 1654x^2y^2].$$

For $W_{2,yy}$ we have triple points at nodes 1 and 3. This implies

$$W_{2,yy} = P_4(x, y)x.$$

Since the function value and all the x-derivatives are zero on the segment defined by nodes 1 and 2, $P_4(x, y)$ must be of the form $yP_3(x, y)$. Hence,

$$W_{2,yy} = P_3(x, y)xy.$$

The factor xy ensures that $W_{2,yy}$ has a simple point at each of the nodes 2 and 3, and a double point at node 1. However, $W_{2,yy}$ must have a triple point at node 1. Therefore, $P_3(x, y)$ must have a simple point at node 1 and has no constant term. A similar argument shows that $P_3(x, y)$ must have a double point at node 3 and a simple point at node 2. Taking into account the behavior of the function modulo the coordinate axes, we find that $P_3(x, y)$ must be of the form

$$P_3(x, y) = ax(1 - x)^2 + by(1 - y)^2 + xy(cx + dy + e).$$

This factor also has to ensure that $W_{2,yy}$ has a simple point at each of nodes 4, 5, and 6. These conditions, together with the requirement that the second partial derivative of $W_{2,yy}$ with respect to x at $(0, 1)$ must be zero and that the second partial derivative of $W_{2,yy}$ with respect to y at $(1, 0)$ must be unity, enable us to solve for the undetermined coefficients. We finally have

$$W_{2,yy} = \tfrac{1}{20}xy[-6x(1 - x)^2 + 14y(1 - y)^2 + xy(33x + 43y - 37)].$$

Let us now construct U_6. The fact that there are triple points at nodes 1, 2, and 3, and the application of the Maclaurin–Bézout theorem, show us that

$$U_6 = P_2(x, y)xy(1 - x - y).$$

Furthermore, $P_2(x, y)$ must be zero at nodes 1, 2, 3, 4, and 5. Therefore, it is some multiple of the polynomial associated with the conic through these five points. Hence, $U_6 = a(3x^2 + 3xy - 2y^2 - 3x + 2y)xy(1 - x - y)$. The value of a is chosen so that U_6 has unit value at node 6. This implies that

$$U_6 = \tfrac{512}{5}(3x^2 + 3xy - 2y^2 - 3x + 2y)xy(1 - x - y).$$

7.5.3 Interpolation on a Square

We will do two examples for this local domain, namely a second-order Lagrange basis and a cubic Hermite basis.

The node placement for the Lagrange case is shown in Figure 7.34, with vertices at the origin and at unit distances on the axes. The dimension on each

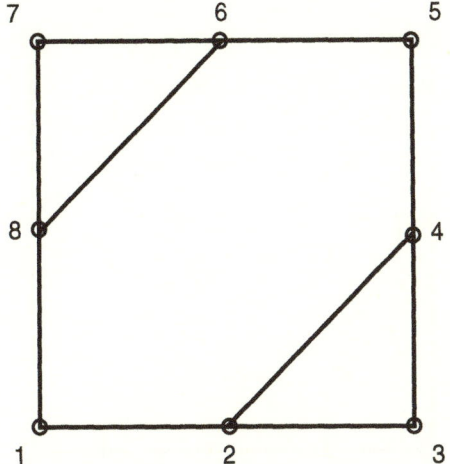

Figure 7.34. Quadratic interpolation on a square.

boundary segment is three, and hence we require at least eight nodes, as shown. In this case we span a larger space than $P_2(x, y)$. Let us proceed by constructing a basis and then deduce exactly what space is spanned. We illustrate the construction of the basis function associated with node 3. Because of the continuity requirement of being identically zero on boundary segments that do not contain the node, $U_3(x, y)$ must be of the form $U_3(x, y) = g(x, y)x(1 - y)$. Also, $g(x, y)$ must be zero at nodes 2 and 4. The simplest function that satisfies these conditions is $g(x, y) = 2x - 2y - 1$. Therefore,

$$U_3(x, y) = (2x - 2y - 1)x(1 - y).$$

This function, though not a polynomial of second degree, is of second order on the boundary segments. Certain cubic terms are included, but neither x^3 nor y^3 occurs. These terms will not occur in any of the basis functions. Because of the biorthonormalization of the basis, the eight basis functions are linearly independent, and hence they must span the eight terms remaining after we have excluded the x^3 and y^3 terms, that is, they span the space $\{1, x, y, x^2, xy, y^2, x^2y, xy^2\}$.

The node placement for the cubic Hermite case is shown in Figure 7.35. Let us construct the basis function $V_{1,x}$. Because the basis function must be zero on the boundary segments defined by nodes 2 and 3 and by nodes 3 and 4, we know that $V_{1,x} = g(x, y)(1 - x)(1 - y)$. The basis function is now guaranteed to have a double point at node 3. It must also have double points at nodes 2 and 4. At node 1 it must be zero, its partial derivative with respect to y must be zero, and its partial derivative with respect to x must be unity. There are

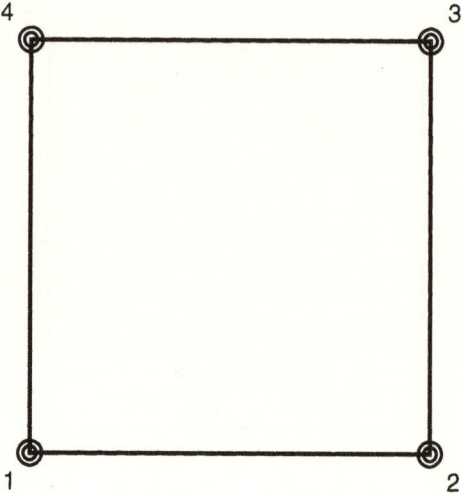

Figure 7.35. Cubic Hermite interpolation on a square.

no x^4 or y^4 terms. If we omit these terms from the general quartic, we are left with thirteen terms. It is common practice to take one more node in the interior of the domain, at which we interpolate only the function value. This will give thirteen basis functions. We use only twelve basis functions by stipulating that the coefficients of x^3y and x^2y^2 must be the same, since this results in a nice symmetry. Finally, we obtain

$$V_{1,x} = x(1 - x - y)(1 - x)(1 - y).$$

Once again, this function is not a cubic polynomial.

7.5.4 Interpolation on a Tetrahedron

The simplices, in any number of dimensions, are particularly easy to deal with. When the local domain R is a tetrahedron with vertices at the origin and at unit distances along the axes (Figure 7.36), we find that $D_{1,3}|R = 4$. The basis functions associated with the vertices are

$$U_1(x, y, z) = 1 - x - y - z,$$

$$U_2(x, y, z) = x,$$

$$U_3(x, y, z) = y,$$

$$U_4(x, y, z) = z.$$

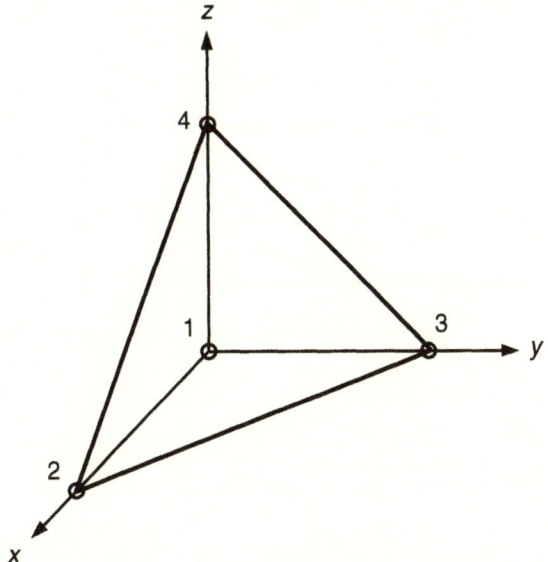

Figure 7.36. Linear interpolation on a tetrahedron.

7.5.5 Interpolation on a Parallelopiped

We consider only a rectangular parallelopiped, because there is no loss of generality in doing so. Consider the problem of finding a second-order basis for this local domain in three dimensions. Then $D_{2,3} = 10$, but we need six nodes on each planar face and three nodes on each edge (line segment) to satisfy the dimension requirements (the interpolant must still be quadratic when restricted to a face or an edge). These requirements determine the node placement for Lagrange interpolation on the unit cube, as shown in Figure 7.37. Let us construct the basis function $U_1(x, y, z)$ associated with node 1, which is at the point $(1, 1, 1)$. This function has to be zero on the three coordinate planes (the boundaries that do not contain the node), that is,

$$U_1(x, y, z) = f(x, y, z)xyz.$$

It also has to be zero at the remaining three nodes not on the coordinate planes, namely 2, 3, and 4. Since these three nodes are not collinear, they define a plane, and the function $U_1(x, y, z) = a(5 - 2x - 2y - 2z)xyz$ now only has to be normalized to unity at $(1, 1, 1)$. Therefore,

$$U_1(x, y, z) = -(5 - 2x - 2y - 2z)xyz.$$

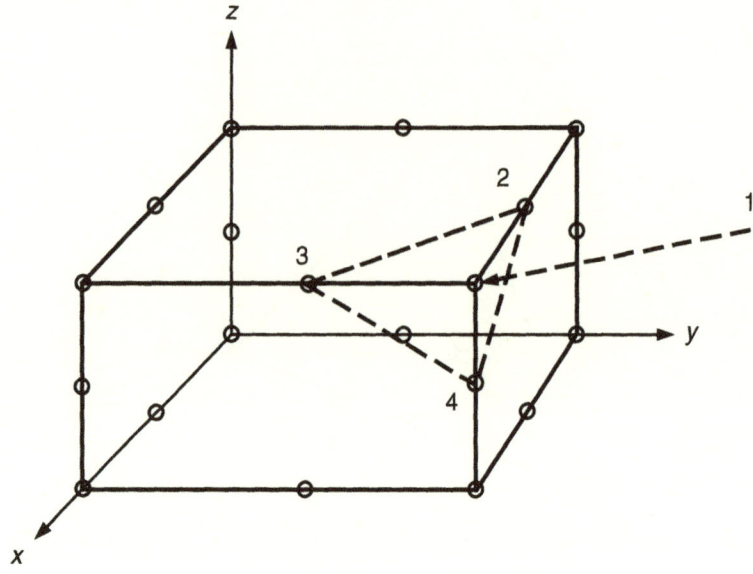

Figure 7.37. Interpolation on a parallelopiped.

7.6 Polynomial Equivalence on an Algebraic Subspace

Before discussing local interpolation on complex domains, we must investigate a property of special collections of curves or surfaces. In what follows we will concentrate on curves, but the ideas can be generalized to surfaces. Let us begin with a simple example.

Consider Figure 7.38. This figure depicts five lines of a pencil of lines through the point $(1, 0)$. The general equation of such a line is

$$y = m(x - 1).$$

Every one of these lines meets the line $y = 0$ at the point $(1, 0)$. Let us refer to these lines as a, b, c, d, and e, and let h denote the line $y = 0$.

Definition 7.7 *We will say that two curves f and g are **equivalent** on a third curve h if*

$$f \bmod h = \alpha(g \bmod h),$$

where α is a scalar. In more compact notation we write this as

$$\beta f = \gamma g \bmod h, \qquad \text{or} \quad f \equiv g \bmod h.$$

In our example of a set of lines through the same point on the line h it is now

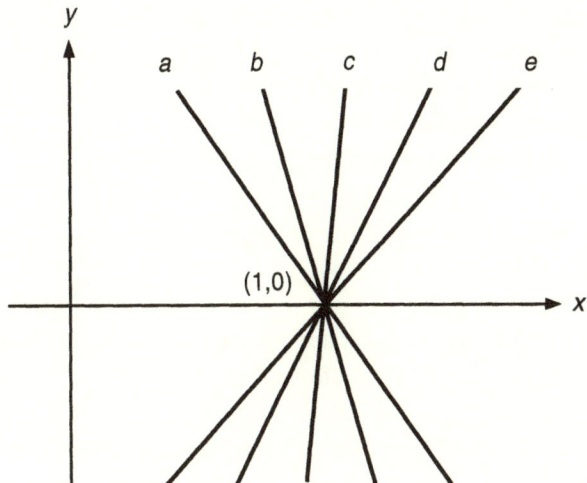

Figure 7.38. A pencil of lines through (1,0).

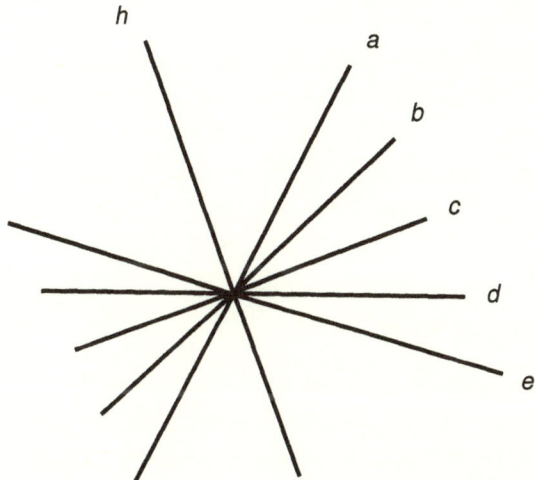

Figure 7.39. A pencil of lines equivalent modulo a line.

clear that all the lines are equivalent modulo the line h. For example,

$$a \equiv b \bmod h, \qquad c \equiv d \bmod h, \quad \text{and} \quad b \equiv e \bmod h.$$

This property is a property of the *intersections* of the collection of curves and has nothing to do with the rather special situation of the line h being the x-axis. Figure 7.39 shows this same situation without any reference to an axis. Shown like this, it is apparent that the line h is simply another member of the pencil and hence $h \equiv a \bmod b$, and so on.

This is a rather trivial result. However, a major generalization of this result, which is just as straightforward to prove, has some very powerful consequences. Let us first look at the following algebraic statement. It is quite acceptable, and perhaps instructive, to think of the symbols in this statement as numbers.

If $x = \alpha y - \beta z$, then $x = 0$ implies $\alpha y = \beta z$ or, equivalently, $y \equiv z \bmod x$.

Theorem 7.2 *If f, g, and h are distinct members of any pencil of curves, then $f \equiv g \bmod h$.*

Proof. Since a pencil can be generated by any two distinct members, we can certainly write $h = \alpha f - \beta g$. Therefore, $f \equiv g \bmod h$. ∎

The proof is a wonderful example of the power of suitable mathematical notation. The algebraic manipulations of the theorem and the preceeding example are identical, but in one case we thought of the symbols as representing numbers, while in the second case the symbols represented algebraic curves of any order. Before we look at an application, we introduce some more notation.

Definition 7.8 *Let a, b, c, ... denote points. The notation $(a; b; c; ...)$ indicates the polynomial associated with the algebraic curve defined by the points $a, b, c, \ldots$. Usually, there will be the appropriate number of points to define a curve of some order uniquely. By $(a; b; c; ...)_p$ we mean the polynomial normalized to have unit value at the point p.*

Example 7.12 *For the unit circle and lines shown in Figure 7.40, consider the equivalence of the lines on the circle.*

We include the axes because we will solve this problem in two different ways. The first method is included to make the point that an understanding of the geometry can be very useful. Let us show that the function

$$W = \frac{(x - y + 1)(x - y - 1)}{x + y + 1}$$

behaves like the function $l = 1 - x - y$ on the curve $x^2 + y^2 - 1 = 0$.

Method 1.

$$W = \frac{(x - y + 1)(x - y - 1)}{x + y + 1}$$
$$= \frac{(x - y + 1)(x - y - 1)(1 - x - y)}{(x + y + 1)(1 - x - y)}$$
$$= \frac{(x^2 - 2xy + y^2 - 1)(1 - x - y)}{1 - x^2 - 2xy - y^2}.$$

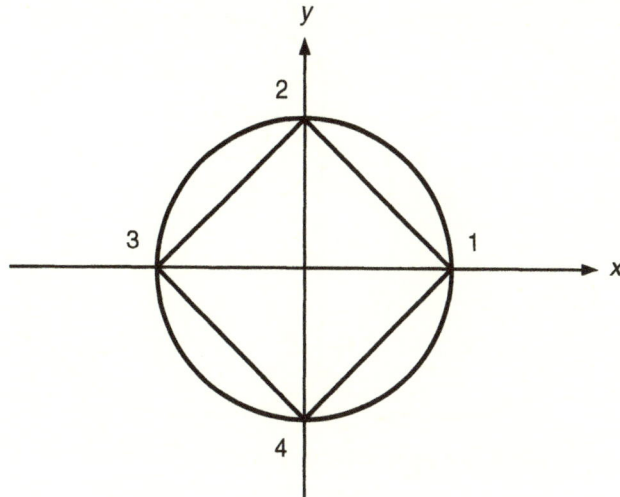

Figure 7.40. Equivalence of lines on a circle.

When $x^2 + y^2 - 1 = 0$, we have $x^2 = 1 - y^2$. Substitution of x^2 yields

$$W \equiv \frac{(-2xy)(1 - x - y)}{(-2xy)} = 1 - x - y.$$

Method 2. Let h denote the circle. The curves $(1; 2)(3; 4) = 0$, $(2; 3)(4; 1) = 0$, and h are three members of the pencil of conics through the four points 1, 2, 3, and 4. Therefore,

$$(2; 3)(4; 1) \equiv (1; 2)(3; 4) \bmod h$$

and

$$\frac{(2; 3)(4; 1)}{(3; 4)} \equiv (1; 2) \bmod h.$$

Since $(1; 2) = \alpha(1 - x - y)$, the required result follows.

The following discussion serves as a warning to be careful when using equivalence. Consider Figure 7.41. Here f and g are circles shown to intersect at the points 1 and 2. We ask the question: is $f \equiv (1; 2) \bmod g$? There should be an immediate reaction that this cannot be the case, since f and g are of a different order from the order of $(1; 2)$ and the three curves are not members of the same pencil. This is true, but we have to remain aware of the context of our discussions, namely the complex projective plane. When we use our techniques to explore the real affine plane, we must be keep this in mind. Two conics meet in four points, and circles, being conics, must do the same. We showed in Chapter 4 that two circles always meet in two specific complex points

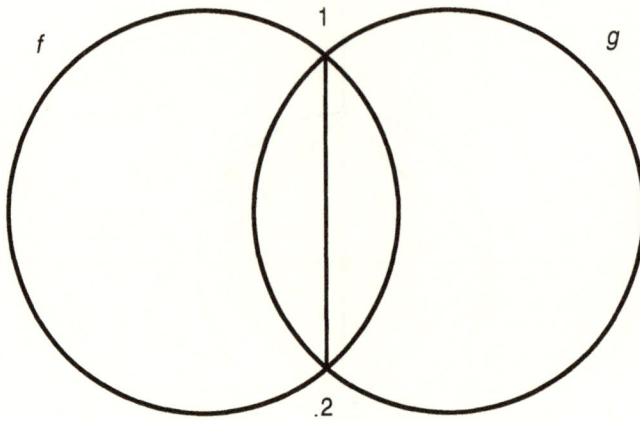

Figure 7.41. Affine equivalence of a circle and a line on a circle.

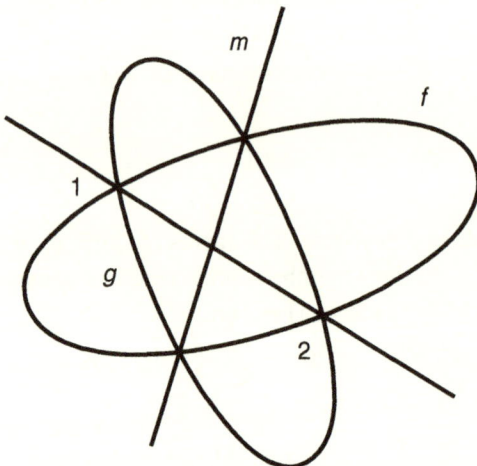

Figure 7.42. Projective equivalence of a pencil.

on the ideal line. Let this line be denoted by m. A schematic complex projective equivalent of Figure 7.41 is shown in Figure 7.42.

It is clear that $f \equiv (1; 2)m \bmod g$. When we move from projective to affine space, the symbol m is replaced by one, and we get the *affine* result $f \equiv (1; 2) \bmod g$. The proof of this result by the manipulation of polynomials in terms of coordinates is left as an exercise (Exercise 7.13). If even one of the circles in the previous example is replaced by some other conic, then the extra intersections will not necessarily be on the ideal line, and the result will not hold, as shown in the next example.

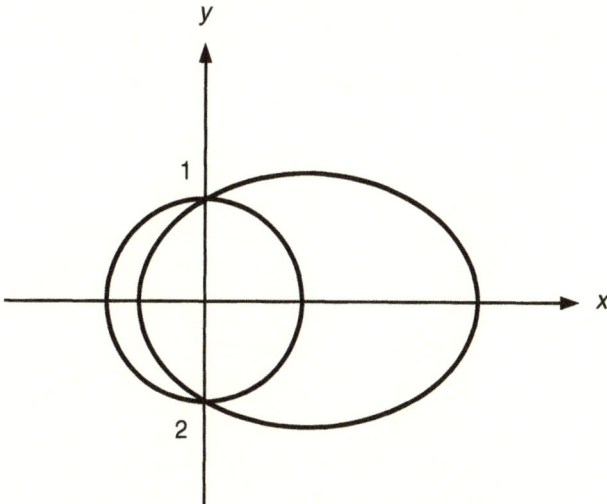

Figure 7.43. Conic and line not equivalent on circle.

Example 7.13 *The circle h and ellipse f intersect in two points 1 and 2 (Figure 7.43), where h is the unit circle with center at the origin and f is the ellipse* $(x - 1)^2 + 2y^2 - 3 = 0$. *Investigate the equivalence of f and* $(1; 2)$ *on h.*

The two curves intersect when, in homogeneous coordinates, $(x - z)^2 + 2(z^2 - x^2) - 3z^2 = 0$, that is, when $x(x + 2z) = 0$. This means that the two conics are members of the pencil through the four points

$$(0, 1, 1), \qquad (0, -1, 1), \qquad (-2, i\sqrt{3}, 1), \quad \text{and} \quad (-2, -i\sqrt{3}, 1).$$

Therefore, $f \equiv (1; 2)(3; 4) \bmod h$, where 3 and 4 denote the two complex intersection points. Since the line $(3; 4)$ is not the ideal line, f and $(1; 2)$ are not equivalent on h.

7.7 Basis Construction for Complex Local Domains

We will continue to considered only local domains where the boundary segments have simple intersections at the vertices of the domain. We also exclude any domains where any boundary segment reenters or touches the domain at any other point. The reason for this will soon become clear. We need one more definition and a theorem before proceeding with constructions.

Definition 7.9 *By the term **external intersections of a local domain** we mean the set of all intersections of the boundary segments other than those points that are vertices of the local domain.*

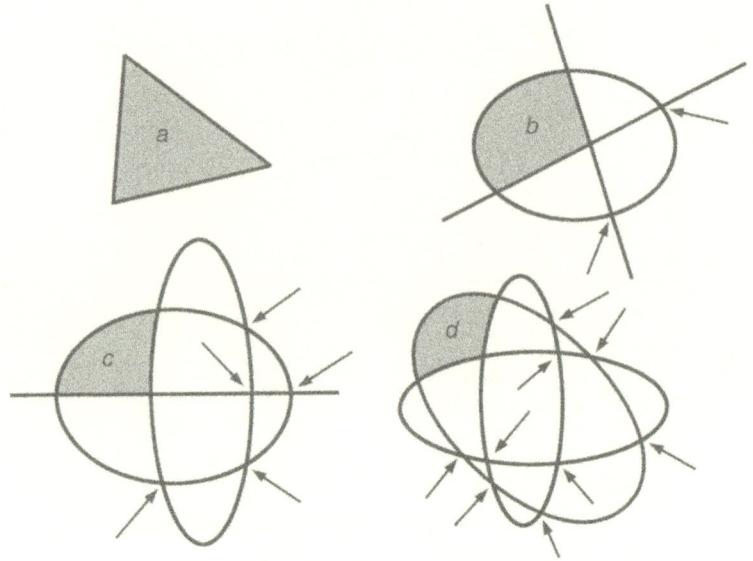

Figure 7.44. External intersections of local domains.

Figure 7.44 shows some local domains. The external intersections are indicated by the arrows. Domain *a* has no external intersections, domain *b* has two, domain *c* has five, and domain *d* has nine.

Theorem 7.3 *For domains bounded by lines or conics, the external intersection points determine a curve of order three less than the sum of the orders of the boundary segments. In symbolic terms this theorem can be stated as follows.*

Let k be the number of boundary segments, and let d_i, $i = 1, \ldots, k$, be the order of the ith boundary segment. Then

$$\sum_{i=1}^{k-1} \sum_{j=i+1}^{k} d_i d_j - k = \frac{1}{2} \sum_{i=1}^{k} d_i \left(\sum_{i=1}^{k} d_i - 3 \right).$$

Proof. We will use induction. Assume that the theorem is true for $k = K$ and consider $k = K + 1$.

$$\sum_{i=1}^{K} \sum_{j=i+1}^{K+1} d_i d_j - (K + 1)$$

$$= \sum_{i=1}^{K-1} \sum_{j=i+1}^{K} d_i d_j - K + d_{K+1} \sum_{i=1}^{K} d_i - 1$$

$$= \frac{1}{2} \sum_{i=1}^{K} d_i \left(\sum_{i=1}^{K} d_i - 3 \right) + d_{K+1} \sum_{i=1}^{K} d_i - 1$$

$$= \frac{1}{2} \sum_{i=1}^{K} d_i \left(\sum_{i=1}^{K+1} d_i - d_{K+1} - 3 \right) + d_{K+1} \sum_{i=1}^{K} d_i - 1$$

$$= \frac{1}{2} \sum_{i=1}^{K} d_i \left(\sum_{i=1}^{K+1} d_i - 3 \right) - \frac{1}{2} d_{K+1} \sum_{i=1}^{K} d_i + d_{K+1} \sum_{i=1}^{K} d_i - 1$$

$$= \frac{1}{2} \sum_{i=1}^{K+1} d_i \left(\sum_{i=1}^{K+1} d_i - 3 \right) - \frac{1}{2} d_{K+1} \left(\sum_{i=1}^{K} d_i + d_{K+1} - 3 \right)$$

$$- \frac{1}{2} d_{K+1} \sum_{i=1}^{K} d_i + d_{K+1} \sum_{i=1}^{K} d_i - 1$$

$$= \frac{1}{2} \sum_{i=1}^{K+1} d_i \left(\sum_{i=1}^{K+1} d_i - 3 \right) - \frac{1}{2} d_{K+1} (d_{K+1} - 3) - 1$$

$$= \frac{1}{2} \sum_{i=1}^{K+1} d_i \left(\sum_{i=1}^{K+1} d_i - 3 \right) - \frac{1}{2} (d_{K+1} - 1)(d_{K+1} - 2)$$

$$= \frac{1}{2} \sum_{i=1}^{K+1} d_i \left(\sum_{i=1}^{K+1} d_i - 3 \right), \qquad \text{provided } d_{K+1} = 1 \quad \text{or} \quad 2.$$

For $K = 2$ there are only two possible local domains to consider – one bounded by a line and a conic, and the other bounded by two conics. In the first case we have $d_1 = 1, d_2 = 2$, so that $d_1 d_2 - 2 = 0$ and

$$\frac{1}{2} \sum_{i=1}^{2} d_i \left(\sum_{i=1}^{2} d_i - 3 \right) = \frac{1}{2}(3)(0) = 0.$$

In the second case we have $d_1 = d_2 = 2$, so that $d_1 d_2 - 2 = 2$ and

$$\frac{1}{2} \sum_{i=1}^{k} d_i \left(\sum_{i=1}^{k} d_i - 3 \right) = \frac{1}{2}(4)(1) = 2.$$

Therefore, the theorem is true for all positive integer values of $k \geq 2$. ∎

Definition 7.10 *The curve defined in Theorem 7.3 is called the **adjoint** curve of the local domain.*

The external intersections of the local domains of Figure 7.44 determine curves of orders 0, 1, 2, and 3, respectively. We are finally ready to construct interpolating bases for some complex local domains. We will illustrate the technique by example. The general procedure for constructing a biorthonormal basis which spans $P_n(x, y)$ on a local domain R will have the following steps:

Step 1. Calculate $D_{n,2} \mid R$, and find the difference $m = D_{n,2} \mid R - \frac{1}{2}(n+1)(n+2)$.

Step 2. Construct m of the basis functions using the adjoint and other curves, as we will show in the following examples.

Step 3. Deduce the remaining $\frac{1}{2}(n+1)(n+2)$ basis functions by requiring that the entire set span $P_n(x, y)$.

Step 4. Show that the basis functions have polynomial behavior on the domain boundary.

Step 5. Show that the basis satisfies the local interpolation property.

Example 7.14 *Construct a linear basis for the local domain shown in Figure 7.45, where f is a conic.*

Step 1. $D_{1,2} \mid R = 4$ and $m = 1$.

Step 2. We construct $W_4(x, y)$. Let h denote the adjoint curve $(a; b)$. Define $W_4(x, y)$ by

$$W_4(x, y) = \frac{(1; 2)_4 (1; 3)_4}{(a; b)_4}.$$

Step 3. Let $N_i(x, y)$, $i = 1, 2, 3$, be the polynomials of degree one such that

$$N_i(x_j, y_j) = \delta_{ij}, \qquad i, j = 1, 2, 3.$$

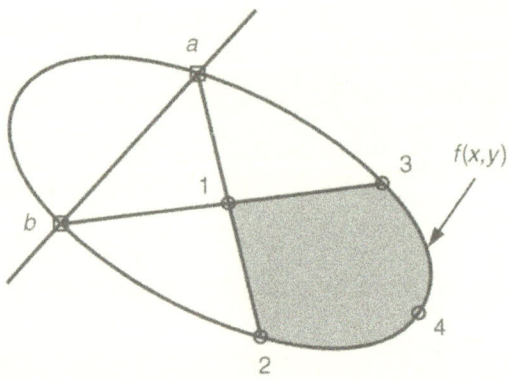

Figure 7.45. A local domain bounded by two lines and a conic.

We require that the biorthonormal basis $\{W_i(x, y), i = 1, 2, 3, 4\}$ span $P_1(x, y)$. In Chapter 1 we saw that this will be achieved if the basis spans any three linearly independent polynomials in $P_1(x, y)$, in particular, if it spans $N_i(x, y), i = 1, 2, 3$, since these functions are independent because of the biorthonormalization. The requirement that the set $\{W_i(x, y)\}$ span the $\{N_i(x, y)\}$ is equivalent to demanding that

$$\sum_{i=1}^{4} N_j(x_i, y_i) W_i(x, y) = N_j(x, y), \qquad j = 1, 2, 3.$$

That is,

$$W_j(x, y) = N_j(x, y) - N_j(x_4, y_4) W_4(x, y), \qquad j = 1, 2, 3.$$

Step 4. The line pairs $(1; 2)(1; 3)$, $(a; b)(2; 3)$, and $f(x, y)$ are all members of the same conic pencil through the points $a, b, 2$, and 3. Therefore, from Theorem 7.2,

$$(1; 2)(1; 3) \equiv (a; b)(2; 3) \bmod f.$$

Hence,

$$\frac{(1; 2)(1; 3)}{(a; b)} \equiv (2; 3) \bmod f,$$

so that $W_4(x, y)$ is linear on the curve. Therefore, the other basis functions also have linear behavior on the curve. Furthermore, $W_4(x, y)$ is identically zero on the line segments, which implies that all the basis functions also have linear behavior on these boundaries.

Step 5. All that remains to be shown is that each of the basis functions is zero on domain boundaries that do not contain the corresponding node. This is obvious (by construction) in the case of $W_4(x, y)$, but not so obvious for the other basis functions. Let us show that $W_2(x, y)$ is identically zero on $(1; 3)$. This function is zero at nodes $1, 3$, and 4 [see the definition of $W_j(x, y), j = 1, 2, 3$]. Also, $W_2(x, y)$ is of the form

$$W_2(x, y) = N_2(x, y) \bmod (1; 3),$$

since $W_4(x, y)$ is identically zero on the line $(1; 3)$. Since $N_2(x, y)$ is linear and zero at two points on the line, it must be identically zero on the line $(1; 3)$. The proofs for the other two basis functions follow along similar lines.

Example 7.15 *Construct a basis spanning $P_2(x, y)$ for the local domain, shown in Figure 7.46, with two conic and one line segment boundaries.*

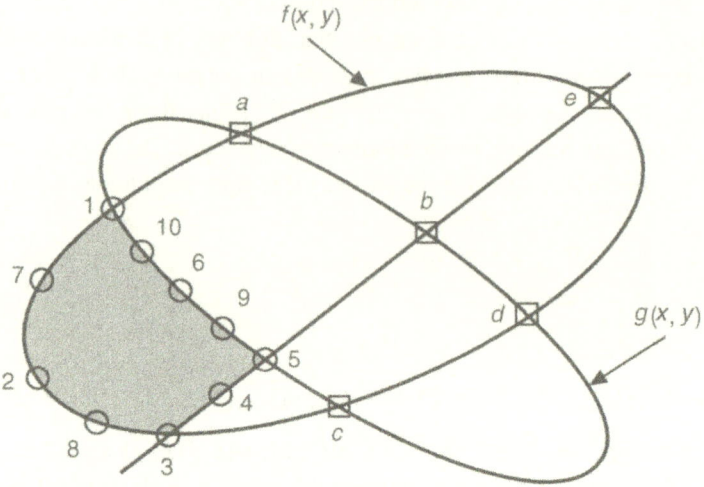

Figure 7.46. A local domain bounded by two conics and a line.

Step 1. $D_{2,2} \mid R = 10$ and $m = 4$.

Step 2. We construct $W_j(x, y)$, $j=7, 8, 9, 10$. Let h denote the polynomial associated with the adjoint curve $(a; b; c; d; e)$, which is a conic. We will construct $W_7(x, y)$ and leave it to the reader to construct $W_j(x, y)$, $j = 8, 9, 10$. Define $W_7(x, y)$ by

$$W_7(x, y) = \frac{g(x, y)(3; 5)_7(2; 8)_7}{g(x_7, y_7)h_7},$$

where h_7 denotes the polynomial h normalized to unity at the point 7.

Step 3. Let $N_i(x, y)$, $i = 1, \ldots, 6$, be the polynomials of degree two such that

$$N_i(x_j, y_j) = \delta_{ij}, \qquad i, j = 1, \ldots, 6.$$

We require that the basis $\{W_i(x, y), i = 1, \ldots, 10\}$ span polynomials of degree two. Following the previous example we define the remaining basis functions by

$$\sum_{i=1}^{10} N_j(x_i, y_i) W_i(x, y) = N_j(x, y), \qquad j = 1, \ldots, 6,$$

that is,

$$W_j(x, y) = N_j(x, y) - \sum_{k=7}^{10} N_j(x_k, y_k) W_k(x, y), \qquad j = 1, \ldots, 6.$$

Step 4. We show that $W_7(x, y)$ behaves like a polynomial of degree two on the boundaries. The argument for the other functions is similar.

The conics g, f, and $(a; d)(1; c)$ are all members of the same pencil of conics. So, also, are h, f, and $(a; d)(c; e)$, and likewise f, $(3; e)(1; c)$, and $(c; e)(3; 1)$. Therefore, by Theorem 7.2,

$$g \equiv (a; d)(1; c) \bmod f,$$

$$h \equiv (a; d)(c; e) \bmod f,$$

$$(3; e)(1; c) \equiv (c; e)(3; 1) \bmod f.$$

We use these equivalences to write $W_7(x, y)$ in the form

$$
\begin{aligned}
W_7(x, y) &= \frac{g(x, y)(3; 5)_7(2; 8)_7}{g(x_7, y_7)h_7} \\
&\equiv \frac{(a; d)(1; c)(3; e)(2; 8)}{(a; d)(c; e)} \bmod f \\
&\equiv \frac{(c; e)(3; 1)(2; 8)}{(c; e)} \bmod f \\
&\equiv (3; 1)(2; 8) \bmod f \\
&= P_2(x, y) \bmod f.
\end{aligned}
$$

Step 5. We know, from the construction, that the basis functions satisfy the property

$$W_i(x_j, y_j) = \delta_{ij}, \qquad i, j = 1, \ldots, 10.$$

Therefore, each basis function is zero at the nodes on boundaries that do not contain the node in question. We must show that this implies that the basis function is identically zero on these boundaries. This is a straightforward consequence of step 4 and the Maclaurin–Bézout theorem, since each basis function behaves like a polynomial of degree two on the boundary and is zero at three points on a line or five on a conic, and is therefore identically zero.

Steps 2 and 4 for the other basis functions are left to the reader (Exercise 7.16).

Example 7.16 *Construct a basis that spans $P_3(x, y)$ on the local domain bounded by two lines and a conic $f(x, y)$ (Figure 7.47), using Hermite interpolation at the vertices.*

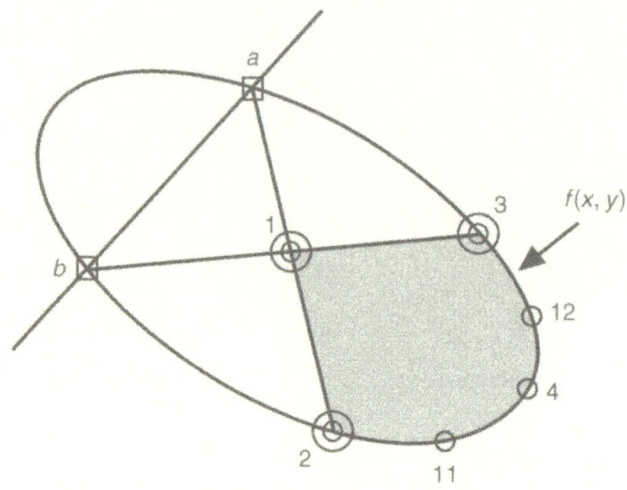

Figure 7.47. Hermite interpolation on a curved local domain.

Step 1. $D_{3,2} \mid R = 12$ and $m = 2$.

Step 2. We construct $W_j(x, y)$, $j = 11, 12$. Let h denote the linear polynomial associated with the adjoint curve $(a; b)$. We construct $W_{11}(x, y)$, and leave it to the reader to construct $W_{12}(x, y)$. Define $W_{11}(x, y)$ by

$$W_{11}(x, y) = \frac{(1; 2)_{11}(1; 3)_{11}(2; 3)_{11}(4; 12)_{11}}{(a; b)_{11}}.$$

The curve associated with this rational function has double points at 1, 2, and 3.

Step 3. Let $N_i(x, y)$, $i = 1, \ldots, 10$, be the polynomials of degree three such that

$$N_i(x_j, y_j) = \delta_{ij}, \qquad j = 1, 2, 3, 4, \quad i = 1, \ldots, 10,$$

$$(N_i)_x(x_j, y_j) = (N_i)_y(x_j, y_j) = 0, \qquad i = 1, 2, 3, 4, \quad j = 1, 2, 3,$$

$$(N_{i+4})_x(x_j, y_j) = \delta_{ij}, \quad (N_{i+4})_y(x_j, y_j) = 0, \qquad i, j = 1, 2, 3,$$

$$(N_{i+7})_y(x_j, y_j) = \delta_{ij}, \quad (N_{i+7})_x(x_j, y_j) = 0, \qquad i, j = 1, 2, 3.$$

We require that the basis $\{W_i(x, y), i = 1, \ldots, 12\}$ span polynomials of degree three. Following our previous examples, we define the remaining basis functions by

$$\sum_{i=1}^{4} N_j(x_i, y_i) W_i(x, y)$$

$$+ \sum_{i=1}^{3} (N_j)_x(x_i, y_i) W_{i+4}(x, y)$$

$$+ \sum_{i=1}^{3} (N_j)_y(x_i, y_i) W_{i+7}(x, y)$$

$$+ N_j(x_{11}, y_{11}) W_{11}(x, y) + N_j(x_{12}, y_{12}) W_{12}(x, y)$$

$$= N_j(x, y), \qquad j = 1, \ldots, 10,$$

that is, for $j = 1, \ldots, 10$,

$$W_j(x, y) = N_j(x, y) - N_j(x_{11}, y_{11}) W_{11}(x, y)$$

$$- N_j(x_{12}, y_{12}) W_{12}(x, y).$$

Step 4. We show that $W_{11}(x, y)$ behaves like a polynomial of degree three on the boundaries. The argument for $W_{12}(x, y)$ is similar. The behavior of the other basis functions then follows from their definition. As in Example 7.14, $(1; 2)(1; 3)$, $(a; b)(2; 3)$, and f are all members of the same conic pencil through the points $a, b, 2$, and 3. Therefore, by Theorem 7.2,

$$(1; 2)(1; 3) \equiv (a; b)(2; 3) \bmod f,$$

or

$$\frac{(1; 2)(1; 3)}{(a; b)} \equiv (2; 3) \bmod f.$$

We use this equivalence to write $W_{11}(x, y)$ in the form

$$W_{11}(x, y) = \frac{(1; 2)_{11}(1; 3)_{11}(2; 3)_{11}(4; 12)_{11}}{(a; b)_{11}}$$

$$\equiv (2; 3)^2 (4; 12) \bmod f$$

$$= P_3(x, y) \bmod f.$$

Step 5. We know from the construction that the basis satisfies all the properties of a cubic Hermite basis. For example, $W_{11}(x, y)$ is unity at (x_{11}, y_{11}) and zero at all the other nodes, and, since it has double points at the vertices, the partial derivatives vanish at those points. We must show that this implies that the basis function is identically zero on these boundaries. Once again, this is a straightforward consequence of step 4 and the Maclaurin–Bézout theorem. Each basis function behaves like a polynomial of degree three on the boundary, and has two double points on a line not associated with it, or two double points and three simple points on a conic if its associated node is not on the conic. It is therefore identically zero on each boundary segment not associated with it.

Steps 2 and 4 for the other basis functions are left to the reader (Exercise 7.17).

Example 7.17 *Investigate an alternative basis for Example 7.16 in which the linear forms* $(4; 12)_{11}$ *and* $(4, 11)_{12}$ *are not used in the construction.*

We only consider a new definition of the function $W_{11}(x, y)$, since the definition of the function $W_{12}(x, y)$ will be similar. Let the given conic intersect the ideal line at the points c and d. This implies that $(4; c)(12; d) \equiv (4; 12)(c; d) \bmod f$, which, in the affine plane and with the symbolic ideal line $(c; d)$ replaced by one, becomes

$$(4; c)(12; d) \equiv (4; 12) \bmod f,$$

where $(4; c)$ denotes the linear form corresponding to the line through node 4 in the direction determined by c, and a similar meaning is attached to $(12; d)$. Then

$$W_{11}^*(x, y) = \frac{(1; 2)_{11}(1; 3)_{11}(2; 3)_{11}(4; c)_{11}(12; d)_{11}}{(a; b)_{11}}$$

still has exactly the same behavior on the boundaries as the function $W_{11}(x, y)$ and can be used, together with $W_{12}^*(x, y)$, in the construction of the basis. The advantage of the new basis functions is that, if the nodes 4, 11, and 12 are nearly collinear, there is no numerical instability involved in the normalization.

7.8 Mappings to Rational Surfaces

Most of Chapter 1 was concerned with the problem where we were given data $\{f_i\}$ associated with points $\{x_i\}$ and constructed basis functions $\{W_i(x)\}$ that enabled us to form the interpolant

$$I(f; x) = \sum_{i=1}^{n} f_i W_i(x).$$

We pointed out that this function has limited use as a general curve. However, there is a way that we can use the same basis functions to produce a curve with enhanced properties. We consider a two-dimensional data set of points given in the form $\{(x_i, y_i)\}$ and then write, formally,

$$x = \sum_{i=1}^{n} x_i W_i(t), \qquad y = \sum_{i=1}^{n} y_i W_i(t).$$

The choice of the independent variable t is flexible, but there must be some association between this variable and the data points. This association is often defined by setting $t = i$ when $(x, y) = (x_i, y_i)$, that is, we identify the variable with the index. Regardless of the way we make this association, we have constructed a curve that passes through all the data points. If the basis functions $\{W_i(t)\}$ are rational, then this curve is a rational curve. Most of the present chapter has been devoted to the construction of interpolants on two-dimensional local domains. These interpolants are surfaces. The techniques we developed provide a way to produce the required functions in a systematic way. These bases are also functions, and the surfaces they form are therefore as restricted as the function curves. We can, however, use exactly the same approach as we did in Chapter 1 to produce more general surfaces. We consider the data set $\{(x_i, y_i, z_i)\}$ and write, formally,

$$x = \sum_{i=1}^{n} x_i W_i(s, t), \qquad y = \sum_{i=1}^{n} y_i W_i(s, t), \qquad z = \sum_{i=1}^{n} z_i W_i(s, t).$$

If the functions $\{W_i(s, t)\}$ are rational, then the surfaces will be rational.

Example 7.18 *Use the linear Lagrange basis for the unit square in the (s, t) plane with nodes at the vertices to form an interpolating surface.*

We construct the Lagrange basis for the unit square in the (s, t) plane with nodes at the vertices (Figure 7.48), namely

$$W_1 = (1 - s)(1 - t),$$
$$W_2 = s(1 - t),$$
$$W_3 = st,$$
$$W_4 = t(1 - s).$$

The basis is bilinear, and spans a four-dimensional subspace of $P_2(s, t)$.

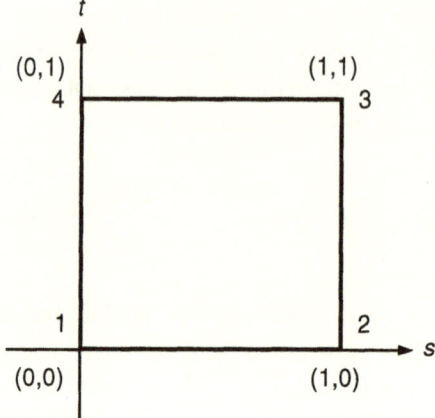

Figure 7.48. Linear interpolation over the unit square.

Assume that we want to interpolate the points $(0, 0, 0)$, $(1, 0, 0)$, $(0, 1, 0)$ and $(1, 1, 1)$. We therefore use the transformation

$$x = \sum_{i=1}^{4} x_i W_i(s, t) = W_2 + W_3 = s,$$

$$y = \sum_{i=1}^{4} y_i W_i(s, t) = W_3 + W_4 = t,$$

$$z = \sum_{i=1}^{4} z_i W_i(s, t) = W_3 = st,$$

to map the vertices of the square to the given three-dimensional points. The square is mapped to a patch of the hyperbolic paraboloid $z = xy$ shown in Figure 7.49. The edges of the square are mapped to straight lines in space. However, not all lines are mapped onto lines, as we can see by considering the image of the line segment $s = t$, $0 \leq s, t \leq 1$, which is mapped onto the parabolic segment

$$x = p, \quad y = p, \quad z = p^2, \qquad 0 \leq p \leq 1.$$

Example 7.19 *Use a cubic Hermite basis on a triangular local domain to interpolate three-dimensional data, thus obtaining a cubic polynomial parametrization of the interpolating surface.*

The construction of the cubic Hermite basis for a triangular local domain is required in Exercise 7.7. The basis functions corresponding to the node

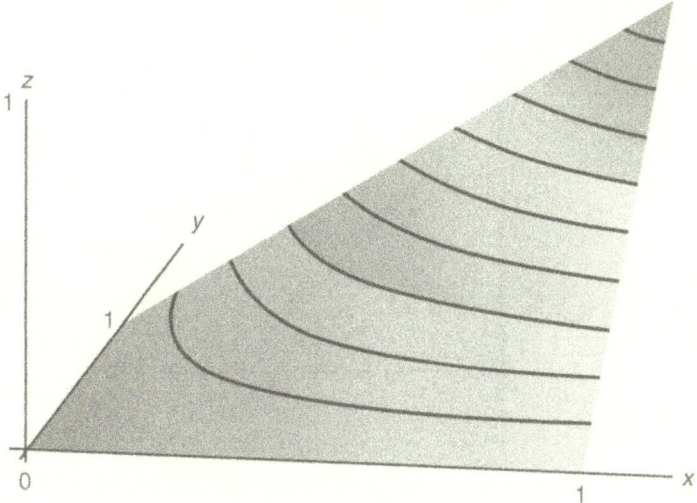

Figure 7.49. The surface $z = xy$.

configuration in Figure 7.50, with interior node 10 at the centroid $(\frac{1}{3}, \frac{1}{3})$, are

$$U_1 = (s + t - 1)(2s^2 + 11st + 2t^2 - s - t - 1),$$

$$U_2 = -s(2s^2 - 7st - 7t^2 - 3s + 7t),$$

$$U_3 = -t(2t^2 - 7st - 7s^2 - 3t + 7s),$$

$$U_{10} = 27st(1 - s - t),$$

$$V_{1,s} = s(1 - s - t)(1 - s - 2t),$$

$$V_{1,t} = t(1 - s - t)(1 - t - 2s),$$

$$V_{2,s} = s(s^2 - 2st - 2t^2 - s + 2t),$$

$$V_{2,t} = st(2s + t - 1),$$

$$V_{3,s} = st(2t + s - 1),$$

$$V_{3,t} = t(t^2 - 2st - 2s^2 - t + 2s),$$

where $V_{i,s}$ denotes the basis function with partial derivative with respect to s normalized to unity at the ith node and zero with respect to the other operators. The functions $V_{i,t}$ are defined analogously.

We map the vertices of the triangle to the points $(0, 0, 1)$, $(1, 0, 0)$, and $(0, 1, 0)$, respectively, and use zero values for all the partial derivatives. The centroid is mapped to the point $(1, 1, 1)$. The parametrized form of the interpolating

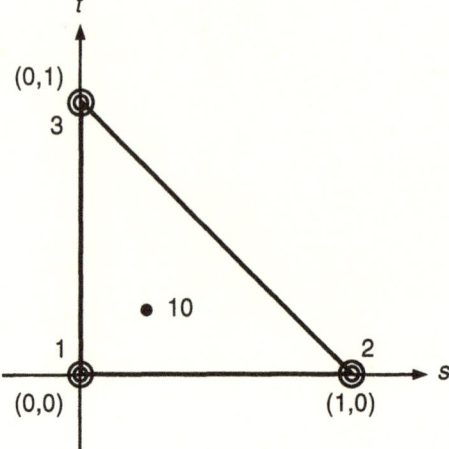

Figure 7.50. Node configuration for the cubic Hermite basis.

surface is given by

$$x = U_2 + U_{10} = -s(2s^2 + 20st + 20t^2 - 3s - 20t),$$
$$y = U_3 + U_{10} = -t(2t^2 + 20st + 20s^2 - 3t - 20s),$$
$$z = U_1 + U_{10} = (s + t - 1)(2s^2 - 16st + 2t^2 - s - t - 1).$$

Straight lines are, in general, mapped to polynomial cubic curves. The surface is illustrated in Figure 7.51.

The mappings described in this section take any line in the (s, t) plane into a rational curve, since any line satisfies a condition of the form $as + bt + c = 0$. This equation can be used to eliminate either s or t, resulting in a mapping of the form

$$x = \sum_{i=1}^{n} x_i W_i^*(t) = X(t),$$

$$y = \sum_{i=1}^{n} y_i W_i^*(t) = Y(t),$$

$$z = \sum_{i=1}^{n} z_i W_i^*(t) = Z(t).$$

These curves are, in general, space curves. This will not be the case when the mapping is second-order. Using homogeneous coordinates, a second-order

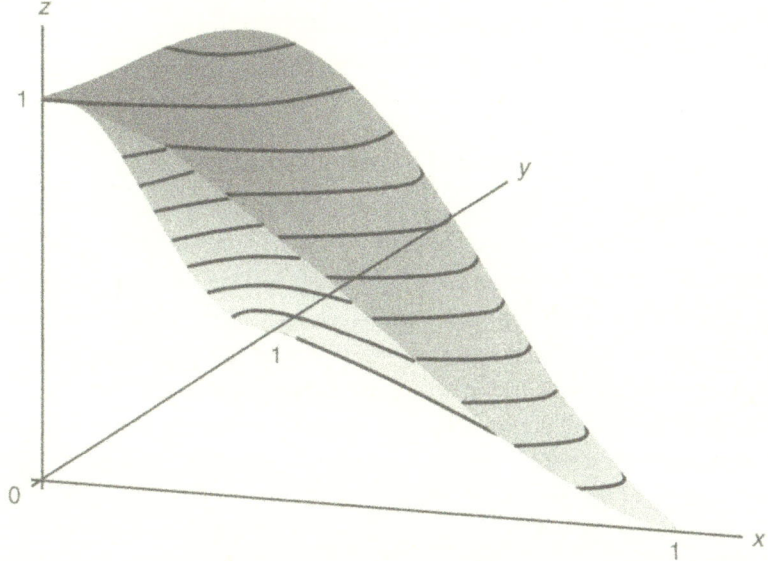

Figure 7.51. A surface given by Hermite interpolation on a triangular patch.

mapping is of the form

$$x = x_2 t^2 + x_1 t + x_0,$$

$$y = y_2 t^2 + y_1 t + y_0,$$

$$z = z_2 t^2 + z_1 t + z_0,$$

$$w = w_2 t^2 + w_1 t + w_0.$$

The homogeneous coordinates x, y, z, and w are all members of the three-dimensional vector space spanned by $\{t^2, t, 1\}$. Since any four vectors in a three-dimensional vector space must be dependent, we know that there must exist scalars a, b, c, and d such that

$$ax + by + cz + dw = 0.$$

This is simply the equation of a plane in homogeneous coordinates, and hence the curve is a planar curve with a second-degree parametrization – a conic.

An alternative argument is to note that the original expressions for x, y, and z must be either linearly dependent or linearly independent. If they are linearly dependent, there is a linear relationship of the form $ax + by + cz = 0$, and hence the curve lies in a plane through the origin. If they are linearly independent, they can be used as a basis for the space, and any other vector in the space can be represented as a linear combination of these vectors. In particular, the vector 1 can be written as a linear combination of x, y, and z, so that $1 = ax + by + cz$, which shows that the curve is a planar curve.

7.9 Order of a Rational Surface

We have developed the ability to generate surfaces with known interpolatory properties. In almost all cases these surfaces will be rational surfaces. We have seen that all rational planar curves are algebraic, and we have investigated the relationship between rational curves and algebraic curves. We now investigate some properties of rational surfaces. In a similar fashion to our definition of a planar real algebraic curve, we define a three-dimensional real algebraic surface as the zeros, in complex projective space, of an irreducible homogeneous polynomial in four variables with real coefficients. Therefore, a surface is given by an equation of the form

$$f_n(x, y, z, w) = 0,$$

where f_n is a homogeneous polynomial of degree n. The *order* of the surface is the number of times an arbitrary line will meet the surface.

There are several of ways we can think about a line. One way is to think of it as defined by one point and a direction. If the point (in affine coordinates) is given by (a, b, c) and the direction cosines by (l, m, n), then the line is given in parametric form by the equations

$$x = a + lt, \qquad y = b + mt, \qquad z = c + nt.$$

When we are given two points (x_1, y_1, z_1) and (x_2, y_2, z_2) rather than one point and a direction, the equations have the form

$$x = x_1 + t(x_2 - x_1), \qquad y = y_1 + t(y_2 - y_1), \qquad z = z_1 + t(z_2 - z_1).$$

The equation of a plane (in nonhomogeneous form) is $ax + by + cz + d = 0$. An alternative way of specifying the line is to say that it is the intersection of two planes. Therefore, a line can be represented by a pair of distinct linear equations

$$a_1 x + b_1 y + c_1 z + d_1 = 0,$$
$$a_2 x + b_2 y + c_2 z + d_2 = 0,$$

which must be satisfied simultaneously.

We can write the nonhomogeneous equation of the surface in the form

$$F_n + F_{n-1} + \cdots + F_0 = 0,$$

where F_i is a homogeneous polynomial of degree i. Consider an arbitrary line given in parametric form. We find the intersections of this line with the surface by substituting the parametric expressions for x, y, and z in the equation of the surface. This leads to an equation of the form

$$F_n^*(t) + F_{n-1}^*(t) + \cdots + F_0^*(t) = 0,$$

where F_i^* is a nonhomogeneous polynomial of degree i in the single variable t. This is a polynomial of degree n in t. The coefficients of this polynomial will be real, since the coefficients of the polynomial that determines the surface are real. The n roots of this polynomial, real or complex, correspond to the points where the line meets the surface. There are n such roots, and hence the order of the surface is n, the same as the degree of the polynomial that determines the surface.

In the case of a rational surface, we consider a line as being determined by the intersection of two planes. If the surface is given by

$$x = X_n(s, t), \qquad y = Y_n(s, t), \qquad z = Z_n(s, t),$$

or, in homogeneous form,

$$x = X_n(r, s, t), \qquad y = Y_n(r, s, t),$$
$$z = Z_n(r, s, t), \qquad w = W_n(r, s, t),$$

then the intersections of this surface with a line determined by two planes will be given by a pair of equations of the form

$$a_1 X_n(s, t) + b_1 Y_n(s, t) + c_1 Z_n(s, t) + d_1 = 0,$$
$$a_2 X_n(s, t) + b_2 Y_n(s, t) + c_2 Z_n(s, t) + d_2 = 0,$$

or, in homogeneous form,

$$a_1 X_n(r, s, t) + b_1 Y_n(r, s, t) + c_1 Z_n(r, s, t) + d_1 W_n(r, s, t) = 0,$$
$$a_2 X_n(r, s, t) + b_2 Y_n(r, s, t) + c_2 Z_n(r, s, t) + d_2 W_n(r, s, t) = 0.$$

We recognize these equations as representing two algebraic curves, each of order n. We know, from the Maclaurin–Bézout theorem, that they will meet in n^2 points and hence the line meets the surface in n^2 points. Therefore, a rational surface with parametrization of degree n is an algebraic surface of order n^2. (In Exercise 7.23 you are asked to show that it *is* an algebraic surface.)

This result may seem disconcerting. Consider surfaces with parametrizations of degrees 1 and 2. The corresponding algebraic surfaces are surfaces of orders 1 and 4. Any second-order surface has a homogeneous equation of the form $f_2(x, y, z, w) = 0$. We know that any line will meet this surface in precisely two points. Therefore, any line through a given point on the surface will meet it again in one other point. Take the set of all lines through the given point in parametric form. We see that the coordinates of the other point in which each of these lines meets the surface will be given explicitly as a rational parametrization of the direction of the line. This is exactly the same argument as we used to show that all second-order plane curves (conics) must be rational. Therefore, we have shown that all second order surfaces, called *quadrics*, are rational. There is, therefore, an apparent paradox, for the degree of the parametrization must be an integer, yet there is no integer such that its square is 2. Before resolving the paradox, let us do an example.

Example 7.20 *Produce a homogeneous parametrization of the quadric*

$$x^2 - 3xy + 2y^2 + 2z^2 - 3zw = 0.$$

The quadric passes through the origin. The set of all lines through the origin can be expressed in the homogeneous parametrized form

$$x = rm, \qquad y = sm, \qquad z = tm, \qquad w = n,$$

where the ratio m/n represents the affine parameter of the line. The intersections of this line with the quadric is obtained by substituting the parametric line equations into the equation of the surface, thus obtaining the parameter values corresponding to the intersections. This substitution yields

$$(r^2 - 3rs + 2s^2 + 2t^2)m^2 - 3tmn = 0.$$

Since the parameter value $m = 0$ corresponds to the origin, the remaining solution

$$m = \frac{3t}{r^2 - 3rs + 2s^2 + 2t^2} n$$

corresponds to the remaining (variable) intersection of the set of lines with the surface. This yields the homogeneous parametrization of the quadric

$$x = 3rt, \qquad y = 3st, \qquad z = 3t^2, \qquad w = r^2 - 3rs + 2s^2 + 2t^2,$$

where we have used the basic property of homogeneous coordinates, namely that the points (x, y, z, w) and $(\alpha x, \alpha y, \alpha z, \alpha w)$, $\alpha \neq 0$, are identical.

Let us return to the paradox. Remember that a point in the projective plane is defined as a triple, *not all zero*, of the form (x, y, z). In three-space, a point is an ordered four-tuple, *not all zero*, of the form (x, y, z, w). In the plane, there is no point with homogeneous coordinates $(0, 0, 0)$, and in three-space there is no point with homogeneous coordinates $(0, 0, 0, 0)$. Therefore, if there are values r_i, s_i, and t_i of the parameters such that

$$X_n(r_i, s_i, t_i) = 0, \qquad Y_n(r_i, s_i, t_i) = 0,$$
$$Z_n(r_i, s_i, t_i) = 0, \qquad W_n(r_i, s_i, t_i) = 0,$$

then these values do *not* correspond to any point on the rational surface

$$x = X_n(r, s, t), \qquad y = Y_n(r, s, t),$$
$$z = Z_n(r, s, t), \qquad w = W_n(r, s, t).$$

Such points in the parameter plane will be called *base points* of the parametrization. Let us assume that there are k such sets of parameter values. Then an arbitrary plane $ax + by + cz + dw = 0$ will correspond to an algebraic curve

$$aX_n(r, s, t) + bY_n(r, s, t) + cZ_n(r, s, t) + dW_n(r, s, t) = 0$$

in the parameter plane, which passes through the k base points $\{(r_i, s_i, t_i)\}$. Therefore, the intersection of two such arbitrary planes will correspond to the intersection of two nth order curves that are known to have these k points in common. Although these triples $\{(r_i, s_i, t_i)\}$ do correspond to points in the parameter plane, they do not correspond to points in the space of the surface, that is, in projective three-space. Hence, the surface is not a surface of order n^2, but a surface of order $n^2 - k$.

Example 7.21 *Investigate the order of the algebraic surface with parametrization*

$$x = t(r + s - t),$$
$$y = s(r - s + t),$$
$$z = (r + s - t)(r - s + t),$$
$$w = (r - s)(r + 2s) - t(r - 2s).$$

Since the parametrization is of second order, the surface cannot be of order greater than four. The line pairs

$$t(r + s - t) = 0, \qquad s(r - s + t) = 0$$

in the parameter plane intersect in the four points $(1, 0, 1)$, $(0, 1, 1)$, $(1, 1, 0)$, and $(1, 0, 0)$. We confirm by substitution that the first three points also lie on the line pair $(r + s - t)(r - s + t) = 0$, as well as on the conic $(r - s)(r + 2s) - t(r - 2s) = 0$. The surface is of order $n^2 - k = 4 - 3 = 1$, that is, it is a plane. In fact, $x - y - z + w = 0$, which is the homogeneous equation of the relevant plane.

Example 7.22 *Show that the parameter-plane curves defined by the parametrization found in Example 7.20 have two points in common.*

We have to solve the set of homogeneous polynomial equations of degree two

$$rt = 0, \qquad st = 0, \qquad t^2 = 0, \quad \text{and} \quad r^2 - 3rs + 2s^2 + 2t^2 = 0,$$

simultaneously, to find the intersections of these curves (if any). Since the third equation implies $t = 0$, the last equation becomes $(r - s)(r - 2s) = 0$, so that the two common points are $(1, 1, 0)$ and $(2, 1, 0)$. Both these points lie on the ideal line in the parameter plane. The surface is of order $n^2 - k = 4 - 2 = 2$, as expected, since we started with a quadric.

7.10 Degrees of Freedom, Tangent Planes, and Multiple Points

7.10.1 Degrees of Freedom

The general homogeneous polynomial of degree n in $d + 1$ variables has the same number of terms as the nonhomogeneous polynomial in d variables, namely $(n + d)!/n!d!$. Hence, the algebraic surface of order n has $\frac{1}{6}(n + 3)$ $(n + 2)(n + 1)$ terms. As in the case of algebraic curves, the number of degrees of freedom will be one less than this, since $\alpha f(x, y, z) = 0$ is the same surface as $f(x, y, z) = 0$. Hence, the algebraic surface of order n has

$$\tfrac{1}{6}n(n^2 + 6n + 11)$$

degrees of freedom and will, in general, be determined by this number of linear conditions on the coefficients of the defining polynomial. The general second-order surface, the quadric, is given in nonhomogeneous form by

$$a_{2,0,0}x^2 + a_{0,2,0}y^2 + a_{0,0,2}z^2 + a_{1,1,0}xy + a_{0,1,1}yz$$

$$+ a_{1,0,1}xz + a_{1,0,0}x + a_{0,1,0}y + a_{0,0,1}z + a_{0,0,0} = 0.$$

It has ten terms and nine degrees of freedom. There is therefore, in general, a unique quadric through nine points. Similarly, there is a unique quartic through thirty-four points.

7.10.2 The Tangent Plane

Any planar section of an algebraic surface of order n meets the surface in a curve of order n. There are two arguments that justify this assertion. The *algebraic* argument is that, since a linear transformation of coordinates will not change the order and could be used to map any plane to the particular plane given by $w = 0$, we may assume, without loss of generality, that the plane has $w = 0$ as its equation. Substituting $w = 0$ in the homogeneous equation for the surface $f_n(x, y, z, w) = 0$ yields $f_n(x, y, z, 0) = 0$, which is an algebraic curve of order n. The *synthetic* argument is that if the surface is of order n, then every line must meet the surface in n points. In particular, every line in the planar section must meet the intersection of the surface with the plane in n points. Since the intersection of the plane with the surface is some curve and every line in the plane of this curve meets the curve in n points, the curve must be a curve of order n.

Let (x_0, y_0, z_0, w_0) be a point on a surface of order n. If (x, y, z, w) is some other point, then a variable point on the line between these points has homogeneous coordinates of the form $(\lambda x - x_0, \lambda y - y_0, \lambda z - z_0, \lambda w - w_0)$. The line determined by these two points will meet the surface where

$$f_n(\lambda x - x_0, \lambda y - y_0, \lambda z - z_0, \lambda w - w_0) = 0.$$

The parameter in this equation is λ, and the equation is of degree n. The n roots of the equation will give us the n values of λ, which in turn will give us the n points where the line meets the surface. We can rewrite this equation in terms of any other basis for the one-variable polynomial space. If we write it as a polynomial in λ, we have

$$\frac{\lambda^n}{n!} \left(x^n \frac{\partial^n f_n(x_0, y_0, z_0, w_0)}{\partial x^n} + \cdots + w^n \frac{\partial^n f_n(x_0, y_0, z_0, w_0)}{\partial w^n} \right) + \cdots$$
$$+ \frac{\lambda^2}{2!} \left(x^2 \frac{\partial^2 f_n(x_0, y_0, z_0, w_0)}{\partial x^2} + \cdots + w^2 \frac{\partial^2 f_n(x_0, y_0, z_0, w_0)}{\partial w^2} \right)$$
$$+ \lambda \left(x \frac{\partial f_n(x_0, y_0, z_0, w_0)}{\partial x} + \cdots + w \frac{\partial f_n(x_0, y_0, z_0, w_0)}{\partial w} \right)$$
$$+ f_n(x_0, y_0, z_0, w_0) = 0. \tag{7.3}$$

Since the point (x_0, y_0, z_0, w_0) lies on the surface, $f_n(x_0, y_0, z_0, w_0) = 0$. Hence, $\lambda = 0$ is one solution to this equation (the solution corresponding to the chosen point on the surface). In general, there will only be a single solution corresponding to $\lambda = 0$. This means that an arbitrary line through a point on the surface will, in general, meet the surface once at that point. However, if

$$x \frac{\partial f_n(x_0, y_0, z_0, w_0)}{\partial x} + \cdots + w \frac{\partial f_n(x_0, y_0, z_0, w_0)}{\partial w} = 0, \tag{7.4}$$

then every line will have at least double contact at the point (x_0, y_0, z_0, w_0) corresponding to $\lambda = 0$. The condition (7.4) is of the form $ax + by + cz +$

$dw = 0$ and is therefore a plane. It is defined to be the *tangent plane* to the surface at the point (x_0, y_0, z_0, w_0). We used the same approach to define tangents to curves.

Example 7.23 *Find the tangent plane at the point* $(2, 2, 1, 2)$ *to the quadric of Example 7.20.*

Since we have

$$f_2(x, y, z, w) = x^2 + 2y^2 + 2z^2 - 3xy - zw = 0$$

with

$$\frac{\partial f_2}{\partial x} = 2x - 3y, \qquad \frac{\partial f_2}{\partial y} = 4y - 3x, \qquad \frac{\partial f_2}{\partial z} = 4z - w, \qquad \frac{\partial f_2}{\partial w} = -z,$$

the tangent plane at $(2, 2, 1, 2)$ is given by $x(-2) + y(2) + z(2) + w(-1) = 0$, or $2x - 2y - 2z + w = 0$.

Example 7.24 *Find the tangent plane to the cubic surface* $2xyz + 2w^3 - w$ $(x^2 + y^2 + z^2) = 0$ *at the point* $(1, -1, -2, 1)$.

In this case we have

$$f_3(x, y, z, w) = 2xyz + 2w^3 - w(x^2 + y^2 + z^2) = 0$$

with

$$\frac{\partial f_3}{\partial x} = 2(yz - wx), \qquad \frac{\partial f_3}{\partial y} = 2(xz - wy),$$

$$\frac{\partial f_3}{\partial z} = 2(xy - wz), \qquad \frac{\partial f_3}{\partial w} = 6w^2 - (x^2 + y^2 + z^2),$$

and the required tangent plane is $x - y + z = 0$.

Since any plane cuts the surface in a curve of the same order, and since the tangent plane has double contact with the surface at the point of tangency, we see that the tangent plane cuts the surface in a curve that has a double point at the point of tangency. This, in turn, shows that tangent planes to cubic surfaces cut the surfaces in rational cubics.

Example 7.25 *Show that the tangent plane to the cubic surface*

$$x^3 + y^3 - xyw + zw^2 = 0$$

at the point $(0, 0, 0, 1)$ *cuts the surface in a rational cubic, and produce a parametrization of the cubic.*

We follow the steps for the construction of the tangent plane as illustrated in the previous two examples, and find that the tangent plane at the given point is the plane $z = 0$. This plane cuts the surface in the curve $x^3 + y^3 - xyw = 0$, which is a cubic curve in the plane $z = 0$ with a double point at the point of contact $(0, 0, 0, 1)$. We parametrize the curve by intersecting it with those lines through the double point that lie in the tangent plane, namely (in homogeneous form)

$$x = \alpha t, \qquad y = \beta t, \qquad z = 0, \qquad w = s.$$

The intersections are the solutions of

$$t^2(\alpha^3 t + \beta^3 t - \alpha\beta s) = 0.$$

The repeated solution $t = 0$ corresponds to the double point at the point of contact of the tangent plane with the surface, while the remaining solution

$$(\alpha^3 + \beta^3)t = \alpha\beta s$$

corresponds to the variable point on the curve. This yields the third-order parametrization

$$x = \alpha^2\beta,$$
$$y = \alpha\beta^2,$$
$$z = 0,$$
$$w = \alpha^3 + \beta^3.$$

In general, there will be two tangents to the intersection curve of a tangent plane with a surface at the point of tangency, which meet the curve, and hence the surface, in more than two coincident points. The argument is that, since the point is a double point of the curve, every line will meet the curve there in at least two points, but two lines, the tangent lines, will meet it in more than two points. Since these lines lie in the plane of the curve, they must also lie in the tangent plane, and hence these two special lines meet the surface with at least triple contact. These lines are called the *inflectional* or *principal* tangents to the surface at the point of tangency. From (7.3) we see that the inflectional tangents are given by the pair of equations

$$x\frac{\partial f_n(x_0, y_0, z_0, w_0)}{\partial x} + \cdots + w\frac{\partial f_n(x_0, y_0, z_0, w_0)}{\partial w} = 0,$$

$$x^2\frac{\partial^2 f_n(x_0, y_0, z_0, w_0)}{\partial x^2} + \cdots + w^2\frac{\partial^2 f_n(x_0, y_0, z_0, w_0)}{\partial w^2} = 0.$$

The first of these equations is the condition for the tangent plane, and the second is the condition for $\lambda = 0$ to be at least a triple root.

Example 7.26 *Find the inflectional tangents to the cubic surface of Example 7.25 at the point* $(0, 0, 0, 1)$.

These tangents must be the tangents to the curve of intersection

$$x^3 + y^3 - xyw = 0$$

in the tangent plane $z = 0$, which, from Chapters 5 and 6, are the axes $x = 0$ and $y = 0$. Using the conditions described above, we see that the inflectional tangents are given by the pair of equations

$$z = 0, \qquad 2xy + 4zw = 0,$$

which represent the intersection of a plane and a quadric surface, in this case the degenerate conic $xy = 0$.

7.10.3 Multiple Points

We have shown that the general equation (7.3) for the intersections of a pencil of lines with a surface can be used to determine the tangent plane at a point on the surface. All lines in the plane (7.4) intersect the surface at least twice in the point (x_0, y_0, z_0, w_0). However, if all the coefficients in this equation are zero, then all lines through the point, not only those in the tangent plane, will intersect the surface at least twice in the point. Such a point is called a *double point* of the surface. The conditions that have to be satisfied at a double point are

$$\frac{\partial f_n}{\partial x} = \frac{\partial f_n}{\partial y} = \frac{\partial f_n}{\partial z} = \frac{\partial f_n}{\partial w} = 0,$$

where all the partial derivatives are calculated at the point. This definition is a direct generalization of the definition of a double point of a curve (Section 5.2). Points of higher multiplicity are defined similarly.

Definition 7.11 *A point $P(x_0, y_0, z_0, w_0)$ of a surface $f_n(x, y, z, w) = 0$ is said to be a **point of multiplicity** r if all the partial derivatives of f_n, up to and including those of order $r - 1$, are zero at P, but there is it least one partial derivative of order r whose value at P is nonzero.*

7.11 The Steiner Surface: Preliminaries

7.11.1 Invariants

We have concentrated, throughout our study, on the properties of curves that are projective invariants. For example, the collinearity of points, and the number and type of multiple points, are projective invariants. So far these invariants have always been functions of a single curve. There are, however, invariants that are functions of more than one geometrical entity. One particular invariant of two conics is important for our discussion of the Steiner surface.

Let f be a polynomial in a set of indeterminates $x = (x_1, x_2, \ldots, x_n)$, and let ϕ be a polynomial function of the coefficients of f. Let T be an $n \times n$ nonsingular matrix, and set $\underline{x} = T\underline{x}^*$, where the underscore denotes a column vector of indeterminates. Then f becomes a polynomial in x^* with coefficients that, in general, will be different from the corresponding coefficients of f. Let ϕ^* denote the function corresponding to ϕ calculated with the new coefficients. If

$$\phi^* = K\phi,$$

where K is independent of the coefficients of the polynomial and depends only on the elements of the matrix T, we say that ϕ is an *invariant* of f.

Let f and g be two conics given by

$$f \equiv \underline{x}'A\underline{x} = 0, \qquad g \equiv \underline{x}'B\underline{x} = 0,$$

where $A = (a_{ij})$ and $B = (b_{ij})$ are symmetric matrices, and the prime indicates the transpose (to avoid confusion between the transformation matrix T and the usual superscript notation for transposition). The pencil of conics through the four points common to these conics is given by $\lambda f + \mu g = 0$ or, in matrix notation, by

$$\underline{x}'(\lambda A + \mu B)\underline{x} = 0.$$

The determinant of the matrix representing a member of this pencil of conics is an invariant since, if

$$A^* = T'AT \quad \text{and} \quad B^* = T'BT,$$

then

$$\det(\lambda A^* + \mu B^*) = \det(\lambda T'AT + \mu T'BT) = \det[T'(\lambda A + \mu B)T]$$
$$= (\det T)^2 \det(\lambda A + \mu B).$$

We have considered this determinant in Section 2.8.1, when constructing a conic through five points, noting that the condition for the conic to be a line pair is that the determinant must be zero. Since this determinant is cubic in the parameters of the pencil, a pencil of conics contains three line pairs in the complex projective plane.

The determinant can be written in the form

$$\det(\lambda A + \mu B) = \lambda^3 \Delta + \lambda^2 \mu \Theta + \lambda \mu^2 \Theta' + \mu^3 \Delta',$$

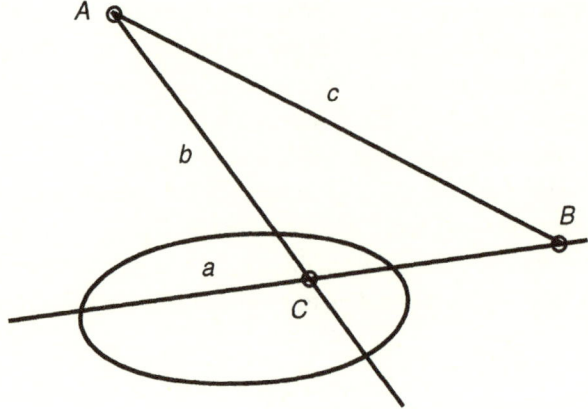

Figure 7.52. A self-polar triangle.

where

$$\Delta = \det A, \qquad \Theta = \sum b_{ij} A_{ij}, \qquad \Theta' = \sum a_{ij} B_{ij}, \qquad \Delta' = \det B,$$

and where A_{ij} (B_{ij}) denotes the cofactor of a_{ij} (b_{ij}). Therefore Δ, Θ, Θ', and Δ' are also invariants. It is the invariant Θ that we will investigate.

7.11.2 The Geometrical Interpretation of $\Theta = 0$

In Figure 7.52, A is an arbitrary point not on the conic and B is any point (not on the conic) on the polar a of A with respect to the conic. If the polar of a point P passes through Q, then the polar of Q passes through P (Section 5.10). Let C be the point conjugate to both A and B, that is, C is on a and on the polar b of B. Then both A and B are on the polar c of C, and the triangle ABC is such that each side is the polar of the opposite vertex with respect to the conic. Such a triangle is called *self-polar* with respect to the conic.

If we take a self-polar triangle as the triangle of reference, then A, B, and C have coordinates $(1, 0, 0)$, $(0, 1, 0)$, and $(0, 0, 1)$, and the sides AB, BC, and CA have equations $z = 0$, $x = 0$, and $y = 0$, respectively. The equation of a conic, referred to a self-polar triangle as the triangle of reference, is $ax^2 + by^2 + cz^2 = 0$, where $abc \neq 0$ for irreducible conics (Exercise 7.29), and the equation of a conic that passes through the vertices of the triangle of reference is $fyz + gzx + hxy = 0$, where $fgh \neq 0$ for irreducible conics (Exercise 7.30). Consider two conics S and S', where S is associated with coefficients a, b, ... and S' with coefficients a', b', ... If S is referred to a self-polar triangle as the triangle of reference, then the invariant Θ is given by

$$\Theta = bca' + cab' + abc'.$$

If this self-polar triangle is inscribed in S', then $a' = b' = c' = 0$ and $\Theta = 0$. It can be shown that the converse is also true, and hence we have the geometrical interpretation of $\Theta = 0$ as the condition that the two conics are such that there is a triangle inscribed in one conic S' that is self-polar to the other conic S. The conic S is said to be *inpolar* to S', and S' to be *outpolar* to S.

This condition is a *linear* condition, either on the coefficients of the point equation of one conic or on the line equation of the other. The reader should verify that the square of any tangent (the repeated tangent line) to a conic S is a degenerate conic that is outpolar to S.

7.12 The Steiner Surface: Some Properties

We are now in a position to investigate the rational surface whose parametrizations are given by polynomials of degree two. This surface is a *Steiner surface* and is given by the equations

$$x_i = C_i(p^*, q^*, r^*), \qquad i = 1, 2, 3, 4,$$

where $\{C_i(p^*, q^*, r^*)\}$ are homogeneous polynomials of degree two. Consider the conics corresponding to these polynomials, that is, the conics in parameter space whose equations are $C_i(p^*, q^*, r^*) = 0$, $i = 1, \ldots, 4$. In Section 7.11 we saw that there is a single linear condition on the coefficients (in line coordinates) of a conic that is inpolar to a given conic. Hence there is web of conics inpolar to two conics, a net inpolar to three conics, and a pencil of conics inpolar to four given conics. In particular, there is a pencil of conics inpolar to the four conics $C_i(p^*, q^*, r^*) = 0$, $i = 1, \ldots, 4$. Let this pencil be given by

$$\lambda S + \mu S' = 0. \tag{7.5}$$

Since the conditions are linear, we know that this pencil is inpolar to any conic of the form

$$\gamma_1 C_1 + \gamma_2 C_2 + \gamma_3 C_3 + \gamma_4 C_4 = 0.$$

The pencil of conics (7.5) can be thought of as the pencil of conics tangential to four lines $L_i(p^*, q^*, r^*) = 0, i = 1, \ldots, 4$. The pencil will be inpolar to each of $L_i^2(p^*, q^*, r^*) = 0, i = 1, \ldots, 4$. Let us assume that at least one of these repeated lines is linearly independent to the set $\{C_i(p^*, q^*, r^*), i = 1, \ldots, 4\}$. If this is the case, then the pencil is inpolar to five linearly independent conics and is uniquely determined. Therefore, it cannot be a pencil. Since this is a contradiction, each of the $L_i^2(p^*, q^*, r^*), i = 1, \ldots, 4$, must belong to the space spanned by the four polynomials $C_i(p^*, q^*, r^*), i = 1, \ldots, 4$. There exist scalars α_{ij}, $i, j = 1, \ldots, 4$, such that

$$\sum_{j=1}^{4} \alpha_{ij} C_j(p^*, q^*, r^*) = L_i^2(p^*, q^*, r^*), \qquad i = 1, \ldots, 4.$$

In other words, there exists scalars α_{ij}, $i = 1, \ldots, 4$, such that

$$\sum_{j=1}^{4} \alpha_{ij} x_j = L_i^2(p^*, q^*, r^*), \qquad i = 1, \ldots, 4.$$

This means that there is a three-space projective transformation of the coordinates x_i, $i = 1, \ldots, 4$, to the coordinates x, y, z, and w, such that

$$x = L_1^2(p^*, q^*, r^*), \qquad y = L_2^2(p^*, q^*, r^*),$$
$$z = L_3^2(p^*, q^*, r^*), \qquad w = L_4^2(p^*, q^*, r^*).$$

The fundamental theorem of projective geometry (Theorem 4.3.2) shows us that we can map any four lines to any four lines. Hence there is a two-space projective transformation of the parameters p^*, q^*, r^* to parameters p, q, r such that

$$L_1^2(p^*, q^*, r^*) = (p + q + r)^2,$$
$$L_2^2(p^*, q^*, r^*) = (p - q - r)^2,$$
$$L_3^2(p^*, q^*, r^*) = (p - q + r)^2,$$
$$L_4^2(p^*, q^*, r^*) = (p + q - r)^2.$$

By the use of two projective changes of coordinates, one in the parameter space and one in the three-space of the surface, the surface given by general second-degree parametrizations can always be written in the form

$$\begin{aligned} x &= (p + q + r)^2, \\ y &= (-p + q + r)^2, \\ z &= (p - q + r)^2, \\ w &= (p + q - r)^2. \end{aligned} \qquad (7.6)$$

From this parametrization it is not so easy to obtain the corresponding quartic equation, but an alternative parametrization simplifies the work. Making the change of coordinates

$$4x^* = x - y - z + w,$$
$$4y^* = x - y + z - w,$$
$$4z^* = x + y - z - w,$$
$$4w^* = x + y + z + w$$

yields

$$x^* = 2pq, \qquad y^* = 2pr, \qquad z^* = 2qr, \qquad w^* = p^2 + q^2 + r^2,$$

from which we can eliminate the parameters to obtain

$$x^{*2}y^{*2} + y^{*2}z^{*2} + z^{*2}x^{*2} - 2x^*y^*z^*w^* = 0. \qquad (7.7)$$

From (7.7) we see that the surface has a triple point at the origin and hence must be rational. This equation also shows us that the surface has the lines

$$x^* = y^* = 0, \qquad y^* = z^* = 0, \qquad z^* = x^* = 0,$$

as double lines (lines composed entirely of double points).

These observations lead to other properties of the Steiner surface. Any plane cuts a surface of order n in a curve of order n, and hence an arbitrary plane cuts the Steiner surface in a quartic. However, since the Steiner surface contains three double lines, an arbitrary plane will also cut these three lines. Therefore, the plane will cut the surface in a quartic that contains three double points. A quartic with three double points must be rational. Hence all planes cut Steiner surfaces in rational quartics. Since a tangent plane to a surface meets the surface, at the point of tangency, in a double point of the intersection curve, a tangent plane to a Steiner surface will, in general, meet the surface in a quartic that has four double points – one at the point of tangency and one each where the plane meets the three double lines. The Maclaurin–Bézout theorem then tells us that this curve must be reducible. A further argument (Exercise 7.33) shows that the quartic will reduce to a pair of conics.

From (7.6) we see that the coordinate planes meet the surface in double conics. For example,

$$x = 0 \quad \Rightarrow \quad y = 4(q + r)^2, \quad z = 4q^2, \quad w = 4r^2,$$

so that $y - z - w = 8r$ and $[(y - z - w)^2 - 4zw)]^2 = 0$.

7.13 The Cubic Surface

As in the case of the Steiner surface, our discussion of cubic surfaces and, in the next section, quadrics will be brief. We use these surfaces to illustrate applications of the properties of curves. We can learn much about a surface by examining the planar algebraic curves that form the intersections of the surface with planes. It is remarkable that the surface given parametrically by second-degree polynomials is a quartic that contains three double lines meeting in a triple point. We will explore similar properties of the cubic surface. In this investigation we use results from our study of curves as well as synthetic arguments. We will first consider the tangential equation of a surface and then proceed to discuss various kinds of tangent planes.

7.13.1 The Tangential Equation of a Surface

In Section 3.5 we noted the duality between points and lines and the dual representation of a curve. The curve can be thought of as a collection of tangent lines, the envelope. We obtain the tangential equation by identifying a line with the tangent line at a variable point and eliminating the variables between the two equations to obtain an algebraic equation in the coefficients of the curve and a variable line. The situation is analogous for surfaces, where the duality is between points and tangent planes.

The equation of the tangent plane to a surface $f_n(x, y, z, w)$ at the point (x_0, y_0, z_0, w_0) is

$$x \frac{\partial f_n(x_0, y_0, z_0, w_0)}{\partial x} + \cdots + w \frac{\partial f_n(x_0, y_0, z_0, w_0)}{\partial w} = 0.$$

A plane is given by $ax + by + cz + dw = 0$. The two planes are the same if

$$\lambda a = \frac{\partial f_n(x_0, y_0, z_0, w_0)}{\partial x},$$

$$\lambda b = \frac{\partial f_n(x_0, y_0, z_0, w_0)}{\partial y},$$

$$\lambda c = \frac{\partial f_n(x_0, y_0, z_0, w_0)}{\partial z},$$

$$\lambda d = \frac{\partial f_n(x_0, y_0, z_0, w_0)}{\partial w}$$

for some nonzero λ, and, since the given point is on the tangent plane, we also have

$$ax_0 + by_0 + cz_0 + dw_0 = 0.$$

To obtain the tangential equation of the surface we eliminate x_0, y_0, z_0, w_0, and λ from these five equations. For a curve, the order of the point equation of the curve is the number of intersections with the curve of an arbitrary line, and the class of the tangential equation is the number of tangents to the curve that can be drawn through an arbitrary point. For surfaces, the order of the point equation of the surface is the number of intersections with the surface of an arbitrary line, and the class of the tangential equation is the number of tangent planes to the surface that can be drawn through an arbitrary line. The order is the same as the degree of the corresponding point equation, and the class is the same as the degree of the corresponding line equation. Just as with curves, the order and class of a surface are not, in general, the same.

7.13.2 Tangent Planes and Tangential Equations

A plane is tangent to a surface if its coordinates satisfy the tangential equation of the surface. As in the case of a tangent line to a curve, this gives a single condition on the plane. A plane is determined by three parameters, and hence there will be a double infinity of planes tangent to a given surface. Each one of these planes is referred to as a *simple tangent plane*. Unless otherwise noted, a tangent plane is considered to be a simple tangent plane. If the tangent plane is tangent to the surface at two points, it is referred to as a double tangent plane or a *bitangent plane*. In general there will be a single infinity of bitangent planes. If the tangent plane is tangent to the surface at three points, it is called a *tritangent plane*. Since three points determine the plane, there can, in general, only be a finite number of tritangents to a surface.

If a planar section of a surface meets the surface in a curve that has a double point, then either the surface has a double point there or the plane is a tangent at that point. If the curve has two double points, then there are three possibilities. The points can both be double points of the surface, the surface can have one double point and the plane be tangent at the other point, or the plane can be a bitangent. Similarly, if the curve has three double points and none of these points is a double point of the surface, then the plane must be a tritangent plane to the surface.

7.13.3 The Twenty-Seven Lines on a Cubic Surface

Assume that there is a line on a cubic surface. Every plane meets the cubic surface in a cubic curve, and planes through this line will be no exception. Hence any plane through this line must meet the surface in a cubic curve that contains the given line. The cubic must reduce to a line and a conic.

Let us assume that the cubic surface has no double points and consider any bitangent plane to this cubic surface. By construction, the cubic curve of intersection contains two double points. By the Maclaurin–Bézout theorem, the cubic must be reducible to a line and a conic. Hence there always exists a line on the nonsingular cubic surface. By the same argument, each of the tritangent planes must meet the cubic surface in a curve that is the product of three lines. Hence we can find tritangent planes by taking any plane through a line on the surface and requiring that the conic we get from an arbitrary plane through this line reduce to a line pair.

Let the line in the surface be given by $x = w = 0$. Then the equation of the surface can be written in the form

$$x\Phi_2(x, y, z, w) + w\Psi_2(x, y, z, w) = 0,$$

where Φ_2 and Ψ_2 are quadratic. Any plane through the line $x = w = 0$ is of the form $w = \mu x$ and will meet the surface where

$$x\Phi_2(x, y, z, \mu x) + \mu x\Psi_2(x, y, z, \mu x)$$

$$= x(\Phi_2(x, y, z, \mu x) + \mu\Psi_2(x, y, z, \mu x)) = 0.$$

Hence, the plane meets the cubic surface in the conic given by

$$\Phi_2(x, y, z, \mu x) + \mu\Psi_2(x, y, z, \mu x) = 0.$$

The condition for this conic to be a line pair is obtained by writing it in quadratic form and equating the determinant of the matrix to zero. This determinant is a polynomial in μ, the roots of which will give us the specific values corresponding to tritangent planes. To determine the degree of this polynomial, we see that the coefficient of x^2 contains, in general, a μ^3 term, the coefficients of xy, xz, and x contain a μ^2 term, and the rest of the coefficients contain terms of order μ. Hence the determinant is of fifth order, and there are, in general, five tritangent planes through any given line that lies on the cubic surface.

Let us consider one of these tritangent planes. We have seen that this plane meets the cubic surface in three lines. Denote these lines by a, b, and c, and let the vertices of the triangle so defined be A, B, and C. We have just seen that through each line in the surface there are five tritangents. Therefore, through each of the lines a, b, and c there must be four more tritangents (apart from the one that contains all three of these lines). Each of these four tritangents will meet the cubic surface in two additional lines. None of these additional lines can pass through A, B, or C, as shown by the following argument.

Let us assume that there are three concurrent lines on the surface. Without loss of generality we can take the lines to be given by

$$x = y = 0, \qquad y = z = 0, \qquad z = x = 0.$$

Then the equation of the cubic surface must be of the form

$$xy\Theta_1 + yz\Phi_1 + zx\Psi_1 = 0,$$

and hence it has a double point at the origin, the point where the three lines meet.

Returning to our main argument, we see that if any of the additional lines were to pass through A, B, or C, there would be three lines through the corresponding points. This would imply that these points are double points. Since this would be contrary to our assumption that the cubic surface is nonsingular, we realize that none of these additional lines can pass through A, B, or C. Since we get two additional lines from each of the four additional tritangent planes through each of the original three lines, we have twenty-four additional lines and hence a total of twenty-seven lines in all. From this result it is now easy to show that there are forty-five tritangent planes (Exercise 7.34). It also follows that the cubic surface must be rational (Exercise 7.35).

7.14 The Quadric Surface

We end this discussion with the quadric surface. Entire books have been written on this surface alone, but we will discuss little more than the fact that it is a rational surface. It provides a nice ending point for our discussion, because we started our study of algebraic geometry with the conic, a second-order curve and, in that sense, the simplest curve that is not a line. We end with the quadric, a second-order surface and the simplest nonplanar algebraic surface. We recall that, if there are two base points of a second-order parametrization of a surface, then the surface must be a quadric. We will illustrate the converse with an example.

Example 7.27 *Produce a second-order parametrization of the quadric*

$$\frac{x^2}{a^2} - \frac{y^2}{b^2} = \frac{2z}{c},$$

and show that this parametrization has two base points.

Since this surface passes through the origin, we will produce the parametrization by passing lines through the origin. The line through the origin can be written in the form

$$x = \alpha t, \qquad y = \beta, \qquad z = \gamma t,$$

and this line meets the surface where

$$t = 0 \quad \text{or} \quad t = \frac{2a^2 b^2 \gamma}{c(b^2 \alpha^2 - a^2 \beta^2)}.$$

The value $t = 0$ corresponds to the origin and the other value will give the parametrization of the variable point on the quadric. In homogeneous coordinates this is given by

$$x = 2a^2 b^2 \alpha \gamma, \qquad y = 2a^2 b^2 \beta \gamma,$$
$$z = 2a^2 b^2 \gamma^2, \qquad w = c(b^2 \alpha^2 - a^2 \beta^2).$$

The parametrization is of second order in the homogeneous parameters α, β, and γ, the coordinate planes being the images of line pairs in parameter space. Since $z = 0$ is the image of $\gamma^2 = 0$, the only possible base points must lie on the line $\gamma = 0$. While any points on this line are on the conics representing the parametrizations of x, y, and z, this line meets the conic corresponding to $w = 0$ at the two points $(a, -b, 0)$ and $(a, b, 0)$. Hence there are precisely two base points.

7.15 Bibliographical Notes

A useful book on the development and applications of splines is *The Theory of Splines and Their Applications* by J. H. Ahlberg, E. N. Nilson, and J. L. Walsh, Academic Press (1967). The review paper "The Steiner Surface Revisited" *Proc. Ray. Soc. Lond. Ser.* A. V 369 (1979) pp. 157–174 by R. J. Y. McLeod gives a comprehensive discussion of the Steiner surface and its use in approximation. The geometry of three dimensions and the properties of general surfaces can be discovered, admired, and enjoyed in the books *A Treatise on the Analytic Geometry of Three Dimensions*, Vols. I, II, by G. Salmon, reprinted by the Chelsea Publishing Company, and *Analytic Geometry of Three Dimensions*, by D. M. Y. Sommerville, reprinted by Cambridge University Press. For those seeking more information on the cubic surface in particular, we recommend A. Henderson's *The Twenty-seven Lines upon the Cubic Surface*.

Exercises

7.1. Construct a surface function that interpolates to the following line segments:

$$z = 0, \qquad 1 - x - y = 0, \qquad 0 \leq x \leq 1,$$
$$x = 0, \qquad 1 - y - z = 0, \qquad 0 \leq y \leq 1,$$
$$y = 0, \qquad 1 - z - x = 0, \qquad 0 \leq z \leq 1.$$

7.2. Construct a surface function that interpolates to the following curve segments in the (x, y) plane:

$$1 - |x| + y = 0, \qquad -1 \leq x \leq 1,$$
$$\text{and} \quad 1 - x^2 - y^2 = 0, y \geq 0.$$

7.3. Construct a surface function that interpolates to the following curve segments:

$$1 - x^2 - y^2 = 0, \qquad y \geq 0, \qquad z = 0,$$
$$1 - x^2 - z^2 = 0, \qquad z \geq 0, \qquad y = 0.$$

7.4. Construct a surface function that interpolates to the following curve segments:

$$z = 0, \qquad f(x, y) = ax^3 + by^3 - (1 + a)x - (1 + b)y + 1,$$

$$x, y \geq 0,$$

$$x = 0, \qquad 1 - y - z = 0, \qquad 0 \leq y \leq 1,$$

$$y = 0, \qquad 1 - z - x = 0, \qquad 0 \leq z \leq 1.$$

7.5. Determine conditions on n such that $D_{n,2}|R = D_{n,2}$, where R comprises four segments, which can be either lines or conics.

7.6. Construct a cubic polynomial basis for the standard triangle, using Lagrange interpolation at the nodes.

7.7. Construct a cubic polynomial basis for the standard triangle, using Hermite interpolation at the vertices and a function evaluation at an interior node.

7.8. Complete the construction of the remaining twelve basis functions for quartic polynomial interpolation on the standard triangle.

7.9. Construct a cubic polynomial basis for the unit square, using Hermite interpolation at the vertices and one function evaluation at the interior node $(\frac{1}{2}, \frac{1}{2})$. Investigate the space spanned by the basis.

7.10. Construct a Lagrange basis for quadratic interpolation on a tetrahedron.

7.11. Complete the construction of the Lagrange basis for quadratic interpolation on a parallelopiped.

7.12. Use both methods as illustrated to show that, in Example 7.12,

$$\frac{xy}{1 + x + y} \equiv 1 - x - y \bmod x^2 + y^2 - 1.$$

7.13. Let two circles f and g intersect in the points 1 and 2. Use coordinate geometry to show that, in the affine plan,

$$f \equiv (1; 2) \bmod g.$$

7.14. Complete step 5 of Example 7.14 for the remaining basis functions.

7.15. Construct a basis that spans $P_2(x, y)$ for the local domain shown in Figure 7.46.

7.16. Complete steps 2 and 4 of Example 7.15 for the remaining basis functions.

7.17. Construct $W_{12}(x, y)$ for Example 7.16, and complete step 4 for the other basis functions.

7.18. Construct a basis that spans $P_1(x, y)$ for the local domain shown in Figure 7.53. The domain is bounded by three conics and a line.

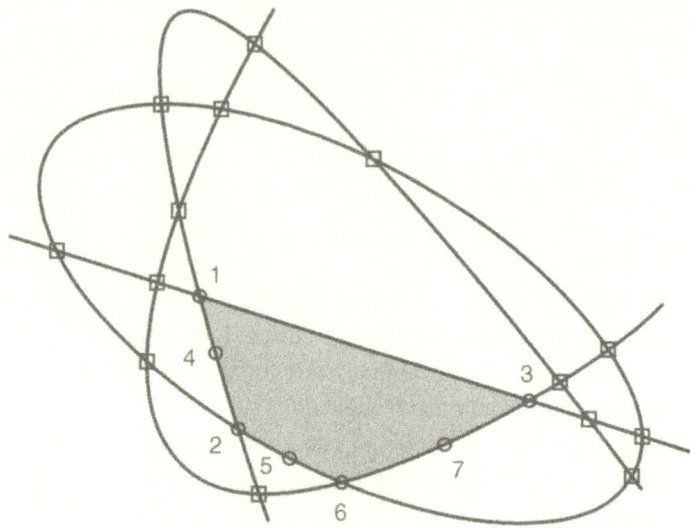

Figure 7.53. A local domain bounded by three conics and a line.

7.19. Use the quadratic Lagrange basis on the standard triangle (nodes at the vertices and at the midpoints of edge segments) to interpolate three-dimensional data, thus obtaining a second-order polynomial parametrization of the interpolating surface. Choose your own data points.

7.20. Repeat the previous exercise, but use the unit square as the local domain in the parameter plane.

(a) Construct two planes which define the affine line

$$x = -1 + 2t, \qquad y = 2 - \tfrac{1}{2}t, \qquad z = 1 + 7t.$$

(b) Find the parametric form of the line defined by the two planes

$$2x - 3y + z = 3, \qquad 5x + y - 4z = 6.$$

7.21. Use homogeneous coordinates and reproduce the discussion about the order of a surface.

7.22. Show that if two points lie in a plane, then the line defined by these two points also lies in the plane.

7.23. Show, by using elimination methods as in Chapter 5, that a rational surface is an algebraic surface.

7.24. Use the quadratic Lagrange basis on the standard triangle (nodes at vertices and midpoints of edge segments), and find the parametrization of the surface obtained by fixing the vertices at $(0, 0, 0)$, $(0, 1, 0)$, and $(1, 0, 0)$ and lifting the midpoint nodes to $(\tfrac{1}{2}, 0, \tfrac{1}{4})$, $(\tfrac{1}{2}, \tfrac{1}{2}, \tfrac{1}{4})$, and $(0, \tfrac{1}{2}, \tfrac{1}{4})$. Find the equation of the surface in algebraic form.

7.25. Determine the order of each of the following rational surfaces. Try to find the algebraic form of the surface equation, to confirm your answer.
(a) $x = 2rt$, $y = 3st$, $z = r^2 + s^2 - t^2$, $w = r^2 + s^2 + t^2$.
(b) $x = rt$, $y = st$, $z = r^2 + rs$, $w = r^2 + s^2 + st$.
(c) $x = (r^2 + 3s^2 - 2rt - 6st)(s - t)$, $y = (r^2 + 3s^2 - 2rt - 6st)(r - t)$,
 $z = (2t - r - s)(r - s)(r - 3t)$, $w = (2t - r - s)(r - s)(s - t)$.
(d) $x = rt^2$, $y = st^2$, $z = t^3$, $w = rst - r^3 - s^3$.

7.26. Find the tangent plane at $(1, 1, 1, 1)$ to the quartic surface

$$x^4 - 6x^2y^2 + y^4 - z^4 + \left(x^2 - y^2\right)w^2 + 6z^2w^2 - w^4 = 0.$$

7.27. Find the cubic curve in which the tangent plane at $(1, -1, 0, 1)$ intersects the surface $2xyz + 2w^3 - w(x^2 + y^2 + z^2) = 0$.

7.28. Consider the surface $x^3 + y^3 - 2xyw + zw^2 = 0$.
(a) Show that the cubic curve in which the tangent plane at $(1, 1, 0, 1)$ intersects the surface is rational by finding its parametrization.
(b) Find the inflectional tangents to the surface at the given point.

7.29. Show that an irreducible conic referred to a self-polar triangle as the triangle of reference has the form

$$ax^2 + by^2 + cz^2 = 0, \qquad abc \neq 0.$$

7.30. Show that an irreducible conic through the vertices of the triangle of reference has the form

$$fyz + gzx + hxy = 0, \qquad fgh \neq 0.$$

7.31. Show that the square of any tangent is a degenerate conic outpolar to the conic.

7.32. Use the fact that the Steiner surface has a triple point, and the corresponding algebraic form of the equation, to produce a parametrization of the surface.

7.33. Show that a tangent plane to a Steiner surface meets the surface in a conic pair.

7.34. Show that there are forty-five tritangent planes to a nonsingular cubic surface.

7.35. Show that a nonsingular cubic surface is rational. (Hint: consider any two of the lines that lie in the surface.)

Index

General topics concerning curves are listed under "curves." Topics relating to special curves are listed under the name of the special curve. For example, things related to conics are listed under "conic," topics relating specifically to ellipses are under "ellipse," those concerning splines under "spline." Much of the text discusses interpolation but only general items are listed under this heading. Most interpolation is listed under more specific headings for example: "basis construction," "construction of curves" or "conic, through four points and touching a line."

For EU product safety concerns, contact us at Calle de José Abascal, 56–1°,
28003 Madrid, Spain or eugpsr@cambridge.org.

www.ingramcontent.com/pod-product-compliance
Ingram Content Group UK Ltd.
Pitfield, Milton Keynes, MK11 3LW, UK
UKHW010853090126
466816UK00011B/204